Hydrogen Storage and Transportation

Jianxin Zou

Hydrogen Storage and Transportation

机械工业出版社
CHINA MACHINE PRESS

Springer

Jianxin Zou
School of Materials Science and Engineering
Shanghai Jiao Tong University
Shanghai, China

ISBN 978-981-96-2877-3 ISBN 978-981-96-2875-9 (eBook)
https://doi.org/10.1007/978-981-96-2875-9

Jointly published with China Machine Press Co.,Ltd.

The print edition is not for sale in China (Mainland). Customers from China (Mainland) please order the print book from: China Machine Press Co.,Ltd..

The original submitted manuscript has been translated into English. The translation was done using artificial intelligence. A subsequent revision was performed by the author(s) to further refine the work and to ensure that the translation is appropriate concerning content and scientific correctness. It may, however, read stylistically different from a conventional translation.

This Springer imprint is published by the registered company Springer Nature Singapore Pte Ltd.
The registered company address is: 152 Beach Road, #21-01/04 Gateway East, Singapore 189721, Singapore

Foreword I

The world is currently experiencing unprecedented changes, with a new round of technological revolutions and industrial transformations intersecting historically with China's economic development goals for high-quality growth. Hydrogen energy development and utilization technologies, represented, for example, by fuel cells, have made significant breakthroughs, providing important solutions for achieving zero-emission energy applications. Therefore, we need to seize the global energy transformation trends and opportunities, accelerating the development of the hydrogen energy industry and promoting China's clean and low-carbon energy transformation.

Globally, major developed countries are placing great emphasis on the development of the hydrogen energy industry. Hydrogen has become an important strategic choice for accelerating energy transition and upgrading, as well as for fostering new economic growth drivers. The key technologies of the global hydrogen energy supply chain are maturing, with rapid growth in fuel cell powered shipments, continuing cost reductions, and noticeable acceleration in hydrogen infrastructure construction. Regional hydrogen supply networks are gradually taking shape.

The "dual-carbon" goals have pointed the way for China's economic and social low-carbon transition and have raised higher standards and stricter requirements for high-emission sectors, such as energy, industry, transportation, and construction. Hydrogen is the key energy storage medium in the future energy system, an essential feedstock for decarbonizing industrial sectors like steel, and the most promising pathway for the low-carbon transformation of heavy-duty trucks, ships, and aviation. It is also a necessary component of zero-carbon buildings and communities. The development of hydrogen energy is closely linked to achieving carbon peak and carbon neutrality goals, and it is a strategic emerging industry and promising industry that drives China's sustainable, high-quality economic development.

In the past 3 years, China's hydrogen industry has developed rapidly under the guidance and support of government policies. Hydrogen, as an invisible gas, is now becoming an integral part of our daily life: hydrogen fuel cell buses shuttle between venues and competition areas of the Beijing Winter Olympics and Paralympics; hydrogen refueling stations are being built on land and at sea; hydrogen-powered dump trucks, transport vehicles, and sanitation trucks are in operation across the country; hydrogen-powered passenger vehicles and bicycles have entered mass production; hydrogen-powered ships are under construction; and hydrogen aircraft have begun to turn people's imagination of hydrogen-powered planes into reality. In March 2022, the National Development and Reform Commission and the National Energy Administration jointly released the "Medium and Long-Term Hydrogen Energy Industry Development Plan (2021–2035)," which outlines that by 2025, China will have mastered core technologies and manufacturing processes, with a fuel cell vehicle stock of about 50,000 units and a number of hydrogen refueling stations deployed; by 2030, a relatively complete hydrogen energy industry technological innovation system and a clean energy hydrogen production and supply system will be built; and by 2035, a diversified hydrogen energy application ecosystem will be established, with a significant increase in the proportion of hydrogen produced from renewable energy in terminal energy consumption. In the future, the hydrogen energy industry will have broader development space in the new pattern, where the domestic large circulation is the mainstay, and domestic and international dual circulations promote each other.

Science and technology are the primary productive forces, and talent is the primary resource. The high-quality development of the hydrogen energy industry depends on the cultivation of a talent system. In July 2021, the Ministry of Education issued the "Carbon Neutrality Technology Innovation Action Plan for Higher Education Institutions," and in April of the following year, the "Work Plan for Strengthening the Talent Cultivation System in Carbon Peak and Carbon Neutrality Higher Education" was released, both of which addressed the technology and talent training requirements for the entire hydrogen energy industry chain, from production to storage, transportation, and utilization. "Hydrogen Energy Science and Engineering" has become a newly approved undergraduate major. The "Medium-Term Hydrogen Energy Industry Development Plan (2021–2035)" also proposes the systematic construction of an innovation system for the hydrogen energy industry, focusing on key areas and links, building industrial innovation support platforms, continuously enhancing core technological capabilities, and promoting professional talent development. In October 2022, the General Office of the Communist Party of China Central Committee and the General Office of the State Council issued the "Opinions on Strengthening the Construction of High-Skilled Talent Teams in the New Era," requiring the establishment of a high-skilled talent training system led by industry and enterprises, based on vocational schools, and supported by the government and society and greater efforts to cultivate urgently needed and scarce skilled talents.

The rapid development of the hydrogen energy industry presents challenges in talent cultivation. The hydrogen industry urgently needs interdisciplinary talents with a solid theoretical foundation, a complete knowledge system, and a focus on practical applications. The "Hydrogen Energy and Fuel Cell Industry Application Talent Training Series" books published here have been compiled by the Hydrogen Energy Center of the China Electric Vehicle 100-Person Association, with contributions from experts and scholars from academia, industry, and enterprises. This series is an educational compilation for training practical personnel in the hydrogen energy industry, filling a gap in the industry and making an important contribution to talent development.

Hydrogen energy is not only an industry related to the international energy landscape and national development trends but also a lifelong career for every practitioner. The success of this career depends not only on individual perseverance but also on seizing the opportunities presented by the times and riding the wave of industrial development. I hope that readers will take this series of books as the starting point for entering the hydrogen energy industry, maintain their original aspirations, forge ahead, and not miss the great opportunities and missions brought by the development of the industry.

Chinese Academy of Engineering
Beijing, China
Royal Academy of Engineering
London, UK
World Electric Vehicle Association
Washington, DC, USA
October 2022

Qingquan Chen

Foreword II

Hydrogen, as a secondary energy source with diverse origins, efficient applications, and environmental benefits, is widely used in transportation, energy storage, industry, and power generation. The development and utilization of hydrogen has become an important direction for the new global energy technology revolution and is also a critical pathway for achieving net-zero emissions worldwide. With the introduction of China's "dual-carbon" strategy goals, hydrogen energy, with its ability to ensure energy security and facilitate deep decarbonization, has become a vital support for China's low-carbon energy transformation and the establishment of a green industrial system, with a clear and steadfast direction for industrial development.

Currently, the hydrogen energy industry is developing rapidly, having progressed from basic research and development to mass production and full industrialization. As the large-scale application and commercialization of hydrogen energy approaches, professionals with a solid theoretical foundation and engineering practical capabilities will become the key drivers for the development of hydrogen industry. Talent cultivation in hydrogen energy is a systematic project that requires strong support from favorable talent policies and an industrial development background, as well as the creation of innovation platforms, research institutions, and numerous enterprises, all working together to build a sound ecosystem for industry–academia–research collaboration.

In July 2021, the Ministry of Education issued the "Carbon Neutrality Technology Innovation Action Plan for Higher Education Institutions," outlining the promotion of carbon-neutral future technology colleges and demonstration energy colleges and encouraging universities to offer general courses on carbon neutrality. In October 2022, the General Office of the Communist Party of China Central Committee and the State Council issued the "Opinions on Strengthening the Development of High-Skilled Talent Teams in the New Era," which emphasized: "Skilled talent is a crucial force supporting China's manufacturing and innovation. Strengthening the development of highly skilled talents at the senior craftsman level is of great significance for consolidating and developing the advanced nature of the working class, enhancing national core competitiveness and technological innovation capacity, alleviating structural employment contradictions, and promoting high-quality development."

In order to implement the decisions of the Party Central Committee and the State Council and to strengthen the development of high-skilled talent teams in the new era, as well as meet the talent needs of the hydrogen energy industry, the Hydrogen Energy Center of the China Electric Vehicle 100-Person Association, in collaboration with the Shanghai Fuel Cell Vehicle Commercialization Promotion Center, Foshan Environmental and Energy Research Institute, Shanghai Hydrogen Energy Utilization Engineering Research Center, Shanghai Intelligent New Energy Vehicle Innovation Platform, Shandong Hydrogen Valley New Energy Technology Research Institute, and other organizations jointly compiled the "Hydrogen Energy and Fuel Cell Industry Application Talent Training Series" books.

This series includes titles such as *Introduction to Hydrogen Energy and Fuel Cell Industries*, *Hydrogen Production Technologies and Processes*, *Hydrogen Storage and Transportation*, *Hydrogen Refueling Station Technical Specifications and Safety Management*, *Hydrogen Fuel Cell Vehicles and Key Components*, *Hydrogen Fuel Cell Vehicle Safety Design*, and *Hydrogen Fuel Cell Vehicle Testing and Maintenance Technologies*. The content of the series covers a complete knowledge system of the entire hydrogen energy and fuel cell industry chain and aims to bridge the gap between theory and engineering practice. With an application-oriented focus, the series emphasizes the practical design of hydrogen energy technologies and activity-based teaching, fostering a deeper integration of the education chain, talent chain, and industry chain. Through learning and training, students or professionals will gain a comprehensive understanding of the hydrogen energy and fuel cell industry's development trends, technological principles, engineering processes, and application solutions and will acquire the fundamental knowledge and practical skills required for working in fields, such as hydrogen production, storage and transportation, refueling station operations, and hydrogen fuel cell vehicle testing and maintenance.

This book focuses on the storage and transportation methods of hydrogen, providing an overview of hydrogen storage and transportation technologies and their development status and detailing the technical principles and current applications of high-pressure gaseous storage, low-temperature liquid storage, hydrogen-rich liquid compound storage, and solid-state storage. Storage and transportation are integral parts of the hydrogen energy supply chain and are key to reducing the cost of hydrogen energy's terminal applications. The purpose of this book is to help students and professionals in the hydrogen energy sector to fully understand hydrogen storage and transportation methods, equipment, and materials, master related knowledge and key technologies, and acquire the basic skills required for working in hydrogen storage and transportation.

While the editorial committee of this series has strived to cover the essential points of the entire industry chain, the rapid development of new technologies means that there are still many shortcomings in the writing process. We welcome valuable feedback and suggestions from readers to continuously refine and improve the content of the book and contribute to the development of highly skilled talent that the industry urgently needs. We would like to express our sincere gratitude to all the collaborating institutions for their strong support and hard work.

We hope that this series can provide assistance to professionals in the hydrogen energy industry, support the cultivation of talent in the sector, and make a modest contribution to the sustainable development of the hydrogen energy industry.

Editorial Committee of the "Hydrogen Energy and Fuel Cell Industry Application Talent Training Series"
Shandong Hydrogen Valley New Energy Technology Research Institute
Jinan, Shandong, China

Zhen Zhang

Preface

The principal methods for hydrogen storage and transportation are gaseous, liquid, and solid-state storage. Currently, gaseous storage is the most widely used method for hydrogen storage. However, with the rapid growth of the hydrogen industry and continuous technological innovations, both liquid and solid-state storage technologies are advancing quickly. This book offers an in-depth exploration of hydrogen storage and transportation methods, combining the latest advancements to analyze the underlying principles and key technologies. Topics covered in this book include an overview of hydrogen storage and transportation technologies, physical and chemical hydrogen storage methods, solid-state hydrogen storage, and the status of hydrogen storage and transportation applications.

This book has important reference value for the training of engineering and technical personnel in colleges and universities, as well as for hydrogen fuel cell research institutions and industries. It is helpful for promoting innovation in the field of hydrogen energy and provides support for the advancement of hydrogen fuel cell vehicle technologies.

Editorial Committee

Editorial Committee

Shanghai, China Jianxin Zou

Contents

1 Overview of Hydrogen Storage and Transportation 1
1.1 Introduction 1
1.2 High-Pressure Gaseous Hydrogen Storage and Transportation 6
1.2.1 High-Pressure Gas Cylinders 6
1.2.2 Hydrogen Transportation by Pipelines 14
1.3 Low-Temperature Liquid Hydrogen Storage and Transportation 16
1.4 Hydrogen Storage and Transportation by Hydrogen-Rich Liquid Compounds 18
1.4.1 Liquid Organic Hydrogen Carriers (LOHCs) 19
1.4.2 Liquid Ammonia 19
1.4.3 Methanol 20
1.5 Solid-State Hydrogen Storage and Transportation 20
1.5.1 Hydrogen Storage Alloys and Metal Hydrides 24
1.5.2 Complex Hydrides 26
1.5.3 Ammonia Boranes and Their Derivatives 27
1.5.4 Physical Adsorptive Materials 27
1.6 Comparison of Various Hydrogen Storage and Transportation Methods 28
1.7 Hydrogen Safety in Storage and Transportation Processes 36
Exercises 42
References 43

2 High-Pressure Gaseous Hydrogen Storage and Transportation 45
2.1 Principle of High-Pressure Hydrogen Storage 46
2.1.1 Principle of Hydrogen Pressurization 46
2.1.2 Hydrogen Pressurization Equipment 47
2.1.3 Principle of Hydrogen Embrittlement of Materials 53

2.2 High-Pressure Hydrogen Storage Equipment ... 59
2.2.1 Classification of High-Pressure Hydrogen Storage Equipment ... 59
2.2.2 Fixed Hydrogen Storage Containers ... 63
2.2.3 Mobile Hydrogen Storage Containers ... 69
2.3 Applications of Gaseous Hydrogen Storage and Transportation Technology ... 72
2.3.1 Applications of High-Pressure Gas Cylinder for Hydrogen Storage and Transportation ... 72
2.3.2 Applications of Hydrogen Transportation by Pipelines ... 78
Exercises ... 84
References ... 85

3 Low-Temperature Liquid Hydrogen Storage and Transportation ... 89
3.1 Principle of Hydrogen Liquefaction ... 89
3.1.1 Ortho-Para Hydrogen Conversion ... 90
3.1.2 Joule-Thompson Effect ... 91
3.1.3 Hydrogen Liquefaction Processes ... 93
3.2 Key Materials and Equipment of Liquid Hydrogen Storage Tanks ... 95
3.2.1 Liquid Hydrogen Storage Tank ... 95
3.2.2 Insulation Methods of Liquid Hydrogen Storage Tanks ... 97
3.3 Applications of Liquid Hydrogen Storage and Transportation Technology ... 103
Exercises ... 110
References ... 111

4 Hydrogen Storage and Transportation Technologies Using Hydrogen-Rich Liquid Compounds ... 113
4.1 Liquid Organic Hydrogen Carriers (LOHCs) ... 114
4.1.1 Hydrogen Storage Principle of LOHCs ... 115
4.1.2 Synthesis and Cracking Processes of LOHCs ... 118
4.1.3 Hydrogen Storage Processes of LOHCs ... 121
4.2 Liquid Ammonia ... 124
4.2.1 Principle of Hydrogen Storage Using Ammonia ... 124
4.2.2 Ammonia Synthesis and Cracking Process ... 126
4.2.3 Current Status of Ammonia Energy Application ... 133
4.3 Methanol ... 138
4.3.1 Principle of Hydrogen Storage Using Methanol ... 138
4.3.2 Methanol Synthesis and Cracking Process ... 140
4.3.3 Current Status of Methanol Energy Applications ... 149

4.4 Comparison of LOHC, Ammonia, and Methanol Hydrogen Storage Technologies . . . 153
4.5 Engineering Applications of Hydrogen-Rich Liquid Compounds . . . 156
Exercises . . . 160
References . . . 161

5 Material-Based Solid Hydrogen Storage and Transportation . . . 165
5.1 Alloys and Metal Hydrides for Hydrogen Storage . . . 166
5.1.1 Principle of Hydrogen Storage in Metals (Alloys) . . . 167
5.1.2 Synthesis of Hydrogen Storage Metals (Alloys) . . . 170
5.1.3 Rare Earth AB_5-Type Hydrogen Storage Alloys . . . 172
5.1.4 Magnesium-Based Hydrogen Storage Alloys . . . 174
5.1.5 Other Metal Alloys . . . 182
5.2 Complex Hydrides for Hydrogen Storage . . . 187
5.2.1 Principle of Hydrogen Storage in Light Metal Borohydrides . . . 187
5.2.2 Principle of Hydrogen Storage in Light Metal Aluminohydrides . . . 199
5.2.3 Principle of Hydrogen Storage in Light Metal Amides . . . 202
5.3 Ammonia Boranes and Their Derivatives for Hydrogen Storage . . 204
5.3.1 Principle of Hydrogen Storage in Ammonia Boranes . . . 204
5.3.2 Principle of Hydrogen Storage in Ammonia Borane Derivatives . . . 206
5.4 Physical Adsorbents for Hydrogen Storage . . . 208
5.4.1 Principle of Hydrogen Storage in Carbon Materials . . . 209
5.4.2 Principle of Hydrogen Storage in MOF Materials . . . 216
5.5 Solid-State Hydrogen Storage and Transportation Technology and Its Applications . . . 224
5.5.1 Hydrogen Absorption and Desorption Reaction Equilibrium Pressure Equation of Hydrogen Storage Alloys . . . 229
5.5.2 Hydrogen Absorption and Desorption Reaction Kinetic Equation of Hydrogen Storage Alloys . . . 230
5.5.3 Heat Transfer Equation . . . 233
5.5.4 Mass Transfer Equation . . . 234
Exercises . . . 235
References . . . 237

Appendixes . . . 245

Chapter 1
Overview of Hydrogen Storage and Transportation

With the purchase of this book, you can use our "SN Flashcards" app to access questions free of charge in order to test your learning and check your understanding of the contents of the book. To use the app, please follow the instructions below:

1. Go to https://flashcards.springernature.com/login
2. Create a user account by entering your e-mail address and assigning a password.
3. Use the following link to access your SN Flashcards set: ▶ www.sn.pub/5lls30 If the link is missing or does not work, please send an e-mail with the subject "SN Flashcards" and the book title to customerservice@springernature.com

1.1 Introduction

Hydrogen (H) is a chemical element, ranking first in the periodic table of elements. Hydrogen mainly appears in a diatomic state (H_2), and under normal circumstances, the elemental form of hydrogen is hydrogen gas. Hydrogen gas is the known gas with the smallest density, composed of diatomic molecules, colorless, tasteless, with reducation properties, which can be produced from water, fossil fuels and other hydrogen-containing substances, and is an important industrial raw material and energy carrier. Hydrogen gas has a low ignition point, a wide explosion range, and a large diffusion coefficient, see Table 1.1. Therefore, hydrogen gas is easy to dissipate after leakage, and it is not easy to form an explosive mist, which is relatively safe and controllable in open spaces. Hydrogen gas has a high calorific value and is an ideal secondary energy carrier due to its clean nature.

J. Zou, *Hydrogen Storage and Transportation*,
https://doi.org/10.1007/978-981-96-2875-9_1

Table 1.1 Comparison of hydrogen gas, gasoline vapor, and natural gas

Technical indicators	Hydrogen gas	Gasoline vapor	Natural gas
Explosion limit (%)	4.1–75	1.4–7.6	5.3–15
Combustion point energy/MJ	0.02	0.2	0.29
Diffusion coefficient/(m^2/s)	6.11×10^5	0.55×10^5	1.61×10^5
Calorific value/(MJ/kg)	142	44	47.5

Hydrogen energy refers to a clean, carbon-free and renewable energy that uses hydrogen as an input, which has a wide range of sources for production and rich application scenarios. As one of the new types of energy, the environmental protection effect of hydrogen energy is excellent. It has multiple advantages such as high calorific value, low production cost, and zero carbon emissions. It can be used for energy storage, power generation, fuel driving of transportation tools, and household fuel. Therefore, hydrogen energy has also become an ideal solution to support the large-scale development of renewable energy and promote the transformation of traditional energy structure. It is an important way to achieve China's "dual carbon" goal and a major way to achieve large-scale deep decarbonization in various fields such as transportation, industry, electricity, and construction in worldwide. It helps to drive the common development of the upstream and downstream of the industrial chain, the joint progress of multiple industries, and provide the driving force for economic development.

At present, many countries and regions around the world have issued development roadmaps for the coming hydrogen industry. For example, in April 2020, the Netherlands officially released a national hydrogen energy policy, whose plan was to build 50 hydrogen refueling stations and put into use 15,000 fuel cell vehicles and 3000 heavy-duty vehicles by 2025; by 2030, 300,000 fuel cell vehicles will be put into use. In June 2020, the German government officially passed the national hydrogen energy strategy, setting an action framework for the future production, transportation, use, and related innovation and investment of clean energy. In June 2020, the French Minister of Transport announced the support for a plan to achieve green hydrogen fuel aircraft by 2035. In July 2020, the European Union released the "EU Hydrogen Strategy" and the "EU Energy System Integration Strategy", hoping to set a new clean energy investment agenda for the EU to achieve carbon neutrality by 2050, while creating job opportunities in related fields, and further stimulate the European Union's economic recovery in the post-pandemic era. The United States, Japan, and South Korea will also continue to support the development of the hydrogen industry based on the existing hydrogen energy development roadmap. In March 2022, the National Development and Reform Commission and the National Energy Administration of China jointly issued the "Medium- and Long-term Plan for the Development of the Hydrogen Energy Industry (2021–2035)", which clearly defined the three strategic positions of hydrogen energy: (i) hydrogen energy is an important part of the future national energy system; (ii) hydrogen energy is an important carrier for the green and low-carbon transformation of energy use terminals; (iii) the hydrogen industry is a strategic emerging one and the key direction for future industrial development. According to the China Hydrogen Alliance's

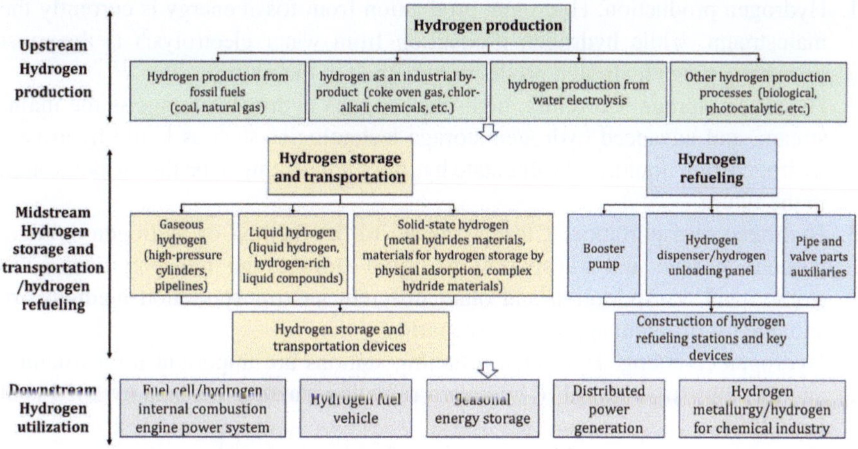

Fig. 1.1 Main links of the hydrogen industry

forecast, the output value of China's hydrogen industry will reach 1 trillion yuan between 2020 and 2025, and the output value will reach 5 trillion yuan between 2026 and 2035.

The hydrogen industry is mainly divided into hydrogen production, storage, transportation, refueling, and utilization, as shown in Fig. 1.1. Hydrogen energy supply system, for the goal of green, efficiency and convenience, China will gradually make break throughs in the production, storage, transportation, and refueling of hydrogen.

Upstream hydrogen production, midstream hydrogen storage and transportation/hydrogen refueling, downstream hydrogen utilization.

Hydrogen production.

Hydrogen production from fossil fuels (coal, natural gas), hydrogen as an industrial by-product (coke oven gas, chlor-alkali chemicals, etc.), hydrogen production from water electrolysis, other hydrogen production processes (biological hydrogen production, photocatalytic hydrogen production, etc.).

Hydrogen storage and transportation.

Gaseous hydrogen storage and transportation (high-pressure gas cylinders, hydrogen transportation by pipelines), liquid hydrogen storage and transportation (liquid hydrogen, hydrogen-rich liquid compounds), solid-state hydrogen storage and transportation (metal hydrides, materials for hydrogen storage by physical adsorption, complex hydrides).

Hydrogen refueling.

Booster pump, hydrogen dispenser/hydrogen unloading panel, pipe and valve auxiliaries.

Construction of hydrogen refueling stations and key devices.

Hydrogen storage and transportation devices.

Fuel cell/hydrogen internal combustion engine power system, hydrogen fuel vehicle, seasonal energy storage, distributed power generation, hydrogen metallurgy/hydrogen for chemical industry.

1. Hydrogen production: Hydrogen production from fossil energy is currently the mainstream, while hydrogen production from water electrolysis is the most promising green hydrogen production method in the future.
2. Hydrogen storage: Currently, high-pressure gas hydrogen storage is the mainstream, and advanced hydrogen storage technologies such as liquid hydrogen, hydrogen-rich liquid, and solid-state hydrides are waiting to be the break through in the future.
3. Hydrogen transportation: Closely related to the method of hydrogen storage, gaseous storage and transportation, liquid storage and transportation, solid storage and transportation and other different gas transportation methods are suitable for different application scenarios.
4. Hydrogen refueling: Hydrogen refueling stations are important infrastructure, and the construction goal of hydrogen refueling stations in China by 2025 is at least 1000.
5. Hydrogen utilization: Hydrogen gas as a fuel is mainly converted into electrical energy or kinetic energy through fuel cells or hydrogen internal combustion engines, and used in fields such as hydrogen fuel vehicles, seasonal energy storage, distributed power generation; or as a raw material, used in hydrogen metallurgy or hydrogen chemical industry and other fields.

In the cross link of the hydrogen industry, the storage and transportation of hydrogen gas is a key issue. Depending on the method of hydrogen storage, the storage and transportation of hydrogen can be divided into high-pressure gaseous hydrogen storage and transportation, liquid hydrogen storage and transportation, and solid-state hydrogen storage and transportation, as shown in Fig. 1.2.

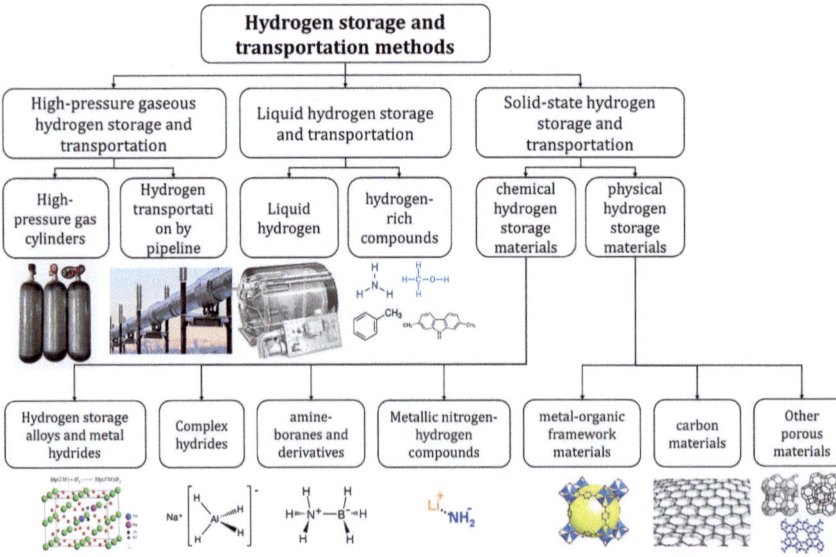

Fig. 1.2 Main methods of hydrogen storage and transportation

Hydrogen storage and transportation methods.

High-pressure gaseous hydrogen storage and transportation, liquid hydrogen storage and transportation, solid-state hydrogen storage and transportation.

High-pressure gas cylinders, hydrogen transportation by pipeline, liquid hydrogen storage and transportation, hydrogen-rich compound storage and transportation, chemical hydrogen storage, physical hydrogen storage.

Hydrogen storage alloys and metal hydrides, complex hydrides, amine-boranes and derivatives, metal-organic framework materials, carbon materials, porous materials.

1. High-pressure gaseous hydrogen storage and transportation refers to the technology of storing and transporting hydrogen in a high-pressure gaseous form. The storage density of hydrogen is directly related to the pressure of hydrogen. High-pressure gaseous hydrogen storage and transportation mainly include high-pressure gas cylinders and hydrogen transportation by pipeline. The former stores hydrogen gas in containers for transportation, while the latter transports it continuously through pipelines.
2. Liquid hydrogen storage and transportation refers to the technology of transporting hydrogen in the form of liquid hydrogen or hydrogen-rich liquid compounds. Liquid hydrogen storage and transportation mainly include liquid hydrogen storage and transportation and hydrogen-rich compound storage and transportation. The former involves cooling hydrogen to −253 °C to liquefy it into liquid hydrogen for storage and transportation, while the latter involves storing hydrogen in hydrogen-rich liquid compounds (such as liquid ammonia, methanol, toluene, dimethylimidazole, etc.) for storage and transportation, and storing and releasing hydrogen through catalytic hydrogenation/dehydrogenation.
3. Solid-state hydrogen storage and transportation refers to the technology where the vast majority of hydrogen is fixed in solid-state hydrogen storage materials for storage and transportation. There are many types of solid hydrogen storage materials, which can generally be divided into hydrogen storage alloys and metal hydrides (such as rare-earth-based, Ti-based, Mg-based, V-based, etc.), complex hydrides (such as $LiAlH_4$, $NaBH_4$, etc.), metal amide hydrides (such as $LiNH_2$, $LiAl(NH_2)_4$, etc.), amine-boranes and derivatives (such as NH_3BH_3, $LiNH_2BH_3$, etc.), metal-organic framework materials (such as ZIF-8, etc.), carbon materials (such as graphene, nanotubes, etc.), and more.

Currently, high-pressure gas cylinder storage and transportation is the most mature one among hydrogen storage and transportation technologies; hydrogen transportation by pipeline and liquid hydrogen have more mature applications in worldwide, but there is a relative lack of engineering applications in China; hydrogen-rich liquid compounds and solid-state hydrogen storage and transportation technologies are in the early stages of industrial development, with only a small number of demonstration applications. According to the "Medium- and Long-term Plan for the Development of Hydrogen Energy Industry (2021-2035)" of China, the future hydrogen storage and transportation method in China is a hydrogen storage and transportation system with characteristics of high-density, lightweight, low-cost and diversity, and various hydrogen storage and transportation methods shall be reasonably selected according to different application scenarios.

1.2 High-Pressure Gaseous Hydrogen Storage and Transportation

High-pressure hydrogen storage refers to storing hydrogen in a high-density gaseous form in a pressure vessel by increasing the pressure, which facilitates the transportation and use of hydrogen. It is currently the most common, direct, and mature method of hydrogen storage. Compared to other methods of hydrogen storage, this method not only has a low cost and fast charging and discharging speed, but it can also achieve hydrogen charging and discharging rate adjustments at room temperature. Moreover, during the storage, transportation, and utilization processes, hydrogen does not undergo phase changes, so the overall energy loss is small, and the number of related equipment required is also limited.

1.2.1 High-Pressure Gas Cylinders

According to the different usage requirements of high-pressure hydrogen storage containers, high-pressure hydrogen storage can be divided into stationary high-pressure hydrogen storage and transportable high-pressure hydrogen storage (lightweight high-pressure hydrogen storage for onboard application and high-pressure hydrogen storage for transport).

1.2.1.1 Stationary High-Pressure Hydrogen Storage

Stationary high-pressure hydrogen storage is mainly used for hydrogen storage in hydrogen refueling stations or green hydrogen production, which is developed to meet the requirements of large-scale and low-cost hydrogen storage. According to statistics from H2stations.org, as of the end of 2020, a total of 553 hydrogen refueling stations have been built worldwide, of which about 430 use high-pressure hydrogen storage technology. By the end of the same year, China has built 118 hydrogen refueling stations (excluding 3 dismantled hydrogen refueling stations), over 85% using the high-pressure hydrogen storage technology. Most of them, such as Anting Hydrogen refueling Station, Zhangjiakou Hydrogen refueling Station, Donghua Energy Hydrogen refueling Station and others use a hydrogen refueling standard of 35 MPa, while a few, such as Changshu Toyota Hydrogen refueling Station, use a hydrogen refueling standard of 70 MPa. Most of the hydrogen refueling stations in other countries and regions (such as the Hydrogen refueling Station Hamburg HafenCity, Germany, and the Honda Hydrogen refueling Station in California, USA) also use a hydrogen refueling standard of 70 MPa.

Hydrogen storage is the most important part of a hydrogen refueling station, and there are generally two types of storage: one is to use a large-volume pressure vessel for storage; the other is to use a small-volume pressure vessel group for storage. The high-pressure hydrogen storage containers used in the hydrogen storage system of the hydrogen refueling station are mainly divided into seamless high-pressure hydrogen storage containers and steel strip staggered high-pressure hydrogen storage containers. The former is mainly a 45 MPa large-volume steel seamless hydrogen storage container, which is designed and manufactured in accordance with the American Society of Mechanical Engineers (ASME) standards and the requirements of TSG 21—2016 "Supervision Regulation on Safety Technology for Stationary Pressure Vessel". The main material is 4130X high-strength structural steel [1], and the nominal volume of a single container is 0.895 m^3. The latter is generally a large-volume multi-layer steel high-pressure hydrogen storage container, such as the 1–20 m^3 50 and 98 MPa pressure vessels designed by Zhejiang University, the specific parameters are shown in Table 1.2. Since the hydrogen refueling station mainly uses the pressure difference between the hydrogen storage container and the on board vehicle hydrogen supply system for hydrogen refueling, its highest design pressure level usually matches the pressure level of the on board hydrogen storage. In addition to using long-tube trailers as 20 MPa mobile hydrogen storage facilities, the highest stationary hydrogen storage design pressure in the 35 MPa hydrogen refueling station is generally 45, 47 and 50 MPa, while the highest stationary hydrogen storage design pressure in the 70 MPa hydrogen refueling station is generally 82, 87.5, 98 and 103 MPa [2]. Under normal circumstances, the hydrogen storage containers used in the hydrogen refueling station use low pressure (20–30 MPa), medium pressure (30–40 MPa), and high pressure (40–75 MPa) three-stage pressure for hydrogen storage, sometimes hydrogen gas tube trailers also serve as a stage of gas storage (10–20 MPa) facilities, forming a 4-stage gas storage method [3].

1.2.1.2 Transportable High-Pressure Hydrogen Storage Vessel

In transportable high-pressure hydrogen storage, high-pressure hydrogen vessels for transportation are mainly used to transport hydrogen from the production site to the usage site or hydrogen refueling station, while lightweight high-pressure hydrogen storage for vehicles is developed to meet the mobile hydrogen supply requirements of hydrogen energy vehicles. In the early stage, tube trailers were often used for transporting high-pressure hydrogen, which consist of several high-pressure containers formed by spinning processing. The hydrogen storage pressure is

Table 1.2 Parameters of large-volume multi-layer steel high-pressure hydrogen storage containers

Specification	1	2	3	4	5	6	7
Design pressure/MPa	98	50	50	50	50	50	50
Volume/m^3	1.0	5.0	7.3	10.0	13.0	15.0	20.0
Inner diameter (mm)	500	1200	1500	1500	1500	1500	1500
Total length/m	5.9	5.5	5.3	6.8	8.5	9.6	12.2
Effective hydrogen storage mass/kg	50	114	210	288	375	432	576

16–21 MPa, and the total hydrogen transport capacity of the whole vehicle generally does not exceed 380 kg. To further increase the single-vehicle hydrogen transport capacity and reduce transportation costs, some research institutions have begun to use winding technology to develop high-pressure hydrogen storage containers for transportation and have successfully developed related products. For example, in 2008, Spencer Composites Corporation successfully developed a low-cost large-capacity high-pressure hydrogen storage container with a full winding structure of glass fiber. Hexagon LinCOln Composites in Norway also successfully developed a large-volume high-pressure hydrogen storage container with a carbon fiber winding structure, with a working pressure of 25 MPa, a single effective volume of 8.5 m^3, and a hydrogen storage capacity of about 150 kg. Subsequently, the company developed a fully wrapped carbon fiber reinforced composite high-pressure hydrogen storage container with a nominal pressure of 25–54 MPa, and applied it to tube trailers, with a single-track hydrogen transportation capacity of 560–720 kg. In China, the 20 MPa large-volume seamless pressure vessel produced by Shijiazhuang Enric Gas Equipment Co., Ltd. is widely used.

To ensure the driving range of hydrogen fuel cell vehicles, lightweight high-pressure hydrogen storage tank for vehicles needs to store as much hydrogen as possible in the limited space of the vehicle, which puts higher requirements on the high-pressure hydrogen storage container for vehicles. In 2003, the U.S. Department of Energy (DOE) proposed that the gravimetric and volumetric hydrogen storage density requirements for lightweight high-pressure hydrogen storage for vehicles are 6 wt% H_2 and 60 kg H_2/m^3, respectively, but considering the technical foundation, cost and other factors of existing demonstration results, DOE subsequently revised this goal. At present, DOE's technical target for on-board hydrogen storage of lightweight fuel cell vehicles is: by 2020, the mass hydrogen storage density (system) will reach 4.5 wt%, and the volume hydrogen storage density (system) will reach 30 g/L; by 2025, the mass density (system) will reach 5.5 wt%, and the volume density (system) will reach 40 g/L; the final mass density (system) should reach 6.5 wt%, and the volume density (system) should reach 50 g/L. To achieve this goal, the on-board high-pressure hydrogen storage container must not only be light, but also have high hydrogen storage pressure. In recent years, many companies have developed high-pressure hydrogen storage containers of various specifications and models, and their design and manufacturing technology of high-pressure hydrogen storage containers are at the world's leading level. The Japan Automobile Research Institute has developed composite high-pressure hydrogen storage tanks with hydrogen storage pressures of 35 and 70 MPa, but the hydrogen storage capacity of the 70 MPa tank is only 60% higher than that of the 35 MPa tank, and its ultimate hydrogen storage capacity and sealing performance need to be further improved and optimized. The number of 70 MPa type IV hydrogen storage tanks has been increased from 2 on the first generation to 3 on the latest Toyota hydrogen fuel cell vehicle "Mirai 2" (as shown in Fig. 1.3), the hydrogen storage density has increased from 5.7 to 6.0 wt%, the hydrogen storage volume has increased from 122.4 to 142.2 L, and the hydrogen storage amount has increased to 5.6 kg. Hyundai's Nexo fuel cell vehicle in South Korea also carries three 70 MPa Hydrogen

Fig. 1.3 " Mirai 2" 70 MPa Type IV hydrogen storage tank layout

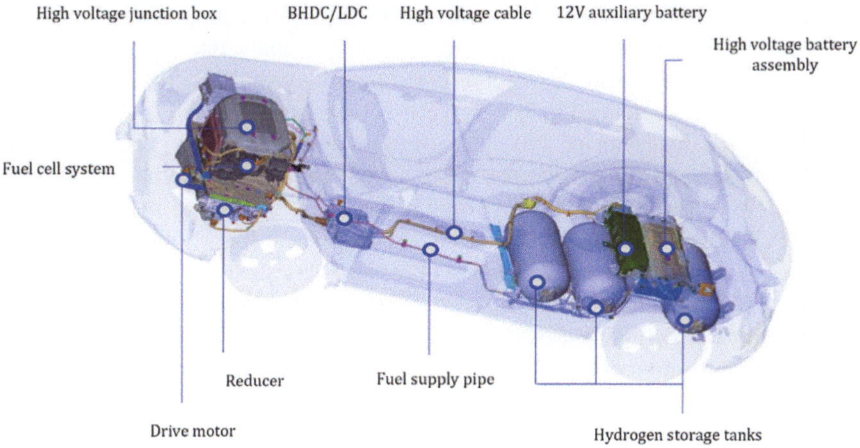

Fig. 1.4 Hyundai Nexo 70 MPa hydrogen storage tank layout

storage tanks (as shown in Fig. 1.4), with a total hydrogen storage volume of 156.6 L, a total hydrogen storage of 6.33 kg, and a hydrogen storage mass density of 5.7 wt%.

⊖ wt% is a commonly used unit in the industry, representing mass fraction.

In China, the research and design of on-board hydrogen storage tanks began during the "Tenth Five-Year Plan" period, which started later than abroad. The Chemical Machinery Research Institute of Zhejiang University was the first in China to successfully trial-produce high-pressure hydrogen storage tanks with a working pressure of 40 MPa and a volume of 0.1–100 L. Tongji University has developed 35 and 70 MPa aluminum alloy liner composite material based hydrogen storage tanks, which have been used in the Roewe 950 fuel cell sedan developed by SAIC. In addition, the 54 and 65 L aluminum liner hydrogen storage tanks developed by Beijing Tianhai Industrial Co., Ltd. and Beijing Chinatank Industrial Co., Ltd. have reached a hydrogen storage pressure of 70 MPa. At present, Faurecia CLD is the only chinese company that has obtained the production license of Fully-Wrapped Carbon Fiber Reinforced Cylinder with a Plastic Liner for on board storage of compressed hydrogen, and the rest are in the exploration stage.

Hydrogen storage tank.

High voltage junction box, BHDC/LDC, high voltage cable, 12 V auxiliary battery, high voltage battery assembly, fuel cell system, reducer, drive motor, fuel supply pipe, hydrogen storage tank.

Ideal gas.

In the practical application of high-pressure hydrogen storage technology, it is worth noting that hydrogen can be considered as an ideal gas at high temperatures and low pressures. Its mass at different temperatures and pressures can be calculated using the ideal gas state equation PV = nRT. However, due to the volume of actual gas molecules and the interaction between molecules, as the temperature decreases and the pressure increases, hydrogen will deviate from the properties of an ideal gas, and the van der Waals equation will be no longer applicable. The deviation of actual gas from ideal gas can be represented thermodynamically by the gas compressibility factor Z, defined as Z = PV/nRT. As can be seen from Fig. 1.5, at 0 °C, the compressibility factor of hydrogen increases with increasing pressure, which means that as the pressure increases, hydrogen becomes more difficult to compress [4]. Therefore, higher hydrogen storage pressure is required to achieve higher hydrogen storage, meet the DOE's technical indicators or higher mass and volume hydrogen storage density targets, which also places higher demands on high pressure vessels. Under the premise of safety, it is necessary to improve the materials and structure of the hydrogen storage container to further increase the hydrogen storage pressure and increase the hydrogen density, while reducing the cost of the hydrogen storage container to meet commercial applications.

High-pressure gaseous transport is the process of pressurizing hydrogen and storing it in large-volume pressure vessels, which are then loaded onto vehicles for transport. The density of gaseous hydrogen at room temperature is extremely low,

Fig. 1.5 Z-P curve of several gases at 50 °C

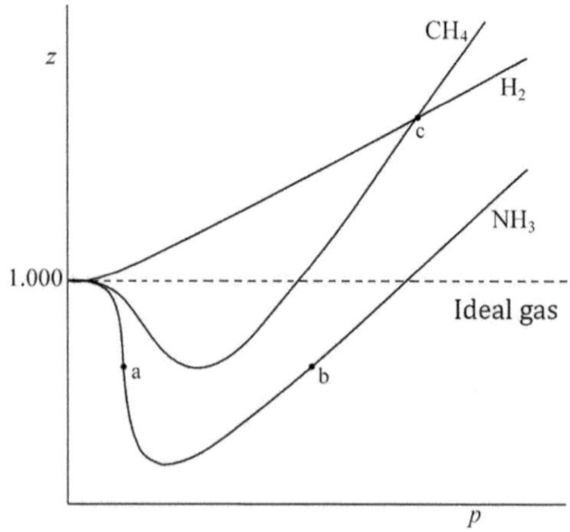

and the mass of hydrogen that can be loaded into high-pressure vessels usually only accounts for 1–2% of the total mass of the transport equipment. Considering the mass loss caused by hydrogen permeation, the longer the time and distance of transport is, the worse the economy is. Therefore, high-pressure vessel transport of hydrogen is only suitable for land transport scenarios with short-distance and small-demand.

1. Transport methods. High-pressure gaseous hydrogen is mainly transported by container and tube trailers. A container is a metal structure frame in which multiple small-volume industrial cylinders are assembled in an upright or lying position. A single cylinder has a volume of about 40 L and a pressure of 15–20 MPa. There are usually 12–28 cylinders of different specifications. Although this method has low space utilization and low mass transport efficiency, and is only suitable for the transport of less than 50 kg of hydrogen, it has the advantage of simpleness and flexibility, and can use trucks and other conventional vehicles to transport multiple containers at once.

Tube trailers transport several cylindrical seamless high-pressure cylinders. There are two types of tube fixation, that is bundled tube trailer and containerized tube trailer. The cylinders of the bundled tube trailer are fixed on the semi-trailer, with both ends fixed by supporting plates. Because it eliminates the weight of the fixed frame, it can load more containers and has higher transport efficiency, but it has higher requirements for the road [5], as shown in Fig. 1.6. The containerized tube trailer connects multiple seamless pressure containers at the head, assembled in a standard container frame, equipped with corresponding control

(a) (b)

a) Cylinder container

b) Containerized tube trailer

Fig. 1.6 Gaseous hydrogen transport methods. (**a**) Cylinder container. (**b**) Containerized tube trailer

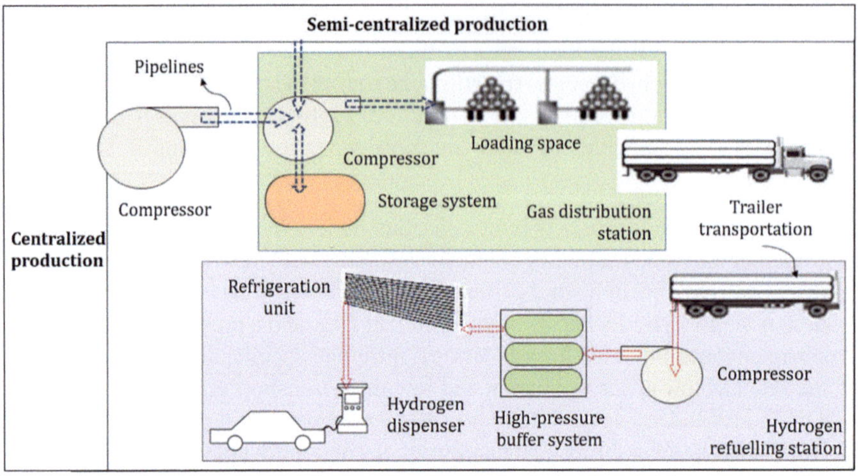

Fig. 1.7 Route of gaseous hydrogen transported by tube trailers [6]

valves, measuring instruments, connecting pipes and safety devices, which are convenient for overall loading and unloading. The unloaded container bundle can be used directly as the gas source for the hydrogen refueling station. The containerized tube trailer has now become the main tool for short-distance hydrogen transport.

2. Transportation route. The route of gaseous hydrogen transported by tube trailers is shown in Fig. 1.7. Hydrogen produced by centralized factories can be transported to tube trailer loading terminals through pipelines. In semi-centralized production processes, the production equipment and gas distribution stations or city gate stations are located at the same place, and the produced hydrogen can be directly injected into tube trailers using the loading space of the gas distribution station, while the surplus hydrogen is stored in a storage system to cope with seasonal demand changes. Tube trailers transport compressed hydrogen to hydrogen refueling stations, where compressors extract hydrogen from the tube trailers and supply it to the high-pressure buffer storage system of the hydrogen refueling station for vehicle refueling.
3. Development and application status. At present, most of the tube trailers in China use large-volume seamless steel pressure vessels made of 4130X (CrMo steel) with a working pressure of 20 MPa, transporting 300–400 kg of hydrogen for each trailer, and the economical distance for hydrogen transportation is within 150 km [7] with a cost of about 2 yuan/kg H_2 [7]. A common tube bundle consists of 9 steel containers with a working pressure of 20 MPa, a diameter of about 0.5 m, and a length of about 10 m, which can be filled with 3500 Nm^3 of hydrogen [8]. One of the reasons why steel containers are widely used is their low manufacturing cost, but steel has a high density and is susceptible to hydrogen embrittlement. If the working pressure is to be further increased, the wall thickness and weight of the container must also increase accordingly, which is limited

by the road carrying capacity and it is difficult to further increase the one-way transportation capacity. At present, the standard regulations guiding high-pressure gas road transportation include GB/T 34542 "Storage and Transportation Systems for Gaseous Hydrogen" series of standards, TSG R0005-2011 "Supervision Regulation on Safety Technology for Transportable Pressure Vessel", GB/T 33145-2016 "Large Capacity Seamless Steel Gas Cylinders", NB/T 10354-2019 "Tube Trailers", etc. There are no corresponding standards for large-volume carbon fiber reinforced pressure vessels used for transportation.

Centralized production, compressor, transportation pipeline, semi-centralized production, compressor, loading space, storage system, gas distribution station, tube trailer transportation to hydrogen refueling station, refrigeration unit, hydrogen dispenser, high-pressure buffer storage system, compressor.

Compared with steel containers, fiber-wound pressure vessels have lower density, thinner wall thickness, and higher strength, which are effective technical means to improve hydrogen transportation efficiency. Zhejiang Rein Gas Equipment Co., Ltd. has developed a 20 MPa large-volume Type II (steel liner, carbon fiber wound) tube bundle box with 9 tubes and put it on the market, with a filling mass of 549 kg. Compared with the original 7-tube Type I tube bundle, the filling mass has increased by 42.6% while the whole vehicle mass has decreased by 14%. Internationally, fiber-wound pressure vessels are widely used, including Type III (aluminum alloy liner) and Type IV (plastic liner) containers, with a maximum working pressure of up to 50 MPa, and a single vehicle can transport 700–1000 kg of hydrogen [9]. The corresponding international guiding regulations include the US Department of Transportation DOT standards, ISO 11120:2015 "Gas Cylinders—Refillable Seamless Steel Tubes of Water Capacity between 150 and 3000 L Design, Construction and Testing", ISO 11515:2013 "Gas Cylinders—Refillable Composite Reinforced Tubes of Water Capacity between 450 and 3000 L Design Construction and Testing", BN EN ISO 11114-4:2017 "Transportable Gas Cylinders—Compatibility of Cylinder and Valve Materials with Gas Contents", etc.

4. Challenges. Current tube trailer transportation of high-pressure hydrogen still faces certain technical difficulties:

    (a) Like the high-pressure containers in the hydrogen refueling station, the long tube containers used by the trailer also need to consider the impact of pressure cycling and filling temperature rise on their strength. Due to the vibration of the vehicle body during transportation, the impact of dynamic loads also needs to be considered, and corresponding vibration reduction measures should be taken.
    (b) The minimum storage pressure and the flow rate of the compressor determine the time required for refueling. A detailed analysis of the gas distribution terminal operation is needed to estimate the required compression and storage pressure as well as the optimal operating pressure.

(c) In order to refuel tube trailers in the shortest possible time, it is necessary to increase the compression capacity of the compressor. However, there is currently a lack of hydrogen compressors on the market that have a large enough flow rate and can operate reliably.

1.2.2 Hydrogen Transportation by Pipelines

High-pressure hydrogen transportation by pipelines is carried out through an underground or overhead seamless steel pipe system. Its high transportation efficiency and low energy consumption make it suitable for long-distance hydrogen transportation. High-pressure pipelines are considered the lowest cost option for supplying hydrogen to large refueling stations (daily demand >1000 kg) in cities (daily demand >150 t) [6]. Currently, the common material for high-pressure hydrogen pipelines is steel, with a hydrogen pressure of 1.0–4.0 MPa, a diameter of 250–500 mm, and the transportation cost per 100 km is only 0.3 yuan/kg H_2 [7]. However, the one-time cost for pipeline construction is enormous.

1. Transportation method. The gas transportation of high-pressure pipelines utilizes the pressure difference at the pipeline inlet and outlet. Due to the viscosity of the gas itself and the friction of the pipeline, pressure drop occurs along the way. To achieve efficient gas transportation over a long distance, while considering the pressure-bearing capacity of the pipeline, it is necessary to limit the minimum and maximum pressure of the gas transported by the pipeline. Usually, a compression station is set up every 80–100 km to recompress the hydrogen to ensure efficient gas transportation. When pipelines of different working pressures need to be connected, a pressure reduction station is set up, and the pressure is reduced through a throttle valve to achieve parallel connection [10].
2. Transportation route. The transportation route is shown in Fig. 1.8. The hydrogen produced centrally is pressurized and sent to the gas distribution station through a long-distance pipeline. The gas distribution network then transports the hydrogen to the refueling station and various user terminals. Hydrogen produced by semi-central plants is directly transported through the gas distribution network. Long-distance pipelines have long transportation distances, high hydrogen pressure, and large diameters, while gas distribution pipelines have short transportation distances, low hydrogen pressure, and small diameters. Surplus hydrogen from centralized production can be stored in underground salt layers, water-bearing layers, oil and gas fields, and other geological structures to cope with seasonal changes in hydrogen demand.

Centralized production, compressor, geological storage, transportation pipeline, gas distribution pipeline, refrigeration unit, hydrogen dispenser, hydrogen refueling station, high-pressure buffer storage system, low-pressure storage system, semi-centralized production.

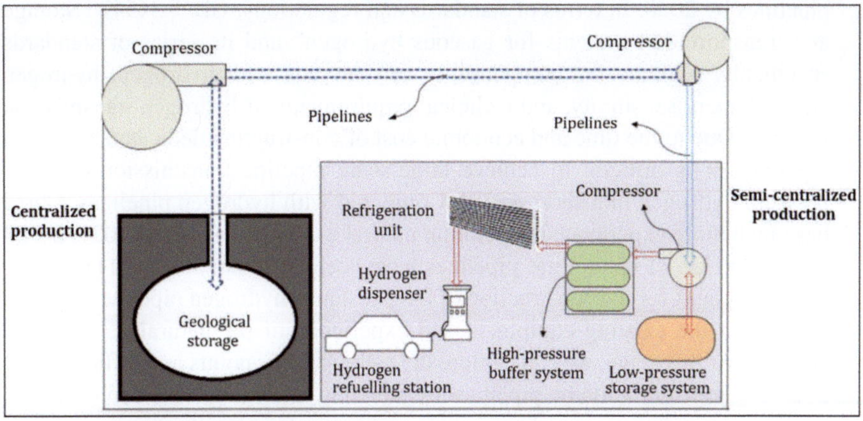

Fig. 1.8 Route of gaseous hydrogen transportation by high-pressure pipelines [6]

3. Current development. As of December 2020, the mileage of hydrogen pipelines in operation in the United States is about 2500 km, 90% of which are located along the Gulf of Mexico in Texas, Louisiana, and Alabama, mainly serving the refineries and synthetic ammonia plants in the region. In Europe, there are about 1600 km of hydrogen pipelines, the longest of which runs through France, Belgium, and the Netherlands, with a total length of 1100 km. The "EU Hydrogen Strategy" and "EU Energy System Integration Strategy" released in 2020 plan to build a pure hydrogen network of 6800 km by 2030 and 23,000 km by 2040, 75% of which will be converted from existing natural gas networks, and 25% will be newly built hydrogen-specific networks. Germany is expected to complete a 130 km green hydrogen pipeline by 2022 and put it into use, with all the hydrogen produced from renewable energy. In line with the rich experience in the construction of long-distance hydrogen pipelines, the design specifications for hydrogen pipelines in Europe and the United States are also relatively complete, such as ASME B31.12:2014 "Hydrogen Piping and Pipelines" of the American Society of Mechanical Engineers, IGC DOC 121/04/E "Hydrogen Transportation Pipelines" standards of the European Industrial Gases Association and others.

The hydrogen transmission pipeline in China is still in the early stages of planning and construction. The total length of hydrogen transportation pipelines is only about 100 km in China, which includes the Baling-Changling hydrogen pipeline (42 km), Jinling-Yangzi hydrogen pipeline (32 km), and Jiyuan-Luoyang hydrogen pipeline (25 km), all of which are used to provide raw hydrogen for the chemical industry. In July 2021, China Petroleum Pipeline Engineering Corporation won the bid for the feasibility study of the Hebei Dingzhou-Gaobeidian long-distance hydrogen pipeline. Its length of 145 km makes it the longest hydrogen pipeline currently planned for construction in China. The "2016: Blue Book on China Hydrogen Energy Industry Infrastructure Development" plans to build more than 3000 km of long-distance hydrogen pipelines by 2030. In terms of standards and regulations, GB/T 34542 "Storage and transportation systems for gaseous hydrogen" and its series of standards specifically regulate the compatibility of materials with hydrogen, hydrogen embrittlement sensitivity, and technical requirements of hydrogen transmission systems. Due to the time and economic cost of constructing dedicated hydrogen pipelines, it is difficult to achieve large-scale pipeline transmission of high-pressure hydrogen in a short period. Compared with hydrogen pipelines, China has much more experience in operating natural gas pipelines. As of 2021, a total of 110,000 km of natural gas pipelines have been built nationwide [11], so it is of great benefit for the construction of long-distance hydrogen pipelines to make full use of the existing equipment and experience of the natural gas pipeline network. At this stage, we can explore the technical key points and difficulties of hydrogen transportation by pipelines by blending a certain proportion of hydrogen into the existing natural gas pipelines or transforming the existing natural gas pipelines into hydrogen pipelines.

4. Challenges. At present, the promotion and application of high-pressure hydrogen transmission pipeline still face certain technical problems, including:

   (a) High pressure, pressure cycling, and hydrogen embrittlement can affect the durability of steel pipeline. Further detailed research on the permeation mechanism of hydrogen in structural materials, appropriate materials, proper forming and heat treatment processes should be considered to extend the service life of the pipeline and reduce maintenance costs.
   (b) Leaks at any point along the pipeline can cause fires and explosions, leading to failures of the entire hydrogen transmission line. Long-distance hydrogen pipeline leak detection relies on low-cost, efficient, and high-precision long-distance sensing technology. Although mature leak detection technologies are already applied to natural gas pipelines, it is still necessary to study their applicability to high-pressure hydrogen pipelines.
   (c) Hydrogen compressors require high reliability and high cost for components, and their lubricants may contaminate the hydrogen. Solutions to this problem include improving the design of specific components such as diaphragms and seals, developing lubricant-free new compression technologies with high power, and researching low-cost hydrogen purification processes, etc.

1.3 Low-Temperature Liquid Hydrogen Storage and Transportation

Low-temperature liquid hydrogen storage and transportation technology refers to the technology for cooling hydrogen gas from normal temperature to −253 °C liquefaction, and storing and transporting liquid hydrogen. The theoretical energy consumption of hydrogen liquefaction is between 4 and 5 kW h/kg, while the actual energy consumption in low-temperature engineering for liquefying hydrogen gas is between 6.5 and 20 kW h/kg, which is related to the scale of hydrogen liquefaction: when the scale of hydrogen liquefaction is 2 t/day or less, the comprehensive energy consumption exceeds 20 kW h/kg, while when the scale of hydrogen liquefaction is 150 t/day, it can be reduced to 6.7 kW h/kg or even lower. The comprehensive energy consumption of a 10–30 t/day hydrogen liquefaction project is between 10 and 14 kW h/kg. This is also an important reason for continuously increasing the scale of hydrogen liquefaction projects, as only large-scale applications of liquid hydrogen are economical. Generally speaking, the economic balance point of a hydrogen liquefaction project is 8–10 t/day, which is related to the available electricity price and hydrogen source.

Liquid hydrogen is mainly transported over long distances by land or sea. The single transport capacity of a liquid hydrogen tank truck is 2.5–3.3 t, which is 6–8 times the single transport capacity of a 20 MPa tube trailer, and the weight of the transport vehicle is reduced by about 30%. Therefore, the economic transport distance of liquid hydrogen can reach more than 1000 km, as shown in Fig. 1.9a. Sea transportation of liquid hydrogen is another important method of liquid hydrogen storage and transport. LH2 Europe and C-Job NavAl Architects have jointly designed and developed a 141 m liquid hydrogen transport ship. This ship is powered by hydrogen fuel cells and will be equipped with three liquid hydrogen tanks with a total capacity of 37,500 m^3 to refuel 400,000 medium-sized hydrogen fuel cell vehicles or 20,000 heavy-duty hydrogen-powered trucks (Fig. 1.9b).

After liquid hydrogen is transported to the hydrogen refueling station, if it is used to refuel fuel cell vehicles with 35 MPa/70 MPa high-pressure hydrogen storage, a liquid hydrogen pump can be used to increase the pressure and then a vaporizer to reheat (using the environment as a heat source, no additional energy consumption is required). Not only is the liquid hydrogen pump that compresses liquid more energy-efficient than the compressor that compresses gas, but the vaporizer can

(a) (b)

Fig. 1.9 Low-temperature liquid hydrogen storage and transport. (**a**) Liquid hydrogen storage tank and its tank truck. (**b**) Liquid hydrogen transport ship choose to reheat to −40 °C, without considering the impact of high-pressure refueling expansion temperature rise. Therefore, even the comprehensive energy consumption of liquid hydrogen refueling for 70 MPa high-pressure vehicles does not exceed 2 kW h/kg, and the energy consumption advantage is obvious. The high-power long-range hydrogen energy heavy trucks requires that on-board hydrogen storage amount reaches 50 kg or more. If high-pressure hydrogen cylinders are used and even they are 70 MPa hydrogen cylinders, it is still necessary to use 5–8 250 L large-volume hydrogen storage cylinders. However, liquid hydrogen cylinders only require one 0.8–1.3 m^3 liquid hydrogen fuel tank. The volume density and weight density of its hydrogen storage system are much higher than those of high-pressure hydrogen storage systems. The power system fuel cell power of heavy trucks exceeds 150 kW, and the thermal management system adopts water cooling. In this system, on board liquid hydrogen fuel also needs to be heated and reheated to room temperature through heat exchange equipment, so the cold energy recovery and utilization of liquid hydrogen can be used to cool the high-power fuel cell, making the design of the heavy truck fuel cell more compact, efficient, and long-lived. At the same time, the superior quality of liquid hydrogen is unmatched by high-pressure hydrogen. All impurity gases except helium will solidify and separate when they encounter liquid hydrogen, so ultra-pure hydrogen can be obtained directly after liquid hydrogen vaporization. This quality can be maintained from upstream liquefaction to the terminal entering the fuel cell stack, which is of great significance for the fuel cell vehicle power system with long life and high performance.

Overall, the application of liquid hydrogen is suitable for large-scale scenarios. Its economic benefits are reflected in the reduction of comprehensive energy consumption, the economy of storage, transportation, and refueling, as well as high-density hydrogen storage and high-quality utilization at the terminal. On board vehicle liquid hydrogen storage is more suitable for heavy trucks, ships, trains, airplanes, etc., and does not have an advantage over high-pressure hydrogen storage in passenger cars and small and medium-sized commercial vehicles.

1.4 Hydrogen Storage and Transportation by Hydrogen-Rich Liquid Compounds

In addition to high-pressure gaseous hydrogen storage and low-temperature liquid hydrogen storage, liquid hydrogen-rich compound storage technology based on chemical hydrogen absorption and release reactions has attracted the attention of the hydrogen industry due to its large hydrogen storage capacity, high energy density, and safe and convenient liquid storage and transportation. It can be used not only for long-cycle seasonal hydrogen storage but also for long-distance hydrogen transmission to solve the problem of uneven energy distribution between regions. The bottleneck of this technology is how to develop a high conversion rate, high selectivity, and stable hydrogen absorption/desorption catalysts. Cyclic utilization and low-cost hydrogen storage and transportation can be achieved by preparing high-performance catalysts and reaction modes with reasonable structural design. This book will mainly introduce the widely studied hydrogen storage technologies, that is, liquid organic hydrogen carriers, liquid ammonia, and methanol.

1.4.1 Liquid Organic Hydrogen Carriers (LOHCs)

Liquid Organic Hydrogen Carriers (LOHC) use the hydrogenation and dehydrogenation reactions between unsaturated liquid aromatic compounds and corresponding saturated organics to achieve hydrogen storage. Common LOHC substances include benzene, toluene, naphthalene, carbazole, N-ethylcarbazole (NEC), dibenzyltoluene (DBT), etc. LOHC hydrogen storage technology has a high hydrogen storage density (5–10 wt%) and can store hydrogen under certain reaction conditions, so it has high safety and is convenient for transportation. Its main disadvantage is the more difficult absorption and release of hydrogen than physical hydrogen storage, side reactions, low stability of cyclic use, required additional reaction equipment and hydrogen separation purification devices, heating requirement of the hydrogen release process and increased cost caused by high energy consumption.

1.4.2 Liquid Ammonia

Liquid ammonia (NH_3), as an important source of nitrogen and fertilizer in agricultural production. It is rich in hydrogen and does not contain carbon and has an extremely high mass hydrogen storage density (17.6 wt%) and volume hydrogen storage density (108 g/L), and its transportation cost is low. Storage and transportation infrastructure and operating norms already exist. Compared with low-temperature liquid hydrogen storage technology, the liquefaction temperature of ammonia under one atmospheric pressure (−33 °C) is much higher, and the liquefaction temperature under 9 atm is 25 °C, Moreover, the "hydrogen-ammonia-hydrogen" storage method has relatively lower energy consumption, implementation difficulty, and transportation difficulty. The "hydrogen economy" that uses ammonia as a chemical hydrogen storage medium is named the "ammonia economy", which is currently attracting more attention. Its main process includes: hydrogen and nitrogen synthesize ammonia under the action of a catalyst. Ammonia is stored and transported in the form of liquid. The decomposition of liquid ammonia under catalysis is to release hydrogen for applications. To realize the entire "ammonia economy" recycling, it is necessary to solve the problems of clean energy-based ammonia decomposition for hydrogen production, hydrogen purification and efficient green ammonia synthesis technology, to avoid the energy consumption problems of electrolytic water hydrogen storage and hydrogen conversion from synthetic ammonia. In addition, the equipment for synthesizing ammonia hydrogen storage and ammonia decomposition for hydrogen production still need to be integrated with the industrial equipment for the terminals.

1.4.3 Methanol

Methanol (CH_3OH) is an important raw material in chemical engineering and biomass energy carrier. As a liquid-phase chemical hydrogen storage material, it has a high hydrogen storage mass density (12.5 wt%) and volume density (99 g/L). Because it can further obtain additional hydrogen from water after the catalytic reforming reaction with water, the hydrogen storage density per unit mass of methanol can be as high as 18.75 wt%. Methanol is a liquid at normal temperature and pressure and easy to store and transport. With abundant and diversified sources, it can come from the traditional chemical industry or be obtained through renewable energy preparation. Another important advantage of methanol hydrogen storage technology is that it does not need to build additional hydrogen refueling stations (high-pressure gas hydrogen or liquid/solid hydrogen storage), but it can simply transform and upgrade from existing gas stations, into joint refueling stations that can refuel both gasoline and diesel and methanol/water solution. The main bottleneck of the hydrogen storage technology based on methanol reforming is the low preparation efficiency, low selectivity, high preparation cost, high energy consumption of green methanol and the low efficiency of the methanol reforming hydrogen production reactor, expensive high-purity hydrogen separation equipment, high operating costs, and limited life.

1.5 Solid-State Hydrogen Storage and Transportation

Solid-state hydrogen storage uses hydrogen storage materials to solidify hydrogen in the material. Compared with gaseous hydrogen storage and liquid hydrogen storage, there is only a very small amount of elemental hydrogen in the traditional sense in the storage and transportation container, so it has the advantages of high safety and low maintenance cost in hydrogen storage and transportation. The common solid-state hydrogen storage materials mainly include metal hydride materials, light metal complex hydrides, physical adsorption materials, etc. Solid-state hydrogen storage material systems can be selected according to different application scenarios, such as the hydrogen charging and discharging pressure of the hydrogen storage material range from 0.01 to 20 MPa, and the hydrogen charging and discharging temperature range from −196 to 600 °C.

From the perspective of hydrogen storage mechanisms, solid-state hydrogen storage materials can be divided into physical hydrogen storage and chemical hydrogen storage. Physical hydrogen storage is a method of adsorbing and desorbing hydrogen gas through the interaction of van der Waals forces between the material and hydrogen molecules, including carbon materials, zeolites, metal organic framework materials, molecular sieves, etc. The chemical hydrogen storage mechanism refers to the storage of hydrogen atoms in the bulk phase through chemical bonds by hydrogen storage materials. Based on different chemical mechanisms, solid-state hydrogen storage materials can be mainly divided into metals (alloys) and their hydrides, complex hydrides, and amine-boranes and their derivatives. Depending on the method of hydrogen release, solid-state hydrogen storage materials can also be divided into pyrolysis hydrogen release materials and hydrolysis hydrogen release materials. Among them, thermal decomposition to release hydrogen gas is a common method of hydrogen release for solid-state hydrogen storage materials, and a small part of hydrogen storage materials can also release hydrogen gas through hydrolysis, mainly including sodium borohydride, amine-borane, and magnesium hydride, etc. Common solid-state hydrogen storage materials are shown in Fig. 1.10.

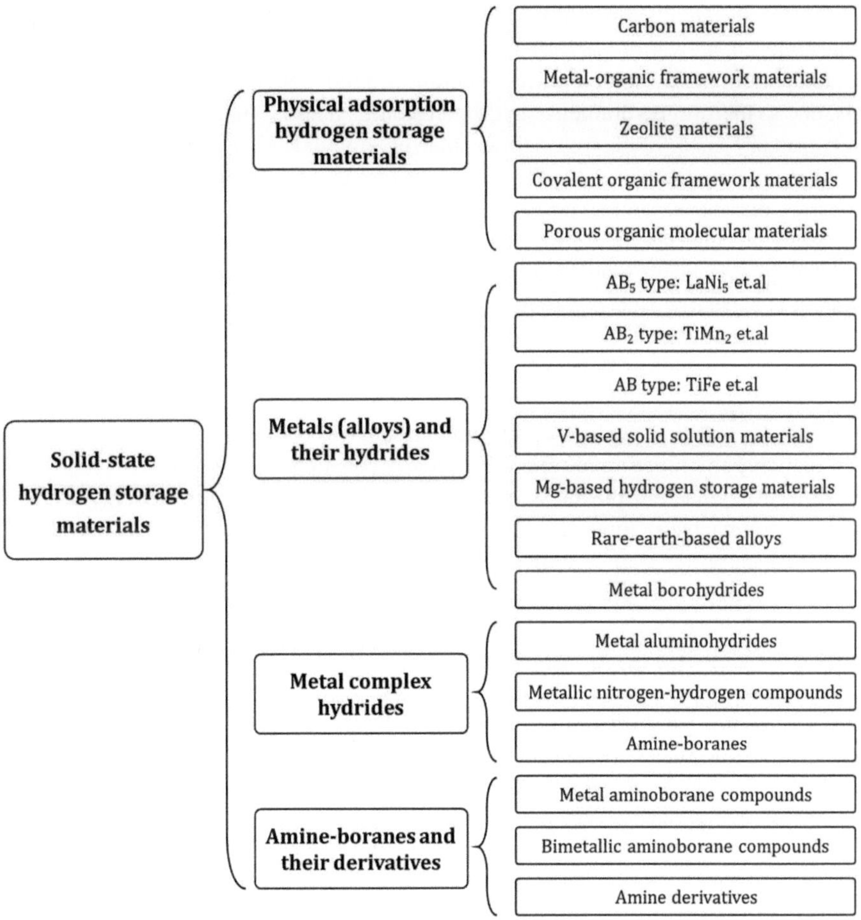

Fig. 1.10 Common solid-state hydrogen storage materials

Solid-state hydrogen storage materials.

Physical adsorption hydrogen storage materials, metals (alloys) and their hydrides, metal complex hydrides, amine-boranes and their derivatives.

Carbon materials, metal-organic framework materials, zeolite materials, covalent organic framework materials, porous organic molecular materials, V-based solid solution materials, magnesium-based hydrogen storage materials, rare-earth-based hydrogen storage alloys, metal borohydrides, metal aluminohydrides, amine-boranes, metal aminoborane compounds, bimetallic aminoborane compounds, amine derivatives.

The use scenarios of solid-state hydrogen storage materials include distributed power generation, hydrogen fuel cell vehicles, standby/emergency power supplies, chemical production, and hydrogen refueling stations, etc. Based on different application scenarios, as well as the use requirements of hydrogen fuel cells, hydrogen metallurgy, hydrogen chemical industry, etc., solid-state hydrogen storage materials should pay attention to aspects such as hydrogen storage density, hydrogen absorption/desorption temperature/rate, cycle life, hydrogen absorption/desorption energy consumption, activation performance, and cost.

In the field of distributed power generation, solid-state hydrogen storage materials, due to their high volumetric hydrogen storage density and high safety, can convert primary energy sources such as solar energy and wind power into hydrogen storage, making them suitable for renewable energy consumption, grid power peak shaving, disaster response, home/building hydrogen utilization systems, and other application scenarios. In 2010, McPhy in France developed the McStore hydrogen storage system using magnesium hydride as the hydrogen storage medium. This system is currently used as a hydrogen storage medium in the INGRID demonstration project in Italy to achieve power regulation. The China Grinm Group Corporation Limited has developed a TiFe hydrogen storage system with a hydrogen storage capacity of 1000 m^3, which will be used in the Hebei Guyuan wind power hydrogen production project in the future, serving as a safe and compact onsite hydrogen buffer, realizing the function of renewable energy consumption.

Emergency/backup power sources have characteristics of high energy storage density, compact structure, easy portability and others. Solid-state material hydrolysis hydrogen production systems can provide excellent hydrogen sources for emergency/backup power sources. The energy density of hydrolysis hydrogen fuel cells can reach 2–3 times that of traditional lithium-ion batteries, thus providing a portable fuel cell power source with the same amount of electricity but only 1/3 to 1/2 the weight of a lithium battery, making it convenient for users to carry. As early as 2001, ProtOnex Technology Corporation developed a sodium borohydride as hydrogen source for fuel cell power source for the U.S. ground forces. In addition, ProtOnex Technology Corporation's unmanned aerial vehicle ProCOre system and South Korea's unmanned reconnaissance aircraft both use sodium borohydride as hydrogen source for fuel cell power sources, which can run for 6–12 h and more than 10 h, respectively. In 2009, NSK Ltd. launched a 50 W portable fuel cell with sodium borohydride as hydrogen source. In China, Shanghai Pearl Hydrogen Energy Technology Co., Ltd. applied hydrogen generated from sodium borohydride hydrolysis from −10 to 40 °C portable power source, with a rated power of 200 W. The hydrogen production density of magnesium hydride hydrolysis can reach 15.2 wt%, and the fuel cell system developed based on hydrolysis has high energy density, high safety, and no pollution, which can be used for underwater equipment, backup power sources, unmanned aerial vehicles, and other special equipment. Shanghai Jiaotong University, Shanghai Magnesium Power Technology Co., Ltd., and Shanghai Yuji Power Technology Co., Ltd. jointly launched a 50–200 W high-energy-density portable fuel cell power source powered by magnesium hydride hydrolysis, which can work at −40 to 50 °C, and the system energy density can reach up to 600 W h/kg. In hydrogen fuel cell vehicles, as early as the 1980s, the Mercedes-Benz 310 TN van was tested in Berlin, using a non-stoichiometric AB_2 type $Ti_{0.98}Zr_{0.02}Cr_{0.05}V_{0.43}Fe_{0.09}Mn_{1.5}$ hydrogen storage alloy, combined with a compressed hydrogen internal combustion engine. Toyota Motor Corporation in Japan used Ti-Mn alloy in fuel cell vehicles in 1996. The external dimensions of its hydrogen storage device are 700 mm × 150 mm × 170 mm, with 100 kg of Ti-Mn series hydrogen storage alloy material and a hydrogen storage capacity of 2 kg, whose maximum traveling distance is 250 km. To further increase the hydrogen storage capacity, Toyota Motor Corporation designed a new type of high-pressure metal hydride hydrogen storage tank in 2005. The tank has adopted a lightweight composite container with a volume of 180 L and a pressure resistance of 35 MPa. The cavity is filled with Ti-Cr-Mn alloy material and a built-in heat exchanger. The total weight of the tank body is 420 kg, and the hydrogen storage capacity reaches 7.3 kg. Currently, the demand for solid-state hydrogen storage continues to increase. In the future, solid-state hydrogen storage materials, are suitable for large and medium-sized buses, medium and heavy-duty truck and other vehicles with long-distance and high-energy-density demand based on their hydrogen storage density and safety advantages.

In hydrogen chemical industry, hydrogen metallurgy replacing coke steelmaking is a revolutionary change in blast furnace technology. In May 2021, HBIS Group Co., Ltd. launched the “first hydrogen-rich gas direct reduction demonstration project in the world” in Zhangjiakou, Hebei. In December 2021, China Baowu Steel Group Corporation started the construction of the first million-ton-level hydrogen gas vertical furnace direct reduction demonstration project and supporting facilities in the world in Baosteel Zhanjiang Iron & Steel Co. Ltd., which can flexibly use coke oven gas, natural gas, and hydrogen in different proportions. In solid-state hydrogen storage materials, Shanghai Jiaotong University and H2 Store (Shanghai) Energy Tech. Co. Ltd. jointly developed China’s first 70 kg-level magnesium-based solid-state hydrogen storage prototype device, which can be used for large-scale storage and transportation of hydrogen in the future. They also cooperated with Baowu Clean Energy Ltd. to develop a “solar power generation-electrolysis” skid-mounted integrated hydrogen energy system called “Hydrogen Quadriga” to verify the feasibility of magnesium-based solid-state hydrogen storage technology in hydrogen energy and hydrogen metallurgy processes.

As for hydrogen refueling stations, since a low-pressure alloy hydrogen storage system has the advantages of low working pressure, high volumetric hydrogen storage density, good safety, no need for high-pressure containers and pressurizing equipment, and safe and stable hydrogen supply, a hydrogen refueling station based on solid-state hydrogen storage alloy was first built in Liaoning by Shenzhen Up Power Technology Co., Ltd. in July 2019. This low-pressure hydrogen refueling station has significant advantages, such as no need for high-pressure compressors and on-site storage tanks, a low initial investment cost of only 3 million yuan, occupying a small area, and recycling the hydrogen storage alloy without damage. However, there are also problems that need to be overcome, such as relatively high prices and large weight of the systems.

1.5.1 Hydrogen Storage Alloys and Metal Hydrides

Metals (alloys) have the characteristic of regular atomic structure arrangement, and the lattice gaps are used to store hydrogen atoms. Under certain temperature and hydrogen pressure conditions, some metal materials can "absorb" a large amount of hydrogen gas, that is, react with hydrogen to form metal hydrides, while releasing heat. When heated, metal hydrides decompose and release the stored hydrogen. The "absorption" and "release" process of hydrogen gas is reversible and can be repeated. The hydrogen absorption reaction of metals can be divided into the following four steps:

1. Under the action of van der Waals forces, hydrogen gas is first adsorbed on the metal surface. Under the action of surface metal atoms, H_2 dissociates into H atoms.
2. H atoms diffuse from the surface to the interior of the metal, entering the gaps in the metal atomic structure.
3. With the continuous increase in the concentration of H atoms in the body phase, a "solid solution" begins to form in the metal lattice.
4. The concentration of hydrogen atoms continues to increase. After the solubility of the "solid solution" is saturated, a chemical reaction occurs, producing β-phase metal hydrides, and the hydrogen refueling reaction is completed.

Since the hydrogen absorption reaction of most metal hydrides is a reversible reaction, the dehydrogenation reaction step is the reverse process of the above steps. The hydrogen storage alloys currently researched and developed include rare earth AB_5 hydrogen storage alloys, Mg-based hydrogen storage alloys, AB_2 type Laves phase hydrogen storage alloys, AB type Ti alloys, V-based solid solution hydrogen storage alloys, and rare earth-magnesium-nickel hydrogen storage alloys, etc.

The existing development technology of hydrogen storage alloys is mature, including induction melting method, mechanical alloying method, hydriding combustion synthesis method, plasma gas-phase method, etc., which are suitable for large-scale industrial production. In the laboratory, in order to obtain nanoscale hydrogen storage materials with superior performance, in-situ impregnation method, melting method, chemical plating method, precursor in-situ synthesis, etc. are common methods for synthesizing hydrogen storage-catalyst-protectant composite materials with high-performance. The rare earth AB_5 hydrogen storage alloy is the earliest applied hydrogen storage material and is currently widely used in nickel-hydrogen batteries. The advantages of $LaNi_5$ rare earth alloy are mild absorption/desorption conditions with fast speed, easy activation, and insensitivity to impurities. However, due to the 24% expansion of the $LaNi_5$ alloy's crystal cell volume after hydrogen absorption, the material is prone to pulverization with poor cyclic performance. In addition, the high cost also limits the large-scale application of $LaNi_5$. Component optimization, structure regulation, and adjustment of stoichiometric ratio are common processes for optimizing AB_5 type rare earth alloy performance. Youon Technology Co., Ltd. has developed a hydrogen-powered bicycle based on AB_5 type hydrogen storage alloy. With 0.5 m^3 of hydrogen gas storage, this bicycle can continue to drive for 70 km, with a maximum speed of 23 km/h. AB_2 type $TiMn_2$, AB type TiFe, and rare earth-magnesium-nickel alloys are also widely used solid-state hydrogen storage alloys, with reversible hydrogen storage amounts all reaching more than 1.5 wt%. The reversible hydrogen storage characteristics of these three types of alloys are with a practical range, relatively low raw material prices, and abundant resources, so they have certain advantages in industrial production. At present, China Grinm Group Corporation Limited has recently developed an onboard vehicle hydrogen storage system based on the $TiMn_2$ hydrogen storage alloy, with a total hydrogen storage amount of 17 kg, which is applied to hydrogen fuel cell buses. The same company has also developed a TiFe hydrogen storage system with a storage amount of 1000 m^3, which is expected to be applied to the Hebei Guyuan wind power hydrogen production project for safe hydrogen storage. Magnesium-based hydrogen storage materials have the advantages of large volume hydrogen storage density, low working pressure, good safety, etc., which can greatly save installation space, reduce land occupation, and are especially suitable for applications with strict safety restrictions on the site, such as fuel cell cogeneration systems for building/park/home, fuel cell backup power sources, distributed hydrogen storage systems, etc. Solid solution alloys mainly refer to solid solution alloys formed by one or several hydrogen-absorbing metal elements dissolving into another metal, without components of a stoichiometric ratio or close to a stoichiometric ratio. Among solid solution alloys, V-based solid solution alloys have good hydrogen storage performance. The reversible capacity of $Ti_{0.32}Cr_{0.43}V_{0.25}$ alloy is 2.3 wt%. The alloy has good cycle stability. After 1000 cycles of hydrogen absorption and desorption, the reversible capacity still remains around 2 wt%. At present, solid-state hydrogen storage application technology is still in continuous development and improvement, and different hydrogen storage and transportation materials can be selected based on different storage and application scenarios in the future.

1.5.2 Complex Hydrides

Complex hydrides are a type of coordination compound formed by an anionic coordination group which is formed by a central atom and a hydrogen atom in the form of a covalent bond, and a metal ion. Its general formula is Ax(ByHz)N. A represents metals (including Li, Mg, Ca, K, Na, Al, Zn, Zr, Hf, etc.). B represents elements forming coordination groups with hydrogen (including B, N, Al, etc.). X, y, z, and n represent the number of atoms. Complex hydrides mainly include metal aluminohydrides, boron hydrides, and amides, which have a higher theoretical mass hydrogen storage density, but complex hydrides will produce impurity gases such as NH_3 during the hydrogen release process, which needs to be purified by adsorption.

The general formula for metal borohydrides is $M(BH_4)_X$, with the negative charge compensated by metal cations such as Li, Na, K, Be, Mg, Ca, etc. Among various borohydrides, $LiBH_4$, $Mg(BH_4)_2$, and $NaBH_4$ are the three borohydrides of concern. The mass hydrogen storage densities of $LiBH_4$, $Mg(BH_4)_2$, and $NaBH_4$ are 18.4, 14.6, and 10.6 wt%, respectively. The hydrogen release reaction is complex, usually requiring high temperature conditions (>300 °C) and multiple steps to release hydrogen, with a slow hydrogen release rate and poor cycle stability. The main methods to improve the hydrogen storage performance of borohydrides include:

1. Thermodynamic regulation, including element substitution and compounding.
2. Kinetic regulation, providing sufficient active reaction points to enhance reaction kinetics through catalyst doping or in-situ generated active substances.
3. Nanosizing, improving the thermodynamic and kinetic performance of the system by reducing particle size, optimizing material structure, and nano-confinement.

Currently, reversible hydrogen absorption and desorption in $LiBH_4$ has been achieved in the laboratory. After 100 cycles, the hydrogen release capacity of the sample is still close to 8.5 wt%, achieving the best cycle stability performance in current borohydride researches. In addition, $NaBH_4$ is a common hydrolysis hydrogen storage material. $NaBH_4$ can directly react with water without a catalyst under normal temperature and neutral conditions, producing hydrogen and sodium metaborate. The hydrolysis of sodium borohydride to produce hydrogen has the advantages of high storage efficiency, easy control of reaction speed, easy storage, and high safety, making it an optimal hydrogen supply material for backup power sources.

In metal aluminohydrides, the 4 H atoms form a $[AlH_4]$—tetrahedron with the Al atom through covalent action, and $[AlH_4]$—combines with metal cations to form coordination compounds, typical representatives of which are $LiAlH_4$ and $NaAlH_4$. The hydrogen storage capacity of metal aluminohydrides is 7.4–10.7 wt%, which is lower than the decomposition temperature of borohydrides. After modification with catalysts, hydrogen release can be completed at 150 °C. Among them, $NaAlH_4$ can reversibly absorb and release about 4.5 wt% of hydrogen at 100 °C, with no by-products, high purity of hydrogen, and catalysts applicable for non-precious metals such as Ni, CO, etc., with relatively low cost. It is very suitable for hydrogen supply materials for on board mid-temperature hydrogen fuel cell (80–200 °C). However, due to the high cost of raw materials, there is currently no practical application of $NaAlH_4$. Metal amides are compounds formed by metal ions (such as Li, Na, Mg, Ca, etc.) coordinating with $[NH_2]$—or $[NH]^2$—ion groups. Their hydrogen storage capacity is also high ($LiNH_2$, about 8.7 wt%), but thermal decomposition easily produces NH_3 and other by-products, thus the reversibility is also poor, usually combined with other hydrides to form a composite hydrogen storage system, such as $2LiH\text{-}Mg(NH_2)_2$, etc.

1.5.3 Ammonia Boranes and Their Derivatives

Ammonia-borane (NH_3BH_3) is one of the most basic B-N-H compounds, with a theoretical hydrogen content as high as 19.6 wt%. In the ammonia borane molecule, the H atom connected to the N atom exhibits positive electricity, and the H atom connected to the B atom is negative. The hydrogen release methods of ammonia-borane can be divided into hydrolysis hydrogen release and thermal hydrogen release. The ammonia-borane molecule contains three types of proton bonds, that is B-H, B-N and N-H, which has high thermal stability. Therefore, the decomposition process requires high energy and high temperature, meantime the decomposition process is prone to produce ethylborane and other by-products. Ammonia-borane is relatively stable in aqueous solution. Hydrolysis is very slow in the absence of a catalyst, and dehydrogenation at room temperature requires the presence of an appropriate catalyst. The efficiency of the dehydrogenation reaction also largely depends on the choice of catalyst, currently applicable catalysts for the hydrolysis reaction of ammonia borane include precious metals, non-precious metal-precious metal alloys, non-precious metal catalysts, supported metal catalysts, etc. The derivatives of ammonia borane mainly include metal aminoborane compounds, bimetallic aminoborane compounds, and amine derivatives, etc. These derivatives have significantly lower hydrogen release temperatures compared to ammonia borane. However, due to the complexity of their hydrogen release reactions and by-products such as ammonia, ethylborane and others, they are still in the research and development stage.

1.5.4 Physical Adsorptive Materials

Physical adsorption hydrogen storage is a method of storing hydrogen that relies on the interaction of van der Waals forces between the material and hydrogen molecules. During the adsorption and desorption process, hydrogen exists in the form of H_2 molecules. Since physical adsorption is usually an exothermic process, and the binding force between hydrogen and the material is weak, physical adsorption materials for hydrogen storage generally have strong hydrogen adsorption capacity under low temperature conditions (usually at the boiling point of liquid nitrogen, 77 K). Currently, representative physical adsorption materials for hydrogen storage include carbon materials, zeolites, metal-organic frameworks, covalent organic frameworks, porous polymers, etc. Research on the adsorption of hydrogen by carbon materials began in the early twentieth century. Carbon materials maintain chemical and thermal stability during the hydrogen absorption and release process, which can improve the reliability of the storage system and is beneficial to their application in the field of hydrogen storage. At the same time, with abundant resources, easy processing, and relatively low cost, carbon is suitable for industrial and large-scale production. Metal-organic framework materials (MOFs) are compounds with metal ions as coordination centers and organic acid anions as ligands. By regulating the pore surface functionalization of MOFs, they can have a large number of strong hydrogen adsorption sites so as to have a higher hydrogen storage capacity. Researchers at the University of California, Berkeley, developed Ni_2 (m-dobdc) and used a packing density (0.366 g/mL) to calculate the volumetric hydrogen storage working capacity. Such a material can reach a volumetric hydrogen storage effective capacity of 23 g/L at 10 MPa and a temperature range of −75 to 25 °C. However, the conditions for hydrogen storage by physical adsorption are still stringent, requiring a low-temperature environment to have a considerable hydrogen storage capacity. At the same time, hydrogen storage and low-temperature preservation also increase the cost of hydrogen storage to a certain extent. Therefore, the research and development of physical hydrogen storage materials shall focus on increasing the working temperature of the materials.

1.6 Comparison of Various Hydrogen Storage and Transportation Methods

Hydrogen storage and transportation requires safety, large capacity, low cost, and convenience. Currently, the comparison of hydrogen storage and transportation technologies with industrial potential is shown in Table 1.3. Comparison of various hydrogen storage and transportation technologies shows that high-pressure gas cylinder hydrogen storage and transportation technology is currently the most mature and widely used one, but there are bottlenecks in hydrogen storage density and safety; Hydrogen transportation by pipelines technology is a method of large scale hydrogen transportation with the lowest cost, but it needs extremely high fixed input costs resulting in limited application when hydrogen energy is not widely popularized; low-temperature liquid hydrogen storage and transportation technology has the advantage of large hydrogen storage density per unit volume, suitable for long-distance hydrogen storage and transportation, but the current storage cost is too high due to the large energy consumption in the liquefaction process and the extremely high insulation performance of the hydrogen storage container, and liquid hydrogen

Table 1.3 Advantages and disadvantages of hydrogen storage and transportation technologies and their current main applications

Hydrogen storage and transportation technology		Advantages	Disadvantages	Application
High-pressure gaseous hydrogen storage and transportation	High-pressure gas cylinders	Mature technology, simple structure, high hydrogen filling and discharging rate, low energy consumption	Low hydrogen storage density, high transportation cost, low safety, high cost of large-capacity ultra-high pressure carbon fiber wrapped cylinders	Small-scale hydrogen storage and transport, on board vehicle hydrogen storage and transport systems, currently the mainstream method of hydrogen storage and transportation
	Hydrogen transportation by pipelines	Large transport volume, low cost and low energy consumption of ultra-long distance hydrogen transportation	Extremely high fixed investment cost	Less hydrogen transportation by pipelines in China, solving the problem of transportation demand
Liquid hydrogen storage and transportation	Low-temperature liquid hydrogen storage and transportation	High hydrogen storage density, low long-distance transport cost, high purity	High energy consumption for hydrogen liquefaction, high requirements for hydrogen storage containers, volatilization	Long-distance hydrogen transportation abroad, heavy truck hydrogen storage systems, rocket propellers
	Hydrogen storage and transportation by hydrogen-rich compounds	High volumetric hydrogen storage density, relatively good safety, long-term storage, low container requirements	High energy consumption, side reactions exist in hydrogen release, stringent operating conditions	Long distance transport of hydrogen across the sea

cannot be stored for a long time with inevitable volatilization; Hydrogen storage and transportation technology by hydrogen-rich liquid compound, has not yet been able to be applied on a large scale commercially due to its energy consumption, hydrogen release purity, catalyst life and other issues; solid-state hydrogen storage and transportation technology is an important future hydrogen storage and transportation method with the advantages of high safety and capacity, but the technology is relatively immature since it is currently in the early stage of industrialization with relatively high cost.

Hydrogen is applied for hydrogen fuel cell vehicles, seasonal energy storage, distributed generation, hydrogen metallurgy, hydrogen for chemical industry, etc. It is necessary to choose suitable and economical methods of hydrogen storage and transportation according to the characteristics of each application scenario. Hydrogen energy used in the field of transportation uses fuel cells as a power generation method. Compared with other conventional energy utilization forms such as gasoline engines and diesel engines, whether it has economic advantages in energy prices is a key technical and economic indicator for the future development of hydrogen energy. Fuel cell systems can achieve fuel to electricity efficiency of 45–55%, and can be combined with power batteries to form a hybrid mode powertrain. Therefore, compared with conventional fuel vehicles, hydrogen energy vehicles have obvious energy efficiency advantages. The "World Hydrogen Energy and Fuel Cell Vehicle Industry Development Report (2018)" released by the China Society of Automotive Engineers analyzed the hydrogen energy consumption of passenger cars and commercial vehicles [12]. By sorting out the publicly announced fuel consumption for comparison, the fuel consumption of hydrogen energy vehicles and other passenger vehicles is shown in Table 1.4. It can be seen from the table that the energy consumption level of hydrogen energy vehicles is lower than that of pure gasoline vehicles, and there is still room for further reduction. Under the assumed hydrogen gas price of 4 yuan/Nm^3, the actual fuel cost of hydrogen energy vehicles is lower than that of gasoline vehicles, higher than that of diesel vehicles, natural gas vehicles and pure electric vehicles. However, in application scenarios such as buses and heavy trucks, lower-priced hydrogen is needed to meet the competition with diesel trucks. At present, the Ministry of Finance and other five departments have issued a notice on carrying out the demonstration application of fuel cell vehicles, giving a guide price for hydrogen = 35 yuan/kg H_2. Therefore, it is necessary to further reduce the costs of hydrogen production, hydrogen storage and transportation, and hydrogen refueling, especially the cost of hydrogen storage and transportation, so that hydrogen energy can reach the economic turning point at the supply end as soon as possible.

The storage and transportation of hydrogen is complex. Existing and under-research technologies that meet industrial application conditions include high-pressure gas cylinders, pipeline hydrogen, liquid hydrogen, hydrogen-rich liquid compounds, solid hydrogen storage and transportation technologies. Considering the actual situation of current industrial applications, the key indicators of hydrogen storage and transportation technologies are shown in Table 1.5. The cost of hydrogen storage and transportation mainly consists of fixed costs and operating costs. The fixed costs include the investment in hydrogen storage equipment,

Table 1.4 An energy consumption table of passenger cars

Fuel consumption	Hydrogen sedan	Diesel vehicle	Gasoline vehicle	Electric vehicle
Fuel retail price/yuan	4	6.21	6.4	0.8
Fuel consumption per hundred kilometers	10.58 Nm^{3}H$_2$	6.00 L	7.7 L	14.92 kW h
Unit cost/(yuan/100 km)	42.29	37.26	49.28	11.94

Table 1.5 Comparison of existing hydrogen storage and transportation technologies

H_2 Storage technologies	High-pressure gas cylinder	Pipeline hydrogen transport	Liquid hydrogen storage and transportation	Hydrogen-rich liquid compounds	Solid-state storage and transportation
Transport temperature (°C)	Room temperature	Room temperature	−253	Room temperature	Room temperature
Transport pressure (MPa)	~20	4–10	<1	Atmospheric pressure	Atmospheric pressure
System gravimetric storage density (wt%)	~1	/	5–9	5–7	4–7.6
System volumetric storage density (g/L)	~18	/	40–60	40–60	50–75
Single vehicle H_2 storage capacity (kg)	300	Continuous	~3000	1300	1400
Transport equipment	Long tube trailer	Pipeline	Liquid hydrogen tanker	Liquid tanker	Metal tanker
Refueling electricity (kWh/kg H_2)	2	2	12–17	Exothermic	Exothermic
Refueling pressure (MPa)	>20	4–10	~1	>1	1–1.5
Discharge electricity (kWh/kg H_2)	/	/	/	~10	~11
H_2 release temperature (°C)	Room temperature	Room temperature	Room temperature	180–400	300–350
Storage efficiency (%)	~90	95	75	85	>90

transportation equipment, and hydrogen release equipment. The operating costs mainly include hydrogen filling power consumption, transportation mileage fee, and hydrogen release power consumption, that is, six cost quadrants, as shown in Table 1.6. According to the characteristics of various hydrogen storage and transportation technologies, the power consumption of high-pressure gas cylinders and pipelines is small, while the power consumption of liquid hydrogen, hydrogen-rich liquid compounds, and solid hydrogen storage during operation is high. This leads to high-pressure gas cylinders being a more economical method for short-distance small-scale transportation; for medium and long-distance transportation, liquid

Table 1.6 Cost quadrant and analysis of hydrogen storage and transportation technology

Fixed cost	**Hydrogen charge end**	**Transportation end**	**Hydrogen discharge end**	**Current status**
High pressure gas hydrogen 20 MPa	Pressurized filling station	High pressure tube trailer	None	Industrialized
Pipeline hydrogen	Pressurizing equipment	Pipeline and booster equipment	None	Short distance
Liquid hydrogen	Hydrogen liquefaction plant	Liquid hydrogen tanker truck	Liquid hydrogen gasification station	Small scale device
Hydrogen-rich liquid compounds	Hydrogen refueling plant	Liquid tanker truck	Dehydrogenation plant	Small scale device
Solid-state hydrogen storage	Charging equipment	Metal truck	Discharging equipment	Small scale device
Operating cost	**Hydrogen charge end**	**Transportation end**	**Hydrogen discharge end**	**Transport capacity**
High pressure gas hydrogen 20 MPa	Operation and charge electricity consumption	Transportation mileage fee	None	About 300 kg
Pipeline hydrogen	Compression equipment electricity consumption	Pipeline boosting power consumption	None	Continuous
Liquid hydrogen	Operation and liquefaction electricity consumption	Transportation mileage fee	None	About 3 t per vehicle
Hydrogen-rich liquid compounds	Operation	Transportation mileage fee	Catalytic hydrogen release power consumption	About 1.3 t per vehicle
Solid-state hydrogen storage	Operation	Transportation mileage fee	Catalytic hydrogen release power consumption	About 1.2 t per vehicle

hydrogen, hydrogen-rich liquid compounds, and solid hydrogen storage have more competitiveness; for ultra-large-scale ultra-long-distance transportation, pipeline hydrogen is currently the best choice. However, these hydrogen storage and transportation technologies also have obvious characteristics: high-pressure gas cylinders have the problems of low hydrogen storage capacity per vehicle and high safety risks; the fixed investment cost of pipeline hydrogen is extremely high, and it requires national unified planning to build a hydrogen pipeline network; the liquefaction energy consumption of liquid hydrogen is too high, and it is easy to evaporate and cannot store hydrogen for a long time; hydrogen-rich liquid compounds have high hydrogen filling and releasing energy consumption, require precious metal catalysts, and the compounds themselves are mostly toxic/sub-toxic substances, which pose environmental pollution, impure hydrogen release requiring

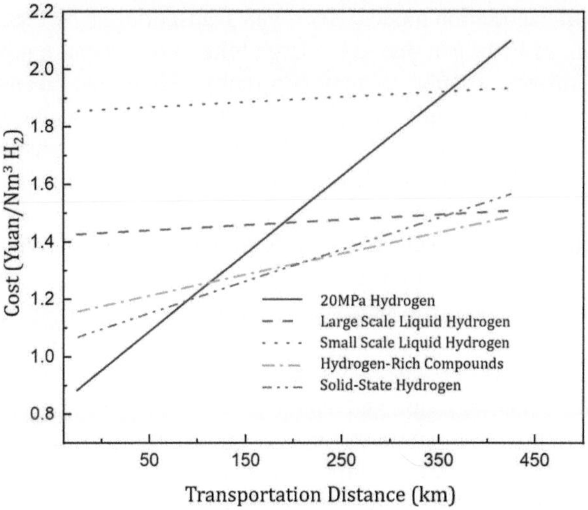

Fig. 1.11 Hydrogen storage and transportation distance and comprehensive cost

purification, and other issues; solid hydrogen storage is safe and can be stored for a long time, but the energy consumption of hydrogen release is high. Therefore, it is still necessary to choose the appropriate hydrogen storage and transportation method according to different application scenarios. According to existing data, under the condition of a supply capacity of 2000 Nm^3/h, the relationship between the cost of hydrogen transportation and the transportation distance is shown in Fig. 1.11.

20 MPa High Pressure Hydrogen, Large Scale Liquid Hydrogen, Small Scale Liquid Hydrogen, Hydrogen-Rich Compounds, Solid-State Hydrogen.

Transportation Distance.

For short-distance, small-scale transportation, high-pressure long-tube trailer transportation remains the main method. As the mileage increases, other storage and transportation methods gradually show their economic advantages. However, due to the currently high initial investment and liquefaction energy consumption of liquid hydrogen, it is more suitable for large-scale long-distance transportation. Under a transportation distance of 100 km, the comprehensive storage and transportation cost of hydrogen by high-pressure long-tube trailer is about 1.2 yuan/Nm^3. As the scale of hydrogen transportation increases, solid-state hydrogen storage and hydrogen-rich liquid compounds begin to show cost advantages for medium and long distances. Organic liquid hydrogen storage, due to its chemical properties of dehydrogenation and the characteristics of fluid transportation, should be more suitable for long-distance sea transportation scenarios. Liquid hydrogen has the unique advantages of high purity and single vehicle transportation volume, and may serve as a large-scale high-quality application in the future. Therefore, for short-distance hydrogen energy distribution scenarios, city hydrogen pipelines, high-pressure gaseous hydrogen, and solid hydrogen will become complementary medium and short-distance transportation modes. Hydrogen Transportation by Pipelines requires a higher amount of hydrogen due to the large initial investment in pipelines and its continuous hydrogen supply system. Currently, Hydrogen Transportation by Pipelines is used for the supply of chemical hydrogen. Large-scale pipeline hydrogen supply in Europe is above 50,000 Nm^3/h, exceeding other transportation methods, and Hydrogen Transportation by Pipelines will become the main mode of trunk line hydrogen transportation in the future.

Seasonal energy storage supports long-term, large-scale, and wide-area energy transfer, and is a key technology to cope with long-term supply breaks in high-proportion renewable energy systems. The cost of fixed energy storage can generally be analyzed from two perspectives: investment cost and full life cycle cost. Researchers [13] analyzed and compared the levelized cost of energy storage (LCOS) of different energy storage technologies in 2020 and 2060, and found that electrochemical energy storage technologies represented by lithium-ion batteries and sodium-ion batteries have a large cost reduction space and are expected to become the lowest-cost short-cycle (hour-level) energy storage technology before 2040. However, as the discharge duration increases, the LCOS costs of all types of energy storage technologies rise, among which the LCOS cost of electrochemical energy storage shows an accelerating upward trend, while the LCOS costs of hydrogen energy, compressed air, and pumped storage increase relatively slowly. In the application scenario of seasonal energy storage, the current LCOS of electrochemical energy storage is more than 6 times that of hydrogen energy, and it is close to 5 times by 2060. Therefore, hydrogen energy will be the main method of seasonal energy storage in the future. Due to the long-term storage characteristics of solid-state hydrogen storage materials and hydrogen-rich liquid compounds, they are potential hydrogen storage methods in the field of seasonal energy storage.

In the future, with the development of hydrogen energy technology and its application in transportation, energy storage, industry, and civil use, a unified social energy system supported by hydrogen energy can be realized. This includes renewable energy input, hydrogen production (electricity to hydrogen, P2H), hydrogen storage (HS), hydrogen energy conversion (hydrogen to electricity H2P), hydrogen to gas (H2G), hydrogen to hydrogen (H2H), hydrogen to heat (H2T), electricity, heat, natural gas, hydrogen transmission networks and load [14], as shown in Fig. 1.12.

Power sources: wind power, photovoltaic power, nuclear power, hydropower, others.

Conversion: electricity, natural gas, hydrogen, heat.

Transmission.

Load: commercial, industrial, transportation, residential, others.

1. Power supply: Composed of solar power, wind power, hydropower, nuclear power, hydrogen power, biomass and other renewable power sources. Among them, wind power and solar power, which have the most abundant resources and the most mature development technology, have uncertainties. To ensure the utilization rate of wind and light, the generator set is coupled with the electrolyzer

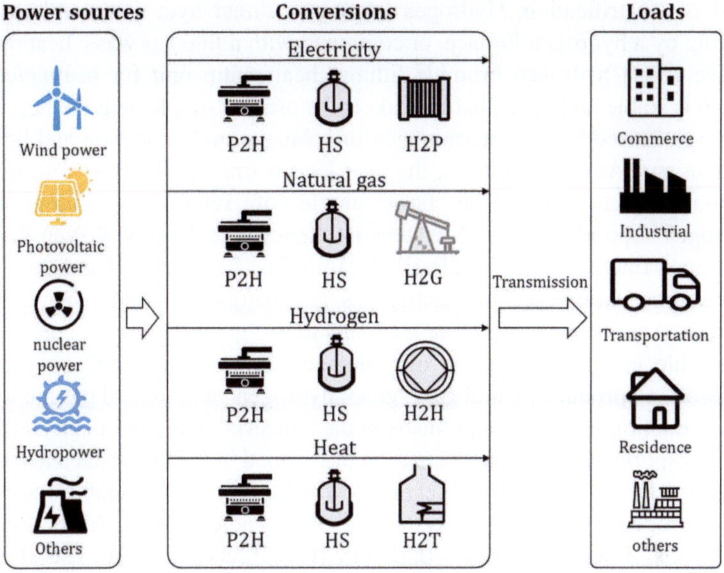

Fig. 1.12 Future energy system with coupled hydrogen and electricity

to convert excess electricity into hydrogen. Stable and controllable power sources like nuclear power plants bear the steady part of the power system load (base load), allowing them to operate as efficiently as possible, saving system fuel consumption, and benefiting safe and economical operation. Hydrogen power is stable and controllable, and hydropower has advantages such as reliable prediction, rapid response, and flexible adjustment, which can be used to fill the power shortage caused by intermittent wind and solar power generation and load fluctuations. In the hydrogen-electricity coupled energy system, the proportion of intermittent power sources such as wind power and solar power is relatively high. Hourly electrochemical energy storage is difficult to meet the stable operation requirements of the power system. By electrolyzing water to produce hydrogen, renewable energy power is converted into hydrogen and stored. When there is a shortage of electricity, the stored hydrogen is used to generate electricity through fuel cells or hydrogen gas turbines to meet the user's demand for electricity. This can achieve daily, monthly, and even seasonal energy storage, effectively solving problems such as renewable energy consumption, fluctuation suppression, and intermittency. The basic mode of reliable operation of the hydrogen-supported power system is: Renewable Energy Power Station—Hydrogen Energy System (P2H-HS-H2P)—Power Grid—Electricity Users.

2. Heat source: Composed of electric heating, hydrogen heating, solar thermal collection, and geothermal and other renewable heat sources. Among them, solar thermal collection is similar to solar power generation and has intermittency; electric heating is reliable, clean, and stable, and is an important means to expand regional electricity consumption, absorb surplus electricity, and improve the

level of electrification. Hydrogen heating is direct hydrogen combustion for heating by a hydrogen furnace, or combined with a flue gas waste heat recovery device and a hydrogen bromide lithium heat pump unit for heating/cooling, which is stable and controllable, and can be used to supplement the heat energy shortage caused by the intermittency of solar thermal collection and heat load fluctuations. At the same time, the heat load is one of the loads with the most seasonal fluctuations. The basic mode of reliable operation of the hydrogen-supported thermal system is: Renewable Energy Power Station—Hydrogen Energy System (P2H-HS-H2T)—Heat Network—Heat Users.

3. Gas source: Composed of electric gas and natural gas and other clean gas sources. Among them, "electric gas" includes: "green hydrogen" produced by renewable energy electrolysis of water and "green hydrogen" coupled with "coal/oil" to produce natural gas; "green hydrogen" and natural gas are used for the production of chemical products in the industrial field, hydrogen metallurgy, and transportation, etc. The fluctuation of gas load in the industrial and transportation fields is affected by market prices. The basic mode of reliable operation of the hydrogen-supported gas system is: Renewable Energy Power Station—Hydrogen Energy System (P2H-HS-H2G/H2H)—Natural Gas/Hydrogen Pipeline Network—Natural Gas/Hydrogen Users.

1.7 Hydrogen Safety in Storage and Transportation Processes

Hydrogen safety is an important issue in hydrogen storage and transportation technology and a prerequisite for large-scale application of hydrogen energy. The typical risk factors in the process of hydrogen storage and transportation are shown in Table 1.7. Corrosion, hydrogen embrittlement, fatigue, and hydrogen permeation of the inner liner and container wall of the hydrogen storage equipment will cause the life of the hydrogen storage equipment to decrease, and in severe cases, it will cause hydrogen leakage accidents. At present, China has successively established relevant standards for the safety of hydrogen storage and transportation systems (GB/T 34542.1–GB/T 34542.3), including the test methods for hydrogen embrittlement sensitivity of metallic materials and the compatibility test methods of metallic materials and hydrogen environment. However, compared with the systematic and completeness of international and American standards (Table 1.8), China's relevant standards still need to be continuously improved.

When hydrogen leaks in a confined space, it is prone to accumulate, forming a flammable hydrogen cloud. Hydrogen has a wide combustion range and low ignition energy. If it is ignited immediately after leakage, it will form a jet flame (hydrogen jet fire). Therefore, during the storage and transportation of hydrogen, it is urgent to continuously develop hydrogen detection technology, remote automatic monitoring, and safety big data analysis technology. In addition, the storage forms

Table 1.7 Typical risk factors during hydrogen storage and transportation process

Typical process	Risk factors
Storage process	Corrosion and hydrogen embrittlement risk: Hydrogen-exposed components, pipelines, and containers are long-term exposed to a hydrogen environment, especially when the stored hydrogen contains corrosive impurities or is under high temperature and pressure, the problems of corrosion and hydrogen embrittlement are more prominent; once hydrogen embrittlement occurs, it will reduce the safety of container storage and eventually lead to hydrogen leakage Fatigue risk: High-pressure hydrogen storage technology has high requirements for high/low pressure fatigue of metal liners; magnesium-based solid hydrogen storage technology involves high/low temperature thermal cycling process, which requires high fatigue life of metal containers Hydrogen permeation risk: In high-pressure hydrogen storage technology, hydrogen will significantly heat up during the rapid filling process, affecting the resin adhesive of composite materials, resulting in delamination, which reduces the load-bearing capacity and safety of the container
Transportation process	Accident risk: Accidents caused by traffic accidents during the transportation process High temperature risk: During the transportation process, the pressure of high-pressure hydrogen gas increases due to the high external environmental temperature, causing hydrogen leakage
Loading and unloading process	Loading and unloading risk: Hydrogen storage tanks are reused multiple times, producing micro-cracks or bumps and friction, causing hydrogen leakage Hydrogen impurity risk: During the process of filling high-pressure hydrogen gas into the tank, impure hydrogen (such as oxygen) will remain in the hydrogen storage tank. If the residual gas is not checked in time after several cycles, the purity of the hydrogen in the storage tank will decrease, causing the hydrogen to become impure and form a flammable mixed gas
Liquefaction process	Vaporization danger: Hydrogen is stored in liquid form at −253 °C. If the surrounding insulation layer is damaged causing the environmental temperature to rise, the liquid hydrogen inside the storage container will quickly vaporize, instantly generating a strong pressure, causing an explosion

Table 1.8 Relevant standards for the safety of hydrogen storage and transportation systems

Some International/U.S. standards	Some domestic standards
International Standard ISO 11114-4:2017 "Transportable Gas Cylinders—Compatibility of Cylinder and Valve Materials with Gas Contents" U.S. Standard ASME BPVC VII .3 KD—10 "Special Requirements for Hydrogen vessel" U.S. Standard ANSI/CSA CHMC 1-2014 "Test Methods for Compatibility of Metallic Materials with High Pressure Hydrogen Gas Environments" U.S. Standard ASTM G142-98 (2016) "Standard Test Method for Determination of Susceptibility of Metals to Embrittlement in Hydrogen Containing Environments at High Pressure, High Temperature, or Both"	GB/T 34542.2—2018 "Hydrogen Storage and Transportation Systems Part 2: Test Methods for Compatibility of Metallic Materials with Hydrogen Environment" GB/T 34542.3—2018 "Hydrogen Storage and Transportation Systems Part 3: Test Methods for Sensitivity of Metallic Materials to Hydrogen Embrittlement"

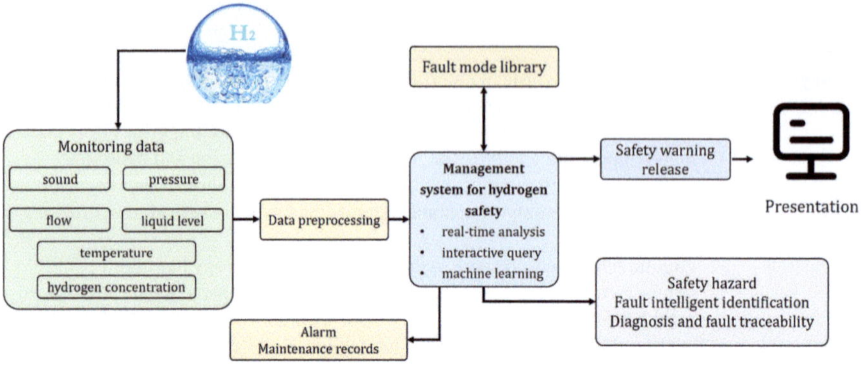

Fig. 1.13 Architecture of the comprehensive management system for hydrogen storage and transportation safety

of hydrogen are diverse and the environment and conditions vary significantly. Safe operation and timely warning cannot be effectively guaranteed, and it is urgent to develop a comprehensive management system for hydrogen storage and transportation safety, as shown in Fig. 1.13.

Monitoring data, sound, hydrogen concentration, flow rate, pressure, liquid level, temperature.

Data preprocessing, alarm, maintenance records.

Fault mode library, comprehensive management system for hydrogen storage and transportation safety, real-time analysis, interactive query, machine learning.

Safety warning release, safety hazard/fault intelligent identification/diagnosis and fault traceability.

High-precision, fast-response, low-cost hydrogen sensors are the key devices for hydrogen safety. A hydrogen sensor is a device that detects hydrogen and generates an electrical signal proportional to the hydrogen concentration. In recent decades, many different types of hydrogen sensors have been commercialized or under development. In 2015, the U.S. Department of Energy (DOE) set challenging performance parameter targets for hydrogen sensors, including concentration range (0.1–10%), operating temperature (−30 to 80 °C), response time (<1.0 s), gas environment (relative humidity 10–98%), service life (>10 years), market price (per unit <40 US dollars), etc. In order to meet the needs of future hydrogen energy development, in addition to reducing the size, cost, and power consumption of sensors, the sensitivity, selectivity, and stability of hydrogen sensors should be improved. Currently, hydrogen sensors mainly include catalytic combustion sensors, electrochemical sensors, resistive sensors, optical sensors, etc.

1. Catalytic combustion sensor

The working principle of the catalytic combustion sensor is that the combustible gas reacts with the oxygen on the surface of the catalytic sensor to release heat. The catalytic combustion sensor consists of two components: a detection piece that reacts with the combustible gas and a compensation piece that does not react

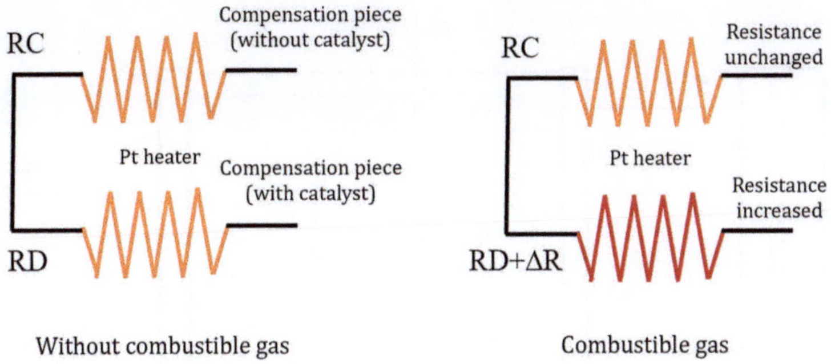

Fig. 1.14 Principle of catalytic combustion sensor

with the combustible gas, as shown in Fig. 1.14. When a combustible gas is present, the combustible gas undergoes catalytic combustion on the surface of the detection piece containing the catalyst (for example: hydrogen and oxygen produce water), and the temperature rise of the detection piece increases the thermistor resistance inside the detection piece. The compensation piece does not undergo catalytic combustion, and its resistance does not change. These components form a Wheatstone bridge circuit. In an atmosphere where there is no combustible gas, the variable resistance can be adjusted to balance the bridge circuit. When the sensor is exposed to a combustible gas, only the resistance of the detection piece rises, so the balance of the bridge circuit is broken, showing an unbalanced voltage that is detected. There is a linear proportional relationship between this unbalanced voltage and the gas concentration, so the gas concentration can be detected by measuring the voltage. This principle can be used to detect any combustible gas including hydrogen. It has the virtues of quick response, high accuracy and a long service life, but needs to be used in an oxygen containing environment, and has relatively poor selectivity with the danger of igniting explosions.

Compensation piece (without catalyst), Pt heater, compensation piece (with catalyst), no combustible gas.

Resistance unchanged, Pt heater, resistance increased, presence of combustible gas.

2. Electrochemical sensor

The working principle of the electrochemical hydrogen sensor is that the hydrogen gas undergoes an electrochemical reaction with the sensing electrode to cause charge transfer or changes in electrical properties. The sensor detects the concentration of hydrogen gas by detecting changes in chemical signals. Electrochemical sensors can be divided into two categories: current type and voltage type.

Current-type hydrogen sensors are more common in commercial applications. They generate a current proportional to the concentration of hydrogen by electrochemically reacting with hydrogen. They are mainly composed of three parts (Fig. 1.15): the first part is the electrode where electron transfer occurs, usually including the working or sensing electrode, counter electrode; the second part is

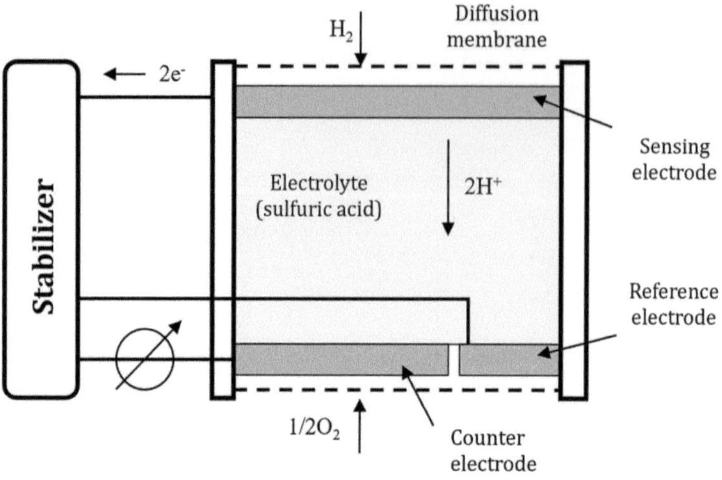

Fig. 1.15 Schematic of the structure of a current-type hydrogen sensor

the electrochemical cell, which contains a solid or liquid electrolyte to allow ion transmission between the electrodes; the third part is the gas permeation layer, which covers the entrance of the sensing electrode and helps to limit the diffusion of hydrogen and oxygen, thus making it the rate-determining step.

Stabilizer, electrolyte (sulfuric acid), diffusion membrane, sensing motor, reference electrode, counter electrode.

The structure of a voltage-type hydrogen sensor is similar to that of a current-type hydrogen sensor. The difference is that the voltage-type hydrogen sensor operates at zero current, and the measured value is the potential difference or electromotive force between the sensing electrode and the reference electrode. These electrodes are usually made of noble elements such as palladium, platinum, gold, or silver. Commonly used solid proton conducting electrolytes include alumina, phosphosilicate glass, and sodium hydride. Electrochemical sensors have the advantages of low power consumption and room temperature operation, as well as good repeatability and anti-interference, but there are still some disadvantages, including a lifespan of only about 2 years, limited temperature range, small selectivity, and sensitivity to environmental conditions (pressure). In contrast, the response of the voltage-type hydrogen sensor is logarithmically related to the hydrogen concentration (with lower accuracy at higher concentrations), while the current-type hydrogen sensor has a linear relationship with the hydrogen concentration (higher sensitivity).

3. Resistance-type sensors

The sensing mechanism of a resistance-type hydrogen sensor is: when the sensor is exposed to hydrogen, the adsorption and permeation of hydrogen will change the resistance of the hydrogen-sensitive material in the sensor, and when the hydrogen is desorbed from the hydrogen-sensitive material, the resistance of the hydrogen-sensitive material will change again. Resistance-type hydrogen sen-

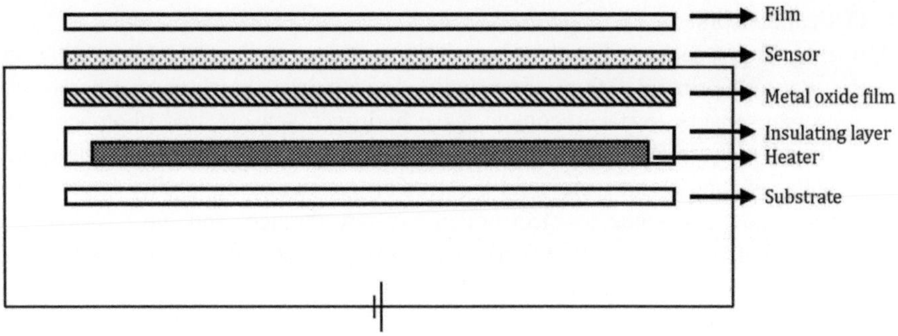

Fig. 1.16 Schematic illustration of the structure of a semiconductor metal oxide type hydrogen sensor

sors are mainly divided into semiconductor metal oxide types and non-semiconductor types (i.e., metal or alloy types).

Semiconductor metal oxide type hydrogen sensors deposit a layer of metal oxide with semiconductor properties (such as doped tin oxide, zinc oxide, tungsten oxide) on a heater and operate at about 500 °C (Fig. 1.16). The working principle is that when oxygen in the environment is adsorbed on the metal oxide layer, the adsorption layer has a high resistivity. When hydrogen diffuses to the sensing layer and reacts with oxygen, it is adsorbed on the surface of the semiconductor metal oxide, and the resistivity of the adsorption layer decreases, and the decreasing value increases with the increase of hydrogen concentration. Most semiconductor metal oxide type hydrogen sensors have high sensitivity, with an average response time between 4 and 20 s, and a hydrogen measurement concentration range from 10^{-5} to 2×10^{-2}. However, the response speed of a single metal oxide is difficult to meet practical needs, and the selectivity of a single metal oxide to hydrogen is poor, lacks sensitivity to hydrogen, and is easily interfered by reductive gases such as CO, CH_4, and alcohol. To improve its selectivity, precious metals (Pt, Pd, Au, etc.) or transition metals (Zn, Mn, Cu, etc.) with good selectivity to hydrogen can be doped to solve this problem. Although semiconductor metal oxide type hydrogen sensors have many advantages, they are only suitable for use in high-temperature working environments and cannot be used at room temperature, which leads to high power consumption. Meantime, sparks can easily be generated during operation, which is not suitable for hydrogen concentration detection in flammable and explosive environments.

Film, sensor, metal oxide film, insulating layer, heater, substrate.

Non-semiconductor type sensors generally use metal hydrides as hydrogen-sensitive materials, especially Pd-based resistive hydrogen sensors, which are widely studied due to their simple process, low cost, high sensitivity, short response time, and operation at room temperature. They are considered to be the most advanced hydrogen sensors at present. At room temperature, Pd reacts reversibly with hydrogen to form palladium hydride PdH_x, which has a higher resistivity than Pd. By detecting the resistance signal of the Pd-based sensor, quantitative detection of hydrogen can be achieved.

4. Optical Sensors

Optical hydrogen sensors detect hydrogen through optical changes. Depending on the working principle, they are usually divided into three types: fiber optic hydrogen sensors, surface acoustic wave hydrogen sensors, and photoacoustic hydrogen sensors. Among them, fiber optic hydrogen sensors have outstanding advantages such as intrinsic safety, corrosion resistance, suitability for remote sensing, and resistance to electromagnetic interference, and have become a research hotspot. Fiber optic hydrogen sensors use a combination of fiber optics and hydrogen-sensitive materials (palladium thin film, tungsten oxide thin film, magnesium-based thin film, etc.). When the hydrogen-sensitive material reacts with hydrogen, the physical properties of the fiber change, causing the optical properties of the transmitted light in the fiber to change. The concentration of hydrogen is detected by measuring changes in physical parameters such as the transmittance and reflectance of the film using fiber optic technology. Fiber optic hydrogen sensors are one of the most promising hydrogen detection sensors. They can be used at room temperature. Also, these sensors use light signals for detection, eliminating the need for heating and avoiding the possibility of explosions. In certain special scenarios, by selecting appropriate wavelengths of light and fibers, long-distance detection can be achieved, which is safer and more practical.

Currently, hydrogen sensors still face many challenges:

(a) The response and recovery time of hydrogen should be further accelerated to less than 1 s.
(b) Further improving the sensitivity to hydrogen at low concentrations is necessary to detect early-stage hydrogen leaks.
(c) Improving the selectivity of the sensor for hydrogen in various gas mixtures, and reducing the impact of other gases are needed.
(d) To ensure the stable operation of the hydrogen sensor under various environmental conditions, the reliability of the hydrogen sensor should be verified at different relative humidity levels (5–98%) and temperatures (−30 to 80 °C).
(e) Hydrogen sensors should operate stably, with high reliability, and no significant signal drift within 6 months.

Exercises

1. If a hydrogen tube trailer is used as a primary gas storage, then the gas storage method used in the hydrogen refueling station can usually be divided into ( ) levels.

   A. 1
   B. 2
   C. 3
   D. 4

2. The deviation of real gas from ideal gas can be represented by the compressibility factor Z in thermodynamics. At 0 °C, as the pressure increases, the trend of the compressibility factor of hydrogen is ( ).

   A. Decreases first then increases
   B. Increases continuously
   C. Increases first then decreases
   D. Decreases continuously

3. The compressibility factor Z of hydrogen at 20 °C and 35 MPa pressure is 1.225. Please calculate the density of hydrogen under this condition. At the same temperature, when the pressure rises to 70 MPa, Z = 1.459, please calculate how much the density of hydrogen has increased compared to 35 MPa.
4. Based on the current technical level, for a transportation distance of 150 km, the lowest cost of hydrogen storage and transportation is ( ).

   A. Liquid hydrogen storage and transportation technology
   B. 20 MPa long tube trailer
   C. Hydrogen-rich compound storage and transportation technology
   D. Solid-state storage and transportation technology

5. According to the different usage requirements of high-pressure hydrogen storage containers, high-pressure hydrogen storage can be divided into two major categories, namely ____ and ____.
6. The hydrogen storage capacity of Mg can reach up to 7.6 wt%, and its reaction formula is $Mg + H_2 = MgH_2$. This reaction releases 74.5 kJ/mol. The heat of H_2, the release of hydrogen is the reverse process of the above reaction, please calculate how much mass of Mg (atomic weight 24.3) is needed to store each kilogram of hydrogen, and how much heat is needed to release each kilogram of hydrogen from MgH_2?
7. Currently, most of the large-volume seamless pressure vessels made of steel used in China operate at a working pressure of 20 MPa, and the material is ____.
8. The main types of hydrogen sensors are ____, ____, ____, and ____.
9. Briefly describe the potential application fields of hydrogen energy.

References

1. Suryan A, Kim HD, Setoguchi T (2013) Comparative study of turbulence models performance for refueling of compressed hydrogen tanks. Int J Hydrog Energy 38(22):9562–9569
2. Zheng J, Ma K, Zhou W et al (2018) High pressure hydrogen storage container for hydrogen refueling station. Pressure Vessel 35(9):35–42
3. Elgowainy A, Mintz M, Kelly B et al (2009) Optimization of compression and storage requirements at hydrogen refueling stations. In: Proceedings of the ASME pressure vessels and piping conference, vol 5, pp 131–136
4. Physical Chemistry Teaching and Research Office, Tianjin University (2017) Physical chemistry, vol I. Higher Education Press, Beijing

5. Wang L (2011) The current situation and safe use of long tube trailers in China. China Special Equip Saf 27(06):51–53
6. Reddi K, Mintz M, Elgowainy A et al (2016) Challenges and opportunities of hydrogen delivery via pipeline, tube-trailer, liquid tanker and methanation-natural gas grid. Hydrogen Sci Eng Mater Process Syst Technol 35:849–874
7. China Hydrogen Energy Alliance (2019) China hydrogen energy and fuel cell industry white paper (2019 edition). China Hydrogen Energy Alliance, Beijing
8. Ma J, Liu S, Zhou W et al (2008) Comparison of hydrogen transportation schemes for hydrogen refueling stations. J Tongji Univ (Nat Sci Ed) 5:615–619
9. Elgowainy A, Reddi K, Sutherland E et al (2014) Tube-trailer consolidation strategy for reducing hydrogen refueling station costs. Int J Hydrog Energy 39(35):20197–20206
10. Gondal IA (2016) Hydrogen transportation by pipelines. Elsevier, Amsterdam, pp 301–322
11. National Energy Administration Petroleum and Natural Gas Department. China Natural Gas Development Report (2021)
12. China Society of Automotive Engineers (2019) World hydrogen energy and fuel cell vehicle industry development report (2018). Social Sciences Academic Press, Beijing
13. Liu J (2022) Economic analysis of energy storage technologies adaptable to renewable energy consumption. Energy Storage Sci Technol 11(1):397–404
14. Zhang H, Yuan T, Tan J et al (2022) Hydrogen energy planning framework for unified energy system. Trans China Electrotech Soc 42(1):83–94

Chapter 2
High-Pressure Gaseous Hydrogen Storage and Transportation

With the purchase of this book, you can use our "SN Flashcards" app to access questions using ▶ www.sn.pub/5lls30 free of charge in order to test your learning and check your understanding of the contents of the book. To use the app, please follow the instructions in Chap. 1.

Gaseous hydrogen storage and transportation technology refers to the technology of storing and transporting hydrogen in the gaseous form. The mainstream methods of gaseous hydrogen storage and transportation mainly include hydrogen storage and transportation by high-pressure cylinders and hydrogen transportation by pipelines. High-pressure cylinder hydrogen storage and transportation refers to the technology of using high-pressure containers for large-scale storage and transportation of hydrogen, while hydrogen transportation by pipelines refers to the technology of using medium-distance and long-distance hydrogen pipelines to transport hydrogen. Among gaseous, liquid, and solid hydrogen storage and transportation technologies, gaseous hydrogen storage and transportation technology is the most common and direct method. By adjusting the pressure reduction valve, hydrogen can be directly released, with a fast and relatively stable release speed. Therefore, this technology is currently the most mature and widely used technology. However, the high energy consumption of high-pressure hydrogen compression, the high requirements for container materials and preparation technology, the cost and stable operation of hydrogen compressors, and the safety of high-pressure hydrogen are key issues to be solved for the application of gaseous hydrogen storage and transportation technology. Moreover, although the pressure and gravimetric storage density of hydrogen have increased a lot with the technological advancement, its volumetric storage density has not been significantly increased. This chapter will

J. Zou, *Hydrogen Storage and Transportation*,
https://doi.org/10.1007/978-981-96-2875-9_2

introduce the principles, preparation, and application of high-pressure gaseous hydrogen storage.

2.1 Principle of High-Pressure Hydrogen Storage

2.1.1 Principle of Hydrogen Pressurization

Hydrogen can be considered as an ideal gas at high temperature and low pressure, and the mass at different temperatures and pressures can be calculated through the ideal gas state equation:

$$PV = nRT \tag{2.1}$$

where P is the gas pressure; V is the gas volume; n is the amount of gas substance; R is the standard gas constant [8.314 J/(mol K)], and T is the thermodynamic temperature. In an ideal state, the volume density of hydrogen is directly proportional to the pressure. However, due to the volume of actual gas molecules and the intermolecular interaction force, as the temperature decreases and the pressure increases, hydrogen increasingly deviates from the properties of an ideal gas, and the van der Waals equation is no longer applicable. Under such conditions, the modified van der Waals equation should be used:

$$P = \frac{nRT}{V - nB} - \frac{An^2}{v^2} \tag{2.2}$$

where A is the dipole interaction force, or the repulsion constant ($a = 2.476 \times 10^{-2}$ $m^6 \cdot Pa/mol^2$); B is the volume occupied by hydrogen molecules ($B = 2.661 \times 10^{-5}$ m^3/mol) [1].

The deviation of actual gas from ideal gas can be represented by the compressibility factor Z in thermodynamics, defined as Z = PV/nRT, and the compressibility factor of hydrogen increases with the increase of pressure.

To correctly describe the properties of real gases, many semi-empirical equations of real gases have been proposed in history, such as the van der Waals equation, Peng-Robinson equation, Redlich-Kwong equation, Beattie-Bridgeman equation, Benedict-Webb-Rubin equation, Martin-Hou equation, etc. The form of these equations is generally obtained by theoretical analysis, and the equations contain two or more constants related to the calculated gas, applicable to different gases. In addition, based on the principle of corresponding states, the corresponding state forms of various equations have been developed, and the constants in the equation no longer vary with different gases, but are common to similar gases within a certain pressure temperature range.

By fitting the real hydrogen property data provided by the National Institute of Standard and Technology (NIST) material property database, a simplified hydrogen state equation can be obtained:

$$Z = \frac{PV}{RT} = 1 + \frac{CP}{T} \quad (2.3)$$

In the formula, C is the coefficient (1.9155 × 10^{-6} K/Pa) [2]. When calculated within the range of 173 K<T<393 K, the maximum relative error is 3.80%; when calculated within the range of 253 K<T<393 K, the maximum relative error is 1.10%.

Figure 2.1 compares the relationship between hydrogen pressure and density at different temperatures calculated by several different state equations.

Density, R-K equation, fitting equation, P-R equation, ideal gas, NIST data, pressure.

Density, R-K equation, fitting equation, P-R equation, ideal gas, NIST data, pressure.

2.1.2 *Hydrogen Pressurization Equipment*

Hydrogen compressors, high-pressure hydrogen storage tanks, and hydrogen refueling machines are the core equipment of the high-pressure gaseous hydrogen station system. After obtaining hydrogen through external hydrogen supply or on-site hydrogen production, the hydrogen station converts it into pressure-stable dry gas after treatment by the pressure regulation drying system, then enters the high-pressure hydrogen storage tank for storage under the delivery of the hydrogen compressor, and finally is refueled by the hydrogen refueling machine for fuel cell vehicles.

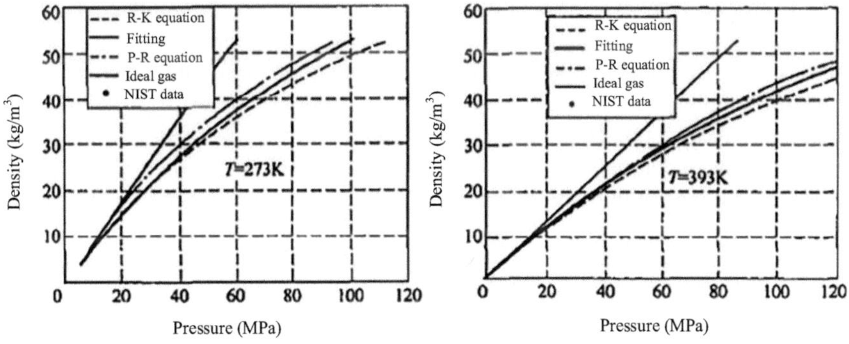

Fig. 2.1 Relationship between pressure and density of hydrogen at T = 273 K and 393 K

High-pressure hydrogen is generally obtained from compressors. The compressor can be regarded as a booster pump, which reduces the pressure on the low-pressure side of the system and increases the pressure on the high-pressure side, causing hydrogen to flow from the low-pressure side to the high-pressure side. In engineering practice, there are mainly two ways to compress hydrogen: one is to directly compress the hydrogen to the pressure required for the hydrogen storage container with a compressor and store it in a large-volume hydrogen storage container; the other is to first compress the hydrogen to a higher pressure for storage, and when refueling is needed, a part of the gas is introduced for filling, and then the hydrogen compressor is started to boost the pressure, so that the hydrogen storage container reaches the required pressure. Hydrogen compressors have various types, such as diaphragm, reciprocating piston, rotary, screw, and turbine. The appropriate type is selected according to the flow rate, suction and exhaust pressure when applied. The piston compressor has a large flow rate, and the single-stage compression ratio is generally 3:1–4:1; the diaphragm compressor dissipates heat quickly, the compression process is close to an isothermal process, and it can have a higher compression ratio, up to 20:1, but because the flow rate is small, it is mainly used in situations where the demand for hydrogen pressure is high but the flow rate is not large.

Generally speaking, for compressors with a pressure below 30 MPa, piston types are usually used, which have been proven to have a higher degree of reliable operation, and can form a compressor composed of multiple stages independently. When the pressure is above 30 MPa and the volumetric flow rate is small, a diaphragm compressor can be chosen. The advantage of the diaphragm type compressor is that it has reliable sealing at high pressure: because the sealing structure of its gas chamber is a diaphragm clamped between the cylinder head and the cylinder body, it is fastened by the main bolt to form a static seal form, which can ensure that the gas will not leak, and the diaphragm chamber is closed, not in contact with any oil droplets, oil mist and other impurities, ensuring that the incoming gas is not contaminated by the outside world during the compression of the gas. This shows its special superiority in situations that require high-purity media.

The structure of the hydrogen compressor includes basic components (such as crankcase, crankshaft, connecting rod, etc.), cylinder components, plunger components, cooler components, safety protection control system, and other auxiliary components. Figure 2.2 is a picture of the HY-C-500-45 type hydrogen diaphragm compressor (with a power of 55 kW).

The hydrogen compressor used in the hydrogen refueling station is a high-purity oil-free hydrogen compressor, which is the core device for pressurizing and injecting the hydrogen source into the gas storage system. The output pressure and gas tightness are its most important performance indicators. Globally, compressors of various types are used. High-purity oil-free hydrogen compressors are mainly divided into metal diaphragm compressors and high-purity oil-free booster compressors.

Fig. 2.2 The HY-C-500-45 type hydrogen diaphragm compressor

2.1.2.1 Diaphragm Compressors

The diaphragm compressor, also known as a membrane compressor, is a special structure volumetric compressor. There is a set of diaphragms in the cylinder, and the space contained between the cylinder head and the diaphragm forms the gas compression chamber. The other side of the diaphragm is the oil pressure chamber. The piston reciprocates in the oil cylinder, and the diaphragm folds back and forth under the action of oil pressure, gas pressure, and its own elastic deformation force, periodically changing the volume of the gas compression chamber to achieve compression and transportation of the gas. This compression method has no secondary pollution, provides very good protection for the compressed gas, and has the characteristics of a large compression ratio, good sealing, so that the compressed gas is not contaminated by lubricating oil and other solid impurities. It is suitable for compressing high-purity, rare and valuable, flammable and explosive, toxic/harmful, and high-pressure gases. However, its cover plate has a special surface, the diaphragm is easy to damage, so the processing and maintenance costs are high. In addition, the exhaust volume of the diaphragm compressor is relatively small due to the limited compression ratio and the volume of the gas chamber.

There are many low-pressure diaphragm compressor manufacturers both home and abroad, but there are not many manufacturers that can provide a pressure of more than 45 MPa. Table 2.1 lists the domestic and foreign companies that can provide diaphragm hydrogen compressors for hydrogen refueling stations.

Currently, diaphragm hydrogen compressors are widely used in hydrogen refueling demonstration stations both at home and abroad, with an exhaust volume of

Table 2.1 Domestic and foreign companies that provide diaphragm hydrogen compressors for hydrogen refueling stations

Country	Brand	Manufacturer	Remarks
China	Tiāngāo	Beijing Tiāngāo Diaphragm Compressor Co., Ltd.	Provided the diaphragm compressor for the first domestically produced hydrogen refueling station (Beijing Green Energy Company) in China, and has provided diaphragm compressors for seven hydrogen refueling stations in China, with the longest continuous operation time. It is the only one in China with experience in developing a 90 MPa hydrogen diaphragm compressor
China	Hengjiu	Jiangsu Hengjiu Machinery Co., Ltd.	Exhaust pressure is 20–45 MPa
China	Zhongding Hengsheng	Beijing Zhongding Hengsheng Gas Equipment Co., Ltd.	Exhaust pressures are 45 MPa and 70 MPa, and the maximum flow rate of a single 45 MPa hydrogen diaphragm compressor reaches 2000 Nm^3/h
China	Jingcheng	Beijing Jingcheng Compressor Co., Ltd.	Signed a hydrogen compressor cooperation agreement with PDC in the United States
United Kingdom	Howden	Howden Burton Corblin	Mainly twin-screw process compressors and single-stage centrifugal blowers, as well as metal diaphragm compressors and piston compressors
United States	PDC	PDC Machines Inc.	Manufacturing technology of hydrogen compressors with a three-layer metal diaphragm structure, with an output pressure limit exceeding 85 MPa. High market share in compressors for hydrogen refueling stations
Germany	HOFER	Andreas Hofer Hochdrucktechnik GmbH	The maximum exhaust pressure of a single machine is 300 MPa
United States	PPI	Pressure Product Industries	The maximum exhaust flow of a single machine is 680 Nm^3/h, and the maximum exhaust pressure is 200 MPa

mainly 20 kg/h or less abroad. The hydrogen diaphragm machines used in domestic hydrogen refueling stations are mainly domestically developed with a 45 MPa exhaust pressure and 41.6 kg/h exhaust volume. The hydrogen diaphragm compressor purchased by the State Energy Group Rugao Hydrogen Refueling Station is from the United States PDC that has an exhaust pressure of 87.5 MPa. For the 70 MPa hydrogen diaphragm compressor used in domestic hydrogen refueling stations, Beijing Tiangao and Tongji University developed a 90 MPa hydrogen prototype diaphragm compressor under the "863 Program" (State High-Tech Development Plan), which has been demonstrated in Dalian Tongxin Hydrogen Refueling Station. The prototype technical indicators meet the operation requirements of the 70 MPa hydrogen refueling station, but the reliability needs to be further improved for commercial application.

The diaphragm compressors equipped in hydrogen refueling stations at home and abroad lack large-scale, high-density, frequent hydrogen refueling application cases. The increasing high-density hydrogen demand in China provides a very good opportunity for the reliability verification of hydrogen diaphragm compressors both at home and abroad.

2.1.2.2 High-Purity Oil-Free Booster Compressors

The high-purity oil-free booster hydrogen compressor is also known as the hydraulic-driven oil-free hydrogen reciprocating piston compressor. The maximum exhaust pressure of the standard design product is 100 MPa. Currently, this type of compressor is used more in skid-mounted hydrogen stations, mainly with an exhaust pressure of 45 MPa and an exhaust volume of 41.6–66.6 kg/h (inlet pressure 12.5 MPa, 500 kg/12 h) or an exhaust pressure of 87.5 MPa and an exhaust volume 41.6–66.6 kg/h (inlet pressure 28 MPa, 500 kg/12 h). Table 2.2 lists the foreign companies that can provide liquid-driven plunger hydrogen compressors for hydrogen refueling stations.

2.1.2.3 Ion Compressors

The basic feature of ion compressors is the use of a certain ion liquid to replace the piston to directly contact the compressed working medium to complete the compression process. Currently, ion compressors are still in the research and promotion stage, and there is no unified technical standard internationally. The structure of ion compressors launched by various compressor manufacturers also has its own merits.

In recent years, the German company Linde has introduced an ion compressor for 90 MPa hydrogen refueling stations, as shown in Fig. 2.3. It uses a hydraulic system to drive the solid piston movement in the cylinder. An ionic liquid is injected into the upper part of the solid piston, separating the piston and the compressed gas.

Table 2.2 Companies related to liquid-driven plunger hydrogen compressors for hydrogen refueling stations

Country	Brand	Manufacturer
Germany	MAXIMATOR	MAXIMATOR GmbH
Germany	HOFER	Andreas Hofer Hochdrucktechnik GmbH
Italy	Idro Meccanica	Idro Meccanica srl
United States	HP	HYDRO-PAC, Inc.
United States	HASKEL	HASKEL International
Netherlands	RESATO	RESATO International B. V.

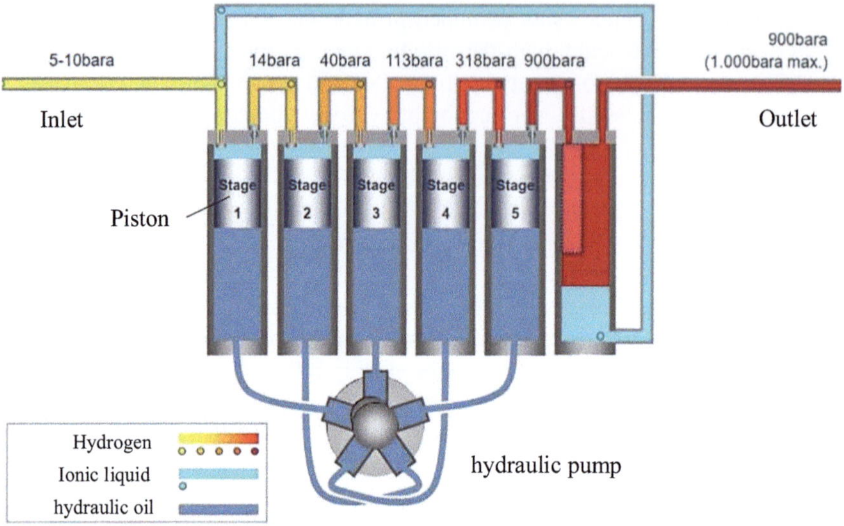

Fig. 2.3 Schematic diagram of Linde ion compressor

This design separates the compression system from the drive system through the solid piston. In the hydraulic drive system, the cam rotates, sequentially driving the hydraulic oil in and out of the cylinder. The hydraulic oil drives the solid piston, which in turn drives the ionic liquid to compress the gas, realizing the intake, compression, exhaust, and expansion process of the compressor. The ion compressor is essentially like a piston machine, so it can achieve a relatively high speed and a large cylinder diameter, achieve large displacement and a high pressure, and has good adaptability to the frequently changing wide conditions of the hydrogen refueling station. Secondly, the ionic liquid in the cylinder not only serves as a medium for compressing gases but also has the functions of sealing and lubrication, solving the problem of lubrication and sealing of traditional piston machines under high pressures. The ionic liquid will not contaminate hydrogen due to its extremely low vapor pressure. At the same time, the presence of the liquid improves the heat dissipation efficiency of the compression system, can achieve an almost isothermal compression process, reduce the exhaust temperature of the compressor, and improve thermal efficiency. The ionic liquid in the cylinder can change the boundary during the compression process, thereby entering and filling spaces, such as valve cavity channels, which can greatly reduce the relative clearance of the cylinder, thereby greatly improving the volume utilization rate of the cylinder, and can achieve a larger single-stage pressure ratio.

Inlet, piston, highest, outlet, hydraulic pump.

Hydrogen, ionic liquid, hydraulic oil.

2.1.3 Principle of Hydrogen Embrittlement of Materials

2.1.3.1 Hydrogen Embrittlement Phenomenon

The hydrogen embrittlement phenomenon of pressure vessels, also known as white spots, refers to the corrosion of the vessel wall by hydrogen, which causes a reduction in the plasticity and strength of the material, leading to cracking or delayed brittle failure. Hydrogen embrittlement is caused by the presence of hydrogen in material defects, usually manifested as a decrease in mechanical properties, such as ductility and toughness.

Under high temperature and high pressure conditions, hydrogen penetrates into the metal in an atomic state, looking for high residual stress areas in the metal where it can accumulate and combine to form hydrogen molecules [2, 3]. In addition, single atomic hydrogen can also form atomic bonds at defect positions, such as vacancies, doping, grain boundaries, and dislocations, thereby causing an increase in internal pressure to produce bubbles or cracks [4]. When the concentration of hydrogen gas in the metal reaches the critical pressure, hydrogen-induced stress cracking will occur. This cracking mechanism can cause ductile steel to undergo brittle fracture under continuous stress loads below the yield stress; materials that are usually easily affected by this mechanism are carbon steel and low alloy steel. In addition, hydrogen embrittlement can also cause a decrease in the ductility of metal materials, and the percentage loss of ductility directly depends on the hydrogen content.

One of the most commonly used materials for hydrogen storage and transportation is steel. Within the steel structure, hydrogen diffuses in the form of protons [5], causing corrosion to the metal surface. To better understand the principle of hydrogen embrittlement, various theories, such as internal pressure theory, hydrogen-enhanced local plasticity theory, and hydrogen-enhanced decohesion embrittlement, have been proposed to explain the phenomenon of hydrogen embrittlement. Among them, the internal pressure theory proposed by Zappfe and Tetelman is widely accepted. The theory points out that on the steel surface, hydrogen atoms are absorbed and gather at different structural positions. At these defect locations, atomic hydrogen can combine and form molecular hydrogen, which increases the cracking caused by local internal pressure, such as carbides, non-metallic inclusions, grain boundaries, dislocations, carbonitrides, and other high-stress concentration areas. When the pressure produced by the hydrogen concentration in these areas reaches the critical value in the metal matrix, cracks will occur [6, 7]. Hydrogen embrittlement depends on environmental factors, material compositions, and the metal surfaces. Among environmental factors, gas pressure, temperature, and other conditions are the most important, as they play a leading role in the susceptibility to the phenomenon. According to Sievert's gas pressure law:

$$K = \frac{[\mathrm{gas}]}{P_{\mathrm{gas}2}^{1/2}} \tag{2.4}$$

where K is the equilibrium constant; [gas] is the mass concentration (wt%) of a specific gas in molten metal; $P_{gas2}^{1/2}$ is the partial pressure of a specific gas formed by diatomic molecules. It can be assumed that the degree of hydrogen embrittlement is directly proportional to the square root of hydrogen pressure. Generally, the sensitivity of hydrogen embrittlement to temperature and pressure largely depends on the material itself. For example, Barthélémy conducted a survey to determine the effect of hydrogen pressure on high yield strength steel and observed that the cracking threshold was about 60 MPa. High yield strength steel, if used for hydrogen storage, carries a high risk. In contrast, in a 70 MPa high-pressure hydrogen gas test, 316-type and A286-type steels showed better performance. Similarly, the hydrogen storage performance of other metal materials (such as aluminum alloys) under high pressures has been explored at this stage. These materials have been proven to resist hydrogen embrittlement, and although some strength loss is observed under high hydrogen pressure, their ductility does not decrease.

Hydrogen embrittlement can occur over a wide temperature range, but for most materials, it is most sensitive to hydrogen embrittlement near room temperature. When it is lower than room temperature, the diffusion rate of hydrogen is too low to fill enough trapping sites, and when it is higher than room temperature, the migration rate of hydrogen increases, leading to less trapping. For some iron-based high-temperature alloys, because their compositions evolve from ordinary stainless steel, they tend to show hydrogen embrittlement in a lower temperature range. At high temperatures with sufficient thermal activation energy, the possibility of hydrogen reaction embrittlement is high, and hydrogen atoms react chemically with certain components or impurities in the grain boundaries.

Sugimoto and Fukai conducted research and calculated several face-centered cubic metals Pt, γ-Fe, Cu, Au, Al, Ag, and Ni, as well as body-centered cubic metal Cr, α-Fe, Mo, and W, as a function of temperature and hydrogen pressure. Regarding the interaction of hydrogen with vacancies, hydrogen can reduce the formation energy of vacancy clusters, promoting the formation of excess vacancies under high temperature and pressure conditions. The vacancy concentration can approach 10–20%, and up to 6 hydrogen atoms can be combined, which leads to an apparent increase in solubility and affects the diffusion rate. Compared with other defect locations, the vacancy density is usually relatively low at lower temperatures; however, rapid dissolution of the metal will produce many local vacancies near the crack surface, or local plastic deformation at the crack tip [8].

Certain operations in the component manufacturing process can cause metals to absorb hydrogen (such as welding, electroplating, pickling, etc.). In order to reduce hydrogen embrittlement in sensitive metals, baking heat treatment is used to expel any hydrogen. Similarly, when the environment allows the cathodic reaction to produce hydrogen, hydrogen can appear as a byproduct of the corrosion process, causing hydrogen to dissociate in atomic form and enter the metal matrix instead of being dispersed in the metal matrix. In this case, cracking may be caused by another failure called stress corrosion cracking. On the other hand, if the reason for hydrogen penetration in the steel matrix is due to the presence of hydrogen sulfide, this phenomenon is called "sulfide stress cracking".

2.1.3.2 Classification of Hydrogen Embrittlement

1. Environmental Hydrogen Embrittlement: When metal is immersed in a hydrogen gas atmosphere (such as a storage tank), hydrogen can be absorbed or adsorbed, which changes the mechanical properties of the material without having to form a second phase. The stress endured by the material largely determines the impact of hydrogen, which also increases at room temperature.
2. Internal Reversible Hydrogen Embrittlement: During material processing, hydrogen enters the matrix, causing structural failure even if the material is not exposed to hydrogen. One of the most notable features is that internal cracks show discontinuous growth. This type of hydrogen embrittlement has been observed in the temperature range of 173–373 K and in an average hydrogen atom of 10^{-7}–10^{-5}. At temperatures close to room temperature, internal reversible hydrogen embrittlement is more severe.
3. Hydrogen Reaction Embrittlement: This occurs when one component of the material reacts chemically with hydrogen to form gaseous hydrogen bubbles (called blistering) or new phases or microstructural elements (such as hydrides), and usually occurs at high temperatures. Under these conditions, this phenomenon occurs through the appearance of blistering or swelling, thereby weakening the material and starting to crack.

The hydrogen diffusion phenomenon necessary for hydrogen embrittlement can occur at low and high temperatures. During the welding process, a heat-affected zone is produced between the filler metal and the parent metal, which is a key point for the occurrence of hydrogen embrittlement, even if the parent metal is not easily affected by this phenomenon. During the pickling process, the metal surface is exposed to an acidic, humid environment, causing corrosion damage, thereby accelerating the occurrence of hydrogen embrittlement.

Hydrogen embrittlement and stress corrosion cracking have many similarities; however, one fundamental difference is the behavior of cathodic protection. Although cathodic protection is one of the most effective corrosion protection methods, it does not reduce the diffusion rate of hydrogen in metals.

2.1.3.3 Hydrogen Embrittlement of Different Materials

1. Stainless steel. In the hydrogen embrittlement process of stainless steel, hydrogen enters into the grain boundaries through the diffusion process, combining with carbon and iron in the alloy to generate methane gas. The generated methane gas does not move, resulting in a significant increase in local pressure, thereby promoting the initiation of cracks. In nuclear power plants where parts are made of metals other than aluminum, the pH of the reactor coolant is kept neutral or alkaline to prevent hydrogen embrittlement.

Atomic hydrogen can be absorbed by the metal lattice at room temperature, diffusing through the grains, and tends to accumulate in inclusions or lattice defects.

Under these conditions, the resulting cracks will be trans-granular. On the other hand, at high temperatures, diffused hydrogen tends to accumulate at grain boundaries, thereby producing intergranular cracks.

When the production of hydrogen gas stops due to changes in environmental conditions and the crack has not yet started, the trapped hydrogen gas diffuses again, thereby restoring the ductility of the material, so hydrogen embrittlement is not a permanent phenomenon.

The problem of hydrogen embrittlement can be solved by different control methods, such as controlling the amount of residual hydrogen, reducing the amount of hydrogen collected during the manufacturing process, searching for hydrogen embrittlement-resistant alloys, developing embrittlement-resistant coatings, and reducing the amount of hydrogen present in the material service environment.

2. Carbon steel. Due to the phenomenon of hydrogen embrittlement, structural components are affected by cracks and brittle fractures under loads lower than the yield strength of the metal. Figure 2.4 shows the absorption of atomic hydrogen in carbon steel at room temperature. The absorption mechanism may involve atomic or molecular forms of hydrogen. After hydrogen absorption, hydrogen diffuses through the metal block and gets captured in the grain boundaries to form bubbles. These bubbles exert internal pressure on the metal grains, which increases over time and reduces ductility and strength.

H atoms, H_2 molecules.

During the machining or assembly operations of structural components, hydrogen diffuses at room temperature in a hydrogen-rich environment, such as pickling used to clean steel surfaces to remove oxides, and electroplating processes. Also, hydrogen embrittlement occurs when metals are exposed to acid or corrosion. Figure 2.5 shows the fracture of a galvanized bolt steel, where individual grains undergo intergranular fracture.

Intergranular fracture.

3. Aluminum and aluminum alloys. Dry hydrogen gas environments have little effect on aluminum and its alloys. The main problem with hydrogen gas is its contact with moisture and the air gaps produced through the melting, casting,

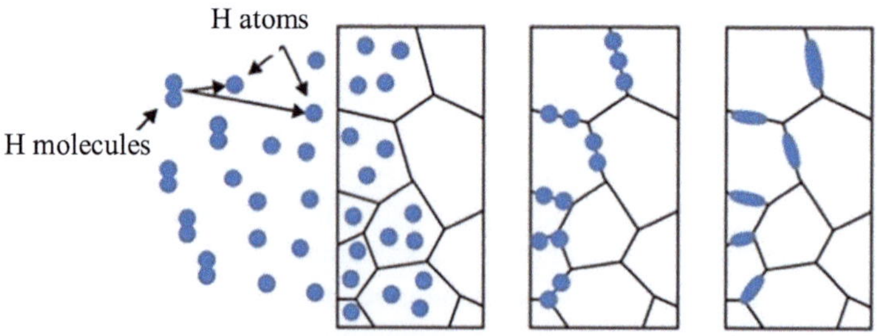

Fig. 2.4 Hydrogen atoms absorbed by carbon steel alloy [1]

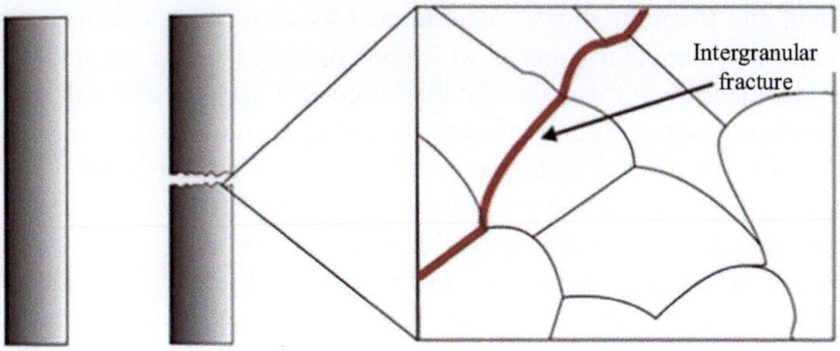

Fig. 2.5 Hydrogen embrittlement failure [1]

and solidification processes of the foundry. These gaps are material defects that affect the mechanical properties of cast aluminum and forged aluminum, such as ductility and fracture toughness. During the cooling process from the melt, hydrogen diffuses into precipitates and casting defects, and cracks are generated due to the reduced solubility of hydrogen in solid metal at lower temperatures. Dry hydrogen gas at room temperature can withstand pressures up to 69 MPa, and no significant hydrogen embrittlement effect occurs in aluminum alloys. However, when high-strength aluminum alloys are electrochemically charged with hydrogen, their ductility decreases. The main mechanism of aluminum alloy embrittlement on water media may be sulfide stress cracking rather than pure hydrogen embrittlement effects.

4. Copper and copper alloys. Copper and copper alloys are generally not sensitive to hydrogen embrittlement unless they contain oxygen or copper oxide. When annealed in a hydrogen atmosphere, atomic hydrogen diffuses and reacts with copper oxide or oxygen to form water, which becomes high-pressure steam when the temperature exceeds 375 °C. The steam promotes hydrogen damage in the form of cracks and bubbles, reducing the fracture toughness and ductility of copper even without the application of external pressure.
5. Nickel and nickel-based alloys. Nickel and its alloys have acceptable high-temperature strength, oxidation resistance, and heat corrosion resistance. However, not all nickel-based alloys have good oxidation grades, and a chemically corrosive environment does not necessarily mean they are also immune to hydrogen embrittlement. As an element, pure nickel is severely embrittled by hydrogen. Therefore, most binary alloys with a rich nickel composition, such as nickel-copper, nickel-iron, nickel-cobalt, and nickel-tungsten, become very brittle due to hydrogen in the nickel-rich region. The same observation has been made in some nickel-rich alloy systems. For example, a nickel-rich alloy known as k-Monel 1 is known to be embrittled by hydrogen under high pressure but is not affected by electrolytic charging. Hydrogen environment sensitivity includes crack initiation on the metal surface, where hydrogen adsorption at the crack tip leads to an increased crack propagation rate, resulting in embrittlement.

6. Titanium and titanium alloys. Titanium and its alloys have excellent corrosion resistance in water environments, which is caused by the thin and stable titanium oxide film naturally formed under oxidizing conditions in air and water. However, under extreme cathodic charging with an applied current, some of these titanium alloys have experienced hydrogen embrittlement in aqueous media. Under low to medium cathodic charging conditions, the naturally formed titanium oxide film on titanium can effectively inhibit hydrogen absorption. However, at high cathodic charging current densities, this protective film will rupture, lose its protective effect on titanium alloys, and allow hydrogen atoms to penetrate the titanium bulk. In environments where the electrolyte (such as seawater) is near neutral and in contact with metals (such as zinc, aluminum, and magnesium) and the temperature is above 80 °C, hydrogen absorption and hydride formation are accelerated. In a dry hydrogen gas environment, with increasing temperature and pressure, titanium and its alloys easily absorb hydrogen. A moderate amount of titanium hydride precipitation does not harm most applications, especially within a hydrogen concentration range of 4×10^{-5} to 8×10^{-5}. However, when the temperature exceeds 250 °C, an excess of titanium hydride forms rapidly. This type of hydrogen embrittlement is a type of hydrogen reaction embrittlement; however, during high-temperature processes, such as welding or heat treatment in the presence of hydrogen, it is also considered by some industries to be a type of internal hydrogen embrittlement.

2.1.3.4 Hydrogen Embrittlement Prevention

The evaluation of a material's sensitivity to hydrogen embrittlement can be determined by high-temperature vacuum hydrogen determination technology to measure the hydrogen content in the metal, the permeation rate of hydrogen in the metal, and the hydrogen embrittlement coefficient (0–1) determined by cross-sectional shrinkage. The smaller the measured value, the less sensitive it is to hydrogen embrittlement.

There are many preventive measures for hydrogen embrittlement, and at this stage they include but are not limited to the following:

1. Replace lower-strength steels and alloys to reduce the risk of hydrogen embrittlement, but special circumstances must be considered to ensure that the material can withstand the load applied throughout the process. If high-strength steels and alloys are the best material choice, some heat treatments can be performed to reduce the hardness and residual stress that may lead to embrittlement.
2. Carry out vacuum dehydrogenation, which is commonly used in industry to remove hydrogen from steel.
3. Perform post-weld heat treatment. Post-weld heat treatment is usually carried out shortly after welding to avoid residual stress and adjust the temperature changes between operating conditions and welding procedures.

2.2 High-Pressure Hydrogen Storage Equipment

2.2.1 Classification of High-Pressure Hydrogen Storage Equipment

High-pressure hydrogen storage equipment is an important material guarantee for the implementation of high-pressure hydrogen storage technologies. It can accommodate high-pressure hydrogen gases with its sufficiently high design strength and realize the storage, transportation, and utilization of hydrogen gases. To better define and describe high-pressure hydrogen storage equipment, it can be classified according to regulatory procedures or manufacturing materials at this stage.

2.2.1.1 Classification According to Regulatory Procedures

According to the different usage requirements of high-pressure hydrogen storage equipment, combined with the two major regulatory procedures, TSG 21—2016 "Supervision Regulation on Safety Technology for Stationary Pressure Vessel" and TSG R0005—2011 "Safety Technical Supervision Regulations for Mobile Pressure Vessels", it can be divided into fixed pressure vessels and mobile pressure vessels. To ensure the safe use of pressure vessels, prevent and reduce accidents, protect people's lives and property, and promote economic and social development, the design, manufacture, installation, etc., of such pressure vessels should meet the requirements in the relevant regulatory procedures.

The "Supervision Regulation on Safety Technology for Stationary Pressure Vessel" defines fixed pressure vessels as pressure vessels installed at fixed locations for use, which mainly applies to pressure vessels that meet the following conditions under this definition:

1. The working pressure is greater than or equal to 0.1 MPa.
2. The volume is greater than or equal to 0.03 m^3 and the inner diameter (the maximum geometric size of the inner boundary of the non-circular cross-section) is greater than or equal to 150 mm.
3. The medium is gas, liquefied gas, and liquids whose maximum working temperature is higher than or equal to their standard boiling point.

This regulation also further divides fixed pressure vessels, for example, according to the design pressure of the pressure vessel, it is divided into low pressure (code L, pressure range 0.1 MPa = P < 1.6 MPa), medium pressure (code M, pressure range 1.6 MPa = P < 10.0 MPa), high pressure (code H, pressure range 10.0 MPa = P < 100.0 MPa) and ultra-high pressure (code U, pressure range 100.0 MPa = P) four pressure levels; according to the degree of danger, it is divided into I, II, III categories (ultra-high pressure vessels belong to the third category), and this type of pressure vessel classification is equivalent to the first, second, and third categories of pressure vessels in the special equipment catalog. In high-pressure hydrogen storage, this type of high-pressure hydrogen storage equipment (i.e., fixed pressure vessels) is usually used for hydrogen storage in hydrogen refueling stations.

The "Safety Technical Supervision Regulations for Mobile Pressure Vessels" defines mobile pressure vessels as transport equipment permanently connected by a tank or large-volume seamless steel gas cylinder and a running device or frame, including railway tank cars, automobile tank cars, long tube trailers, tank containers, and tube bundle containers, etc. It mainly applies to pressure vessels that meet the following conditions under this definition:

1. It has the function of filling and unloading medium, and participates in railway, highway, or waterway transportation.
2. The working pressure of the tank body is greater than or equal to 0.1 MPa, and the nominal working pressure of the gas cylinder is greater than or equal to 0.2 MPa.
3. The tank volume is greater than or equal to 450 L, and the cylinder volume is greater than or equal to 1000 L.
4. The filling medium is gas and the maximum working temperature is higher than or equal to its standard boiling point.

However, it is worth noting that when the tank or cylinder is made of non-metallic materials, this regulatory procedure does not apply. For this issue, the design, manufacture, and installation of new mobile composite pressure vessels made of non-metallic materials at this stage must follow the relevant provisions in the "Vehicle Compressed Hydrogen Plastic Liner Carbon Fiber Fully Wrapped Cylinder". This standard is specifically designed for the design and manufacture of reusable refillable cylinders used as fuel tanks fixed on road vehicles, with a nominal working pressure not exceeding 70 MPa, a nominal volume not exceeding 450 L, a storage medium of compressed hydrogen, a working temperature not lower than −40 °C and not higher than 85 °C. In high-pressure hydrogen storage, such high-pressure hydrogen storage equipment (i.e., mobile pressure vessels) is usually used for hydrogen storage on mobile carriers, such as long tube trailers, tube bundle trucks, and fuel cell vehicles.

2.2.1.2 Classification by Manufacturing Materials

According to the different categories of materials used in the manufacture of high-pressure hydrogen storage equipment, there are four types: Type I (all-metal pressure vessel), Type II (metal liner fiber circumferential winding pressure vessel), Type III (metal liner fiber full winding pressure vessel), Type IV (plastic liner fiber full winding pressure vessel), as shown in Fig. 2.6 and Table 2.3.

Composition, all metal, metal liner + GFRP layer (circumferential winding), metal liner + CFRP layer (full winding), plastic liner + CFRP layer (full winding).

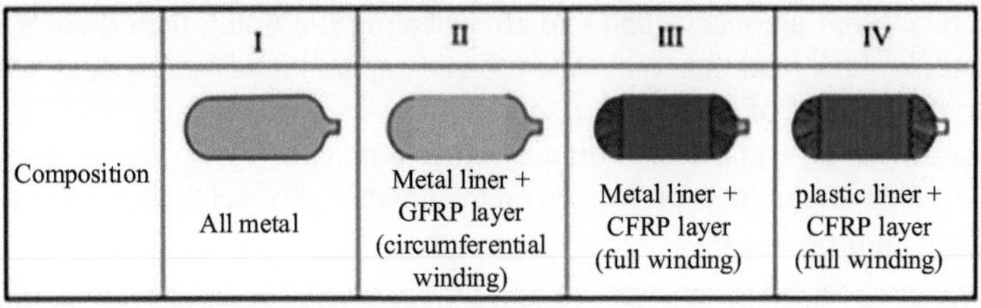

Fig. 2.6 Classification of high-pressure hydrogen storage equipment based on manufacturing materials [9]

Table 2.3 Types and characteristics of high-pressure hydrogen cylinders

Type	Type I cylinder	Type II cylinder	Type III cylinder	Type IV cylinder
Structure and material	Pure metal structure	Metal liner Fiber circumferential winding	Metal liner Fiber full winding	Non-metal liner Fiber full winding
Common working pressure	10–45 MPa	30–45 MPa	30–90 MPa	50–70 MPa
Common materials	30CrMo steel	30CrMo steel Glass fiber reinforced	6061 Aluminum alloy Carbon fiber reinforced	High-density polyethylene Carbon fiber reinforced
Cost	Low	Average	High	High
Gravimetric hydrogen density	Low	Average	High	High
Common uses	Industrial gas cylinders High-pressure long tube trailers Forklift hydrogen storage Hydrogen storage at hydrogen stations	High-pressure long tube trailers Hydrogen storage for hydrogen stations	Vehicle hydrogen storage suitable for various vehicles	Hydrogen storage for passenger cars High-pressure long tube trailers Hydrogen storage at hydrogen stations

The lining materials of Type I, II, and III pressure vessels are usually made of aluminum (6061 or 7060) or steel (stainless steel or chrome-molybdenum steel). This part can be manufactured by three different processes [10]:

1. Form a concave end cap by deep drawing the plate, and then close the other end by extrusion or spinning.

2. First punch out a concave end cap after heating the billet, then draw it into an open tank blank, and then make the top end cap and interface pipe, etc., according to the method of extrusion or spinning.
3. Use seamless pipe material to directly close both ends by extrusion or spinning. Finally, heat treatment is performed on the finished product to obtain the required mechanical properties.

For Type IV pressure vessels, because their lining materials often use high-density polyethylene, polyamide-based polymers, and other high-polymer materials with good fluidity, they can usually be obtained by rotational molding, blow molding, and other molding methods. In Type II, III, and IV pressure vessels, the fiber winding layer is their main reinforcement. By designing and controlling the content, tension, and winding trajectory of high-performance fibers, the performance of high-performance fibers can be fully utilized to ensure the performance of composite material pressure vessels is uniform and stable, and the burst pressure dispersion is small. At present, fibers, such as glass fiber, silicon carbide fiber, alumina fiber, boron fiber, carbon fiber, aramid, and PBO fiber, are all used to manufacture fiber composite material winding gas cylinders, but carbon fiber is gradually becoming the mainstream fiber raw material due to its excellent performance (such as Toray's T300, T700, T1000 in Japan).

From Type I to Type IV pressure vessels, their advantages and disadvantages are different. Mori et al. [9] compared the wall thickness of the materials required for different types of pressure vessels under the same storage conditions, as shown in Fig. 2.7. According to the figure, from Type I to Type IV, the wall thickness of the pressure vessel gradually decreases, and the replacement of the lining material means that the entire pressure vessel becomes lighter and the mass hydrogen density increases. In actual use, although Type I and II pressure vessels have a large mass-to-volume ratio and low mass hydrogen density, they are widely used in industry due to their low cost, for example, hydrogen is often stored in Type I pressure vessels at a pressure of 20–30 MPa (gravimetric storage efficiency is about 1 wt%). Type III and IV pressure vessels are widely used in fuel cell powered vehicles because they use lightweight and high-strength fibers, not only are they lightweight,

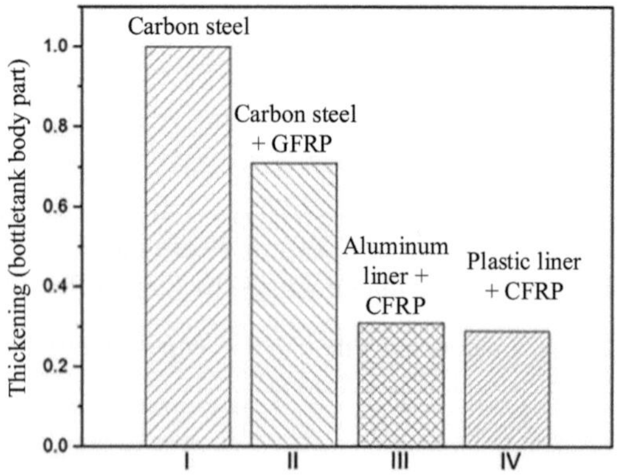

Fig. 2.7 Comparison of the wall thickness of various storage containers under the same volume at 35 MPa

can they also withstand higher pressures to store enough hydrogen in a limited space to ensure their driving range. At present, the main type of hydrogen storage tank used in fuel cell vehicles in China is the 35 MPa Type III hydrogen storage tank, while the more superior performance and larger storage Type IV hydrogen storage tanks are more widely used and mature abroad, and it is still in the research and development stage in China. In recent years, both domestic and foreign companies have proposed a gas tank system that relies solely on carbon fiber winding (i.e., Type V tank), this type of gas tank does not require a liner, which can further reduce the mass of the high-pressure hydrogen storage container, but its safety, pressure resistance level, etc., remains worth further research.

Thickening (tank body part), carbon steel, carbon steel + GFRP, aluminum liner + CFRP, plastic liner + CFRP.

2.2.2 *Fixed Hydrogen Storage Containers*

According to TSG 21-2016 "Safety Technical Supervision Regulations for Stationary Pressure Vessels", stationary pressure vessels are defined as pressure vessels installed in a fixed position. When the working pressure of the vessel is above 10 MPa, it is regarded as a high-pressure vessel. In order to utilize the pressure difference for filling and meet the hydrogenation needs of fuel cell vehicles, the pressure of the fixed hydrogen storage container used in the hydrogenation station is usually higher than the on-board hydrogen storage pressure (35 MPa or 70 MPa). GB 50516-2010 "Technical Specifications for Hydrogen Refueling Stations" stipulates that the hydrogen storage system pressure corresponding to a filling pressure of 35 MPa should not exceed 45 MPa, and the hydrogen storage system pressure corresponding to a filling pressure of 70 MPa should not exceed 90 MPa. The pressure is usually set at the level of 2–3.

Fixed high-pressure hydrogen storage vessels can be divided into seamless high-pressure hydrogen storage vessels, steel-strip staggered high-pressure hydrogen storage vessels, and fiber-wound high-pressure hydrogen storage vessels according to their structural forms. Hydrogen refueling stations are often equipped with different types of fixed containers to meet different hydrogenation needs.

2.2.2.1 Seamless High-Pressure Hydrogen Storage Containers

CrMo steel is commonly used as a material for seamless high-pressure hydrogen storage vessels. The commonly used grade is 4130X. It has the advantages of low cost, excellent comprehensive mechanical properties, and the ability to maintain good plasticity and toughness under high strength levels. The entire container is made of seamless high-strength steel pipes that are forged and closed at both ends to avoid defects, such as cracks, pores, and slag inclusions, caused by welding. The disadvantage of steel hydrogen storage containers is that they are easily affected by hydrogen embrittlement, and the capacity of a single container is limited by standard regulations and material strength. The American Society of Mechanical Engineers standard ASME BPVC-VIII-1-2019 "Boilers and Pressure Vessels" will be integrally forged and the inner diameter of the container is limited to 600 mm, and the water volume of a single container usually does not exceed 2600 L. In order to meet the demand for large-capacity hydrogen storage, multiple containers can be used in parallel, as shown in Fig. 2.8, but this will increase possible hydrogen leakage points and reduce the reliability of equipment operation.

In the initial stage of the construction of hydrogen refueling stations in China, the standards and specifications for high-pressure hydrogen storage equipment have not yet been formulated. Most of the large-volume seamless steel hydrogen storage bottles are designed and manufactured according to the American ASME standards. At present, the relevant domestic specifications are gradually being improved. According to the nominal working pressure of the container, different design and inspection specifications need to be corresponding, see Table 2.4.

Currently, domestic hydrogen refueling stations often use 20 or 25 MPa seamless steel hydrogen storage containers, which require a compressor to achieve a filling pressure above 35 MPa. To better meet the hydrogen refueling needs at 35 and 70 MPa, it is necessary to develop seamless hydrogen storage containers with higher nominal working pressures. Zhejiang Lenney Co., Ltd. has solved the key spinning and heat treatment problems in the manufacturing process of seamless steel containers, and independently developed a 45 MPa large-volume seamless hydrogen storage container stack, which can reduce compressor loss and improve the gas filling rate and system working efficiency. The product was put into use in 2020 at the Handan Hegang Hydrogen Refueling Station and the Daxing Hapelle Hydrogen Refueling Station.

Outer container upper cover, connecting sleeve, connecting rod, inner container upper cover, inner cylinder, outer cylinder, tightening screw cover, outer container lower cover, support leg.

Current research on seamless high-pressure hydrogen storage containers mainly focuses on the strength of the steel and its sensitivity to hydrogen embrittlement. To

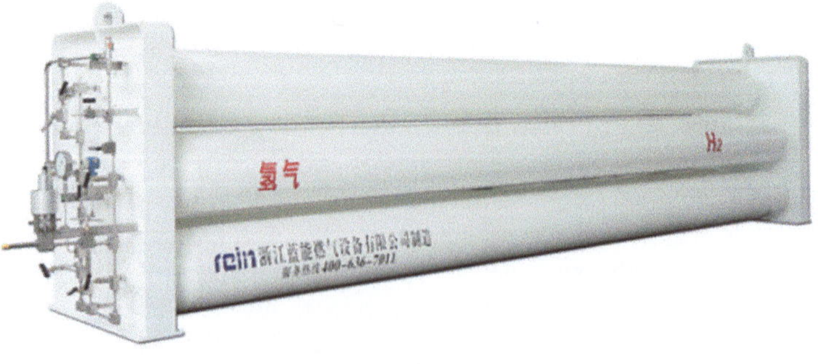

Fig. 2.8 Large-volume steel seamless hydrogen storage container stacks

Table 2.4 Comparison of domestic seamless high-pressure hydrogen storage container design and verification standard

Standard number	T/CATSI 05003—2020	GB/T 33145—2016
Standard name	Special technical requirements for hydrogen storage pressure vessels in hydrogen refueling stations	Large volume seamless steel gas cylinder
Standard type	Group standard	National standard
Working pressure	>41 MPa	10–30 MPa
Design temperature	−40–85 °C	−40–60 °C
Recommended materials	4130X, 30CrMo, S31603	High-quality steel with good low-temperature impact performance
Inspection items	Failure assessment (plastic collapse, brittle fracture, fatigue, local excessive strain, leakage) Performance inspection of material and welding specimens after heat treatment; pressure resistance test; leakage test, etc.	Ultrasonic inspection of steel pipes; metallographic examination Hardness test; non-destructive test Tensile, impact, flattening, cold bending, hydraulic, air tightness, hydraulic burst, fatigue test, etc.
Others	Also suitable for austenitic stainless steel lined hydrogen storage containers and fiber circumferential winding hydrogen storage containers	In addition to being used as fixed containers, it can also be used as long tube trailers or tube bundles

facilitate in-depth research, it is necessary to establish a high-pressure hydrogen gas testing platform and a hydrogen embrittlement database for materials. The Sandia National Laboratory in the United States [11] has been conducting slow tensile tests, fatigue tests, and fracture toughness measurement tests on hydrogen materials since the 1960s, and has formed a database of commonly used hydrogen embrittlement-resistant materials for the selection of materials for high-pressure hydrogen equipment manufacturing. The National Institute of Advanced Industrial Science and Technology in Japan [12] has upgraded the testing pressure of the testing platform to 210 MPa, the highest pressure among all current testing platforms. A team from Zhejiang University [13] independently established China's first high-pressure hydrogen environment material testing platform with a maximum pressure of 140 MPa in 2013, and began to establish a reference database for domestic materials resistant to hydrogen embrittlement, thereby promoting the progress of domestic high-pressure hydrogen storage facilities.

In addition to conducting research related to container material properties, some scholars have also tried to improve the pressure-bearing capacity and reliability of seamless containers by improving the container structure. A new type of jacketed high-pressure hydrogen storage structure [14] is shown in Fig. 2.9. This structure consists of an inner and outer double-layer container. The inner container is a

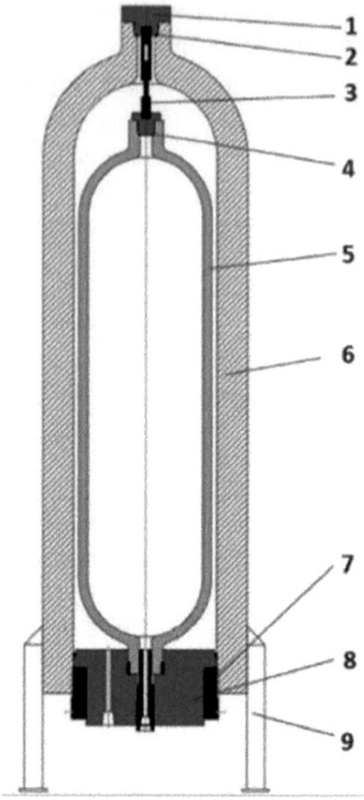

Fig. 2.9 A Jacketed high-pressure hydrogen storage container

large-volume seamless gas cylinder made of austenitic steel, and the outer container is a large-volume seamless gas cylinder made of high-strength steel with a single-sided open mouth, which is supported in the outer container by a support rod and an end plug. The outer container is filled with high-pressure nitrogen. By balancing the gas pressure, the stress level of the inner container can be adjusted, thereby reducing the crack propagation driving force of the inner container. This structure is expected to solve the problem of hydrogen embrittlement of existing single-layer hydrogen storage containers, and overcome the disadvantages of complex manufacturing processes, numerous and dense welds, difficult quality control, long production cycles, and high costs of multi-layer high-pressure hydrogen storage devices.

2.2.2.2 Steel Tape Mis-wound Hydrogen Storage Containers

The steel tape mis-wound container (Multifunctional Steel Layered Vessel, MSLV) is a type of pressure vessel structure first created in China. Since its successful development in 1964, China has accumulated rich experience in the design, manufacture, use, and optimization of this type of container. It has now formed a standard GB/T 26466—2011 "Fixed High-Pressure Hydrogen Storage Steel Tape Mis-wound Container" suitable for high-pressure hydrogen storage. The manufacturing specifications of MSLV have also been listed as a standard case by the American Society of Mechanical Engineers. The structure of MSLV is shown in Fig. 2.10. The

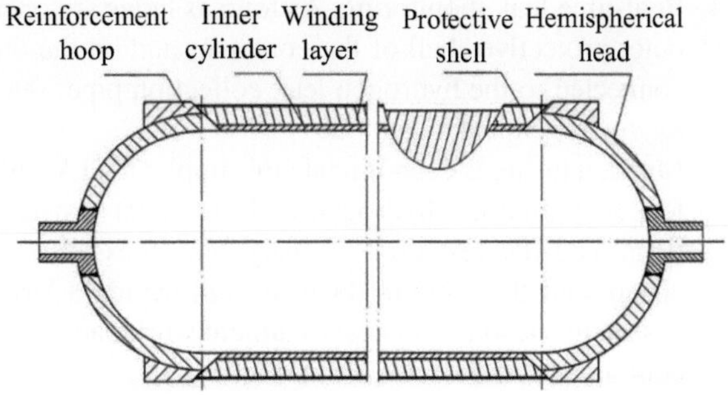

Fig. 2.10 Schematic diagram of the structure of a fixed high-pressure hydrogen storage steel strip mis-wound container

thinner inner cylinder is made by rolling and welding steel plates, and the hemispherical head is formed by stamping steel plates. The container material is austenitic stainless steel with good hydrogen embrittlement resistance, while the base material is Q345R or 16MnDR, and the cladding is 0Cr18Ni9, 00Cr19Ni10 or 00Cr17Ni14Mo2; the steel strip is composed of multiple layers of hot-rolled flat steel strips with a thickness of 4–8 mm and a width of 80–160 mm, and the material is selected from Q345R, 16MnDR or HP345; the reinforcement hoop is first welded from steel plates into short cylinder sections, and then processed into cylindrical and conical surfaces that match the outer hemispherical head; the outermost protective layer is a high-quality thin plate with a thickness of 3–6 mm, which is welded to the outside of the steel strip layer in a bandaging manner. On the outside of the thinner inner cylinder, multiple layers (even number of layers) of flat steel strips are wound at an angle, so that the steel strip and the cylinder body form a certain angle in the circumferential direction, the winding direction of the adjacent layers of steel strips is opposite, and only the two ends of each layer of steel strip are welded to the hemispherical head and the reinforcement hoop, thus forming a steel strip mis-wound high-pressure hydrogen storage container.

Reinforcement hoop, inner cylinder, winding layer, protective shell, hemispherical head.

MSLV has many advantages in the field of hydrogen storage [15]:

1. It is conducive to the manufacture of high-pressure, large-volume containers: The steel strip winding process breaks through the limitations of seamless steel pipe length and wall thickness, ensuring that the container still has excellent strength under large sizes.
2. It has explosion suppression and explosion resistance: The fracture toughness of thin steel plates and narrow thin steel strips is good, and under the action of the pre-tension and friction resistance of the steel strip winding, the container will only leak and fail, and will not undergo overall brittle failure.
3. Defect dispersion: The container has no deep circumferential welds along its entire length, and the winding layer and the head use large bearing area step inclined surface welds, achieving a smooth transition of stress levels, and high structural reliability.

4. Real-time leak monitoring: As long as holes are opened on the upper part of the outer protective shell of the container and on the outer heads at both ends, and connected to the hydrogen leak collection pipe, sensors can be used to monitor the hydrogen leak situation.
5. Manufacturing is economical and simple: MSLV only uses materials with excellent high-pressure hydrogen embrittlement resistance for the hydrogen-facing parts, and the rest uses ordinary high-pressure vessel steel, the steel strip is cheap, and the winding layer design avoids a large amount of welding, non-destructive testing and heat treatment workload, and does not require large heavy equipment and difficult technologies.

Zhejiang University and Juhua Company have cooperated to solve the problems of MSLV application in the smelting of special stainless steel materials for hydrogen, welding process evaluation, and the manufacture of high-pressure threaded sealing structures, and successfully developed a 42 MPa and volume up to 5 m^3 MSLV, and carried out the demonstration applications at China's first hydrogen refueling station located in Zhongguancun; Beijing Feichi Jingli hydrogen refueling station adopted 77, 47 and 42 MPa three-stage MSLV, maintaining a good operation record; In 2017, the Toyota hydrogen refueling station in Changshu, Jiangsu, used a 98 MPa MSLV, which is the highest pressure MSLV currently available. In subsequent research on the 98 MPa container, the Zhejiang University team proposed that by appropriately reducing the design safety factor or using higher strength steel strip materials, the container weight can be effectively reduced while ensuring safety, and designed a 50 MPa and 7.3 m^3 MSLV [16], as shown in Fig. 2.11.

Fig 2.11 A steel strip mis-wound hydrogen storage container developed by Zhejiang University and Juhua Company

2.2.2.3 Fiber-Wound High-Pressure Hydrogen Storage Containers

The fiber-wound high-pressure hydrogen storage container is made of an inner cylinder using materials compatible with hydrogen, and the outer layer is reinforced with fiber, which can overcome the influence of hydrogen material size and thickness on the strength and cost of the container. Compared with MSLV, the fiber-wound container is lighter and avoids welding defects, but the use of carbon fiber materials significantly increases the cost of container manufacturing. The current fiber-wound large-volume high-pressure hydrogen storage containers are still mainly Type II, applicable to the T/CATSI 05003—2020 standard. The 87.5 MPa large-volume hydrogen storage container used in the Dalian Tongxin Hydrogen Refueling Station, which started demonstration operation in 2016 [17], is the first Type III large-volume fixed hydrogen storage container in China. It is a result of the joint development by Shijiazhuang Anruico and Tongji University under the "Twelfth Five-Year" 863 Project. Its water volume reaches 580 L, solving multiple technical problems, such as structural design, molding process, sealing, and hydrogen embrittlement, filling the gap in the field of carbon fiber-wound steel-lined hydrogen storage container technology in China. Zhejiang Lenney obtained the patent for the carbon fiber-wound austenitic stainless steel-lined container in 2021 [18], which is applicable to the design and manufacture of large-volume hydrogen storage containers with a storage pressure above 70 MPa and a volume of 500–2000 L.

2.2.3 Mobile Hydrogen Storage Containers

2.2.3.1 High-Pressure Hydrogen Storage Containers for Transportation

According to TSG R0005—2011 "Safety Technical Supervision Regulations for Mobile Pressure Vessels", mobile pressure vessels are defined as transport equipment permanently connected by a tank body or large-volume seamless steel gas cylinder and a running device or frame, including railway tank cars, automobile tank cars, long tube trailers, tank containers, and tube bundle containers, etc. The equipment currently used for hydrogen transportation is mainly tube bundle containers and long tube trailers, which are only suitable for road transportation within 200 km due to road load and transportation cost limitations. In addition to being used for transportation purposes, this type of container can also directly fill hydrogen into the fuel cell vehicle storage tank through a compressor.

Most of the transportable high-pressure hydrogen storage containers on the market are made of seamless steel tubes consisting of 4130X seamless steel tube materials, with a working pressure of 20–25 MPa. Compared with steel containers, fiber composite material-wound containers have the advantage of being lightweight, with a larger rated storage pressure, and a higher storage capacity. But a higher working pressure also means a thicker carbon fiber layer, a larger container mass, and higher transportation costs. Calculations show that the hydrogen storage pressure that maximizes the gravimetric density of transportation is about 40 MPa [19]. At present, domestic companies, such as CIMC Enric, Zhejiang Lenney, and Luoyang Shuangrui all have the ability to produce Type II 20 MPa containers and tube bundle containers for transportation. In 2021, CIMC Enric successfully mass-produced 30 MPa large-volume carbon fiber-wound storage and transportation containers for shipborne hydrogen fuel, which were certified by the French Bureau Veritas and the British Lloyds and exported to Europe for use as hydrogen fuel storage and transportation devices for maritime and inland waterway navigation. Internationally, Hexagon Lincoln has developed a 25 MPa Type IV tube bundle vehicle, with a single trip hydrogen transport capacity of up to 885 kg; Linde has successfully developed a 50 MPa mobile fiber-wound hydrogen storage container, with a smaller single container volume, loaded vertically in the tube bundle container, increasing the single trip hydrogen transport capacity to 1100 kg.

2.2.3.2 Lightweight High-Pressure Hydrogen Storage Containers for Vehicles

The design and manufacturing requirements for automotive lightweight high-pressure hydrogen storage containers are higher than those for high-pressure hydrogen storage containers for transportation, mainly covering the following requirements:

1. A small volume and reasonable layout, not occupying too much cabin space.
2. Lightweight, excessive weight of the container will lead to increased driving energy consumption and reduced driving range.
3. A higher hydrogen storage density, needs to meet the vehicle's hundreds of kilometers of driving mileage requirements, therefore needs to load hydrogen gas at pressures of 35 MPa or even 70 MPa.
4. Better compatibility with high-pressure gaseous hydrogen, safer and more reliable, not prone to rupture and harm passengers in the event of a car accident.
5. Good economy, only when the purchase cost of fuel cell vehicles reaches the same level as gasoline vehicles, can it be possible to achieve large-scale promotion.

Referring to the development goals set by the DOE, the final gravimetric and volumetric energy density of the on-board hydrogen storage system need to reach 2.2 kW h/kg (6.5 wt%) and 1.7 kW h/L (50 gH_2/L), respectively, the system cost drops to 8 dollars/kW h, and the cost of high-strength carbon fiber for hydrogen storage tanks drops to 13 dollars/kg. Considering all these factors, the high safety, high hydrogen storage density Type III and Type IV on-board hydrogen storage tanks will be the important development direction for on-board hydrogen storage containers at the present and in the future, and also the research hotspot of major vehicles and parts companies and institutions.

In China, Type III tanks are mainly used as on-board high-pressure hydrogen storage containers. The hydrogen storage tanks loaded on the multi-functional model EU-NIQ7 released by SAIC Maxus in 2020 are shown in Fig. 2.12. Three 70 MPa Type III hydrogen storage tanks are equipped in the middle of the chassis, which can be filled with 6.4 kg of high-pressure hydrogen in 3 min, providing a driving range of 605 km under NEDC conditions, and the gas tanks remain intact after the whole vehicle collision test. The main manufacturers of Type III tanks in China include Tianhai Industry, Sinoma Science & Technology, Sylinda Anke, and Beijing Ketaike, etc.

Type IV tanks are the mainstream on-board hydrogen storage containers internationally. Taking the 2021 Toyota Mirai second generation as an example, its hydrogen storage system is equipped with three 70 MPa Type IV hydrogen storage tanks, as shown in Fig. 2.13. In August 2021, the Mirai set off from the TOYOTA Technical Center in California, USA, and drove a total of 1360 km in full hydrogen state, setting a Guinness World Record for the longest driving range of a hydrogen energy vehicle. In terms of the production and manufacturing of Type IV tanks, Norway's Hexagon Composites is one of the leaders in the composite material gas tank industry, capable of mass-producing high hydrogen storage density Type IV hydrogen storage tanks for installation and application by Toyota, Honda, Mercedes-Benz, and other car manufacturers. In addition, French Faurecia, German NPROXX, Japanese JFE, and Korean ILJIN Composite also have the production capacity of high-performance Type IV tanks. In China, the production license for Type IV on-board hydrogen storage tanks is gradually opening up. Sylinda Anke is the first domestic manufacturer to obtain the production license for non-metallic inner liner fiber wound gas tanks and has successfully developed 70 MPa Type IV hydrogen storage tank products and appeared at international exhibitions. Overall, there is still a significant gap between China and abroad on the production and manufacturing technology of Type IV hydrogen storage tanks, and further efforts are needed to achieve commercial applications.

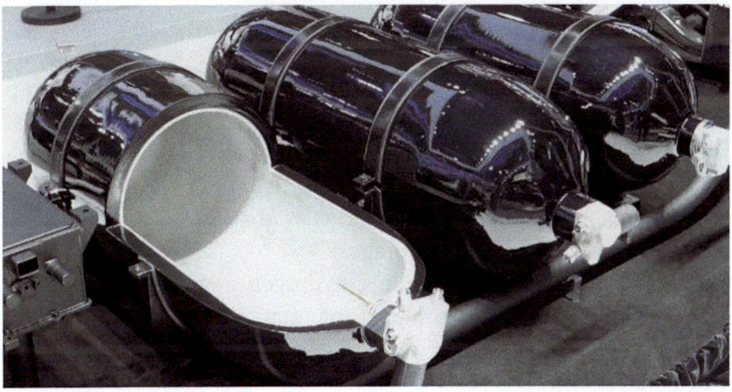

Fig. 2.12 70 MPa carbon fiber wound aluminum alloy gas tank loaded on SAIC Maxus EUNIQ7

Fig. 2.13 Toyota Mirai's power system

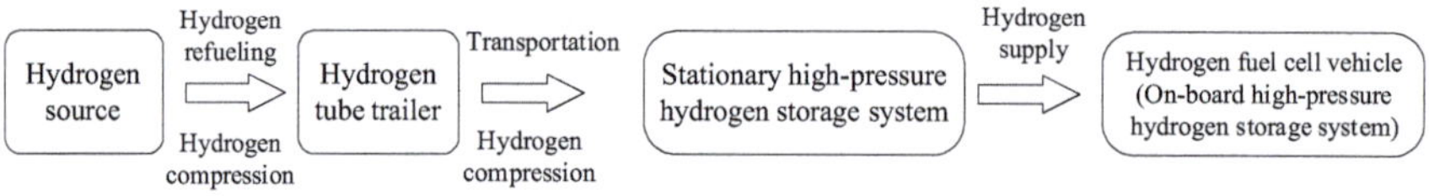

Fig. 2.14 A schematic diagram of the application based on high-pressure gas cylinders for hydrogen storage and transportation

2.3 Applications of Gaseous Hydrogen Storage and Transportation Technology

2.3.1 Applications of High-Pressure Gas Cylinder for Hydrogen Storage and Transportation

Currently, the technology of high-pressure gas cylinders for hydrogen storage and transportation is mainly used in three aspects: high-pressure hydrogen tube trailer transportation, on board high-pressure hydrogen storage systems, and stationary high-pressure hydrogen storage systems. As shown in Fig. 2.14, high-pressure hydrogen tube trailers are mainly used for large-scale hydrogen transportation, on board high-pressure hydrogen storage systems are mainly used for hydrogen fuel cell vehicles, and stationary high-pressure hydrogen storage systems are mainly used for hydrogen refueling stations, stationary/distributed energy storage, and other fields of large-scale hydrogen storage.

Hydrogen source, hydrogen refueling, hydrogen compression, hydrogen tube trailer, transportation, hydrogen compression, stationary high-pressure hydrogen storage system, hydrogen supply, hydrogen fuel cell vehicle (on board high-pressure hydrogen storage system).

2.3.1.1 Applications of High-Pressure Hydrogen Tube Trailer Transportation

High-pressure hydrogen tube trailer transportation refers to compressing and storing hydrogen in several to a dozen of seamless steel cylinders, connecting the cylinders together with piping and valves, and fixing them in the trailer for hydrogen transportation. This is currently the most mature method of hydrogen transportation. The tube trailer was first developed by the American CPI company in 1960, introduced to China in 1987, and successfully manufactured the first tube trailer in China in 1999. In 2002, Shijiazhuang Anruico Gas Machinery Co., Ltd. achieved the mass production of tube trailers. Over the next decade, the design and manufacturing level of tube trailers in China made significant progress, reaching an internationally leading level. However, tube trailers are mostly used for filling natural gas, and the number of high-pressure hydrogen tube trailers has significantly increased since 2011, mainly in the Yangtze River Delta and Pearl River Delta regions. In recent years, with the rise and development of the hydrogen industry, the application of high-pressure hydrogen tube trailers has become more and more widespread.

Tube trailers can be divided into bundled tube trailers and frame tube trailers according to their structural configurations, as shown in Fig. 2.15. Among them, frame tube trailers use a frame to fix the cylinders and are the most numerous, technically mature, and widely used in China; bundled tube trailers directly fix the cylinders on the trailer chassis, reducing the frame weight, and have a higher transportation efficiency compared to frame tube trailers. Due to the flammable and explosive characteristics of hydrogen, China has made special regulations on the manufacture and transportation of tube trailers. The National Energy Administration issued the Chinese energy industry standard NB/T 10354—2019 "Tube Trailer" in December 2019, which clearly stipulates the materials, design, manufacture, test methods, inspection rules, marking, storage, and transportation requirements for transporting hydrogen by tube trailers.

The overall structure of the tube trailer generally consists of three parts: the walking components, large-volume seamless steel cylinders, and their connecting devices. The walking component needs to meet the requirements of total mass and axle load and needs to be modified according to the special structure of the transported cylinders. The external dimensions, total mass, and axle load of the trailer should comply with the relevant requirements of GB 1589—2016 "Dimensions,

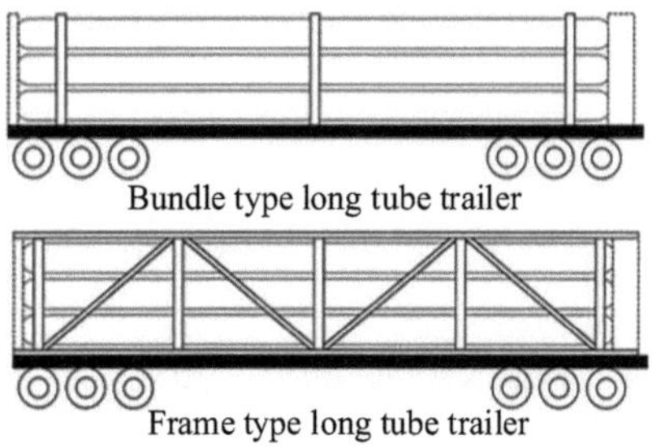

Fig. 2.15 A schematic diagram of the structure of a long tube trailer

Axle Load, and Mass Limits for Motor Vehicles, Trailers, and Vehicle Trains". The length of a single large-volume seamless steel cylinder is usually 5–12 m, with a volume of 2–4.2 m^3, and a working pressure of 15–35 MPa. As the main pressure-bearing component of the tube trailer, the quality of the cylinder is closely related to the safety performance of the tube trailer. China has made detailed requirements for the materials, manufacture, and air tightness test of the cylinders, which should comply with TSG 23—2021 "Cylinder Safety Technical Regulations" [20] and GB/T 33145—2016 "Large Volume Seamless Steel Cylinders" [21] to ensure the quality of the cylinders. The internal state of the cylinder is a shot peening surface, meeting the filling requirements of media with a purity of less than 99.99%. If high-purity hydrogen needs to be stored, the inner surface needs to be ground to remove surface particles, burrs, etc. Grinding is divided into rough grinding and fine grinding processes, using different abrasives, after dozens of hours of grinding, the surface roughness reaches 3S level, achieving a mirror effect, then after high-temperature deionized water cleaning, the cylinder is heated, vacuumed, and replaced, etc., to exhaust the moisture, and finally, the water content of the cylinder medium reaches less than 1 mg/L, avoiding the precipitation of moisture and other impurities on the inner surface [22]. The connection device connects multiple cylinders in parallel, and the connections, such as pipelines and valves, use butt welding to reduce the risk of leakage, and are equipped with safety accessories, such as overpressure relief devices, pipeline safety valves, emergency cut-off devices, and static electricity guiding devices, to ensure the safety of cylinder transportation.

Bundle type long tube trailer, frame type long tube trailer.

Currently, high-pressure long tube trailers in China mainly pressurize hydrogen to about 20 MPa, and the single vehicle hydrogen transport capacity is only 300–400 kg. The technical parameters of a typical 20 MPa high-pressure hydrogen long tube trailer are shown in Table 2.5 [23]. The actual operating filling pressure is generally 19.0–19.5 MPa, the gas is released until the pressure inside the tank reaches 0.6 MPa, each actual hydrogen transport volume is 3750–3920 m^3 (i.e., 334–350 kg), the single vehicle filling time is 1.5–2.5 h, and the unloading time is

Table 2.5 Technical parameters of a typical 20 MPa high-pressure hydrogen long tube trailer

Parameter	Parameter value	Parameter	Parameter value
Nominal working pressure/MPa	20	Cylinder outer diameter × length/mm × mm	559 × 10,975
Ambient working temperature/°C	−40–60	Nominal volume per cylinder/m^3	2.25
Cylinder design thickness/mm	16.5	Number of cylinders	10
Cylinder material (structural steel)	4130	Nominal volume of container bundle/m^3	22.5
Hydrostatic test/MPa	33.4	Filling medium	Hydrogen
Air tightness test pressure/MPa	20	Filling volume (20 MPa, 20 °C) / m^3	3965

1.5–3 h. However, due to the heavy weight of the steel cylinders, the system hydrogen storage density of the 20 MPa high-pressure hydrogen long tube trailer is only 1–2%, which is only suitable for small-scale, short-distance hydrogen transport, and its transport cost significantly increases with distance.

2.3.1.2 Applications of On-board High-Pressure Hydrogen Storage Systems

Lightweight, high pressure, high hydrogen storage mass ratio, and long life are the main features of on-board high-pressure hydrogen cylinders. Currently, on-board high-pressure gaseous hydrogen cylinders mainly include aluminum liner fiber-wrapped cylinders (Type III) and plastic liner fiber-wrapped cylinders (Type IV). The mass of the on-board high-pressure hydrogen cylinder directly affects the driving range of hydrogen fuel cell vehicles. The Type IV high-pressure hydrogen cylinder, with its plastic liner, has a relatively small mass and the potential for lightweighting, making it suitable for passenger vehicles. Countries, such as Japan, France, and the UK, have already achieved mass production of 70 MPa Type IV hydrogen cylinders, realizing the manufacture and commercial application of Type IV hydrogen cylinders, as shown in Table 2.6. In 2014, Toyota mass-produced the world's first hydrogen fuel cell vehicle, Mirai, with a driving range of up to 550 km. Subsequently, passenger vehicle models, such as Honda Clarity in Japan and Hyundai ix35 in Korea, were released, and American manufacturers like General Motors and Ford also deployed hydrogen fuel cell vehicles based on high-pressure hydrogen cylinders.

Since the first application of the 35 MPa aluminum alloy liner fully wrapped hydrogen storage cylinder in new energy vehicles at the Shanghai World Expo in 2010, China has mainly used the 35 MPa Type III cylinder for hydrogen storage. Considering the limitations of vehicle space, 70 MPa is a more economical storage pressure for vehicles, but with the increase in pressure, the difficulty and risk of cylinder manufacturing also increase. In 2017, China implemented the national standard GB/T 35544—2017 "Vehicle Compressed Hydrogen Aluminum Liner

Table 2.6 Main commercial Type IV hydrogen cylinders abroad

Organization name	Technical features	70MPaIV Type Hydrogen storage cylinder application
QUANTUM, USA	35/70 MPa, lifespan 15–20 years, volume 26–994 L, available hydrogen storage 0.8–22.2 kg (70 MPa)	Used by General Motors, USA
TOYOTA, Japan	70 MPa, volume 142.2 L (3 units), 5.6 kg	Used by Toyota, Japan
HEXAGON, Norway	70 MPa, volume 244 L, 9.8 kg	Used by Ford, USA Used by Daimler, Germany
ILJIN, South Korea	70 MPa, volume 156 L (3 units), 6.3 kg	Used by Hyundai, South Korea

Carbon Fiber Fully Wrapped Cylinder" [24], which divides the structure of the Type III hydrogen storage cylinder into the T-type and S-type structure, as shown in Fig. 2.16. The hydrogen storage cylinder is composed of a carbon fiber winding layer, a thermocouple corrosion protection layer, and an aluminum liner. The T-type has a convex bottom structure, and the S-type has a closed structure at both ends. Hydrogen storage cylinders are divided into Class A and Class B according to different working pressures. The nominal working pressure of Class A hydrogen storage cylinders is 35 MPa, and the nominal water volume requirement is 450 L. The nominal working pressure of Class B hydrogen storage cylinders is >35 MPa, and the nominal water volume requirement is 230 L. Each hydrogen storage cylinder that leaves the factory needs to undergo strict water pressure and airtightness tests according to this standard before it can be used. In the future, the development direction of high-pressure hydrogen storage cylinders for vehicles in China is to achieve the domestic production of Type IV high-pressure hydrogen storage cylinders with 70 MPa.

Carbon fiber winding layer, thermocouple corrosion protection layer, aluminum liner.

Carbon fiber winding layer, thermocouple corrosion protection layer, aluminum liner.

2.3.1.3 Applications of Fixed High-Pressure Hydrogen Storage Systems

Fixed high-pressure hydrogen storage systems are mostly used for hydrogen storage in hydrogen refueling stations and energy storage fields. Currently, according to the hydrogen refueling pressure of hydrogen fuel cell vehicles, hydrogen refueling stations are also divided into 35 and 70 MPa. The vast majority of those in use or under construction in China is 35 MPa hydrogen refueling stations. Generally, fixed hydrogen storage containers with a design pressure of 50 MPa are used for the construction of 35 MPa hydrogen refueling stations, and fixed hydrogen storage containers with a design pressure of 98–99 MPa are used for the construction of 70 MPa hydrogen refueling stations [25]. The high-pressure hydrogen storage containers used in foreign 70 MPa hydrogen refueling stations have various forms, including steel single-layer structures, multi-layer metal reinforced structures, seamless liner carbon fiber winding reinforced structures, etc. At this stage, the only

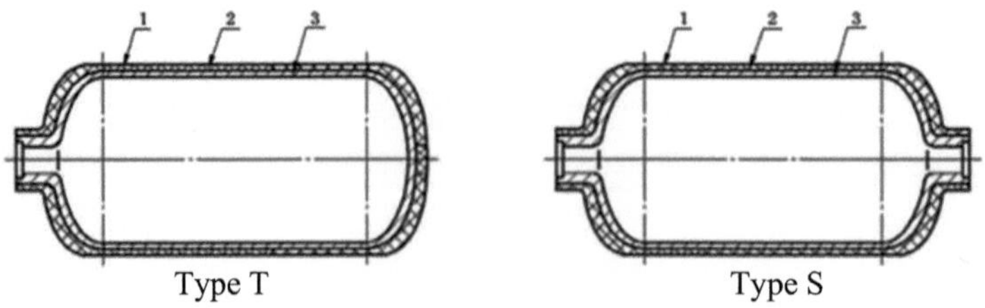

Fig. 2.16 A schematic diagram of the structure of the aluminum liner fiber winding tank

commercialized in China is the steel tape mis-winding hydrogen storage container, which has a heavier weight, higher production cost, lower production efficiency, and greater manufacturing difficulty. The high-pressure storage containers that are mature and widely used abroad are steel seamless tank-type containers and steel seamless liner carbon fiber winding reinforced structures. The more typical ones are the all-steel tank-type container structure of Japan's JFE company and the steel liner carbon fiber winding hydrogen storage container designed by the American FIBA company with a pressure of 100 MPa, which has been used in Korean hydrogen refueling stations. As early as 2012, the National 863 Project "Stationary Large (High) Capacity Hydrogen Storage Container Technology" completed the development of a large-volume hydrogen storage container with a design pressure of 92 MPa and a working pressure of 87.5 MPa and was used for the first 70 MPa hydrogen refueling station demonstration operation in China, and the operation situation was good. This fixed station hydrogen storage container adopts a steel liner and a full carbon fiber winding structure.

The issue of hydrogen embrittlement in high-pressure environments of stationary hydrogen storage containers is key to achieving "leak before burst" in these containers. Hydrogen embrittlement in high-pressure environments refers to the phenomenon where hydrogen enters the metal in a high-pressure hydrogen environment, and under the combined effect of stress and hydrogen, when the local hydrogen concentration reaches a critical value, the ductility and toughness of the metal are reduced, or hydrogen-induced delayed fracture occurs. Hydrogen embrittlement in high-pressure environments involves the dissolution, diffusion, and aggregation of hydrogen, all of which can cause hydrogen-induced cracking failure [25]. Hydrogen embrittlement testing methods for metal materials can be roughly divided into two categories: one is used for preliminary screening of materials, quickly evaluating whether the material can be used to manufacture hydrogen-exposed components, such as disc tests, hydrogen-induced cracking stress intensity factor threshold tests, etc.; the other is used for in-situ testing of material mechanical properties, providing performance data for the design of hydrogen-exposed components or material suitability assessment, such as slow strain rate tensile tests, fatigue crack growth rate tests, fatigue life tests, etc., see Table 2.7. Compared to hydrogen storage cylinders for vehicles, stationary hydrogen storage containers have a high number of pressure fluctuations, up to 103–105 times, and the range of pressure fluctuations is often 20–80% of the design pressure, so stationary hydrogen storage containers also need to consider low-cycle fatigue failure.

The design and manufacture of stationary hydrogen storage containers in China mainly follow the TSG 21—2016 "Safety Technical Supervision Regulations for Stationary Pressure Vessels", GB/T 34542 "Hydrogen Storage and Transportation Systems" series of standards, and steel strip helically wound hydrogen storage containers should also meet the requirements of GB/T 26466—2011 "Stationary High-Pressure Hydrogen Storage Steel Strip Helically Wound Containers". By adhering to the specifications for materials, material manufacturing, and processing stress, it is possible to avoid hydrogen embrittlement failure and fatigue failure of stationary

Table 2.7 Metal material hydrogen embrittlement test methods

Test purpose	Test name	Test method summary	Test standard
Preliminary Material Screening Disc Test	Disc Test	Burst tests are conducted on disc specimens in high-pressure hydrogen and helium environments, with the ratio of burst pressures in helium and hydrogen environments used as the hydrogen embrittlement coefficient to evaluate material hydrogen embrittlement sensitivity	ISO 11114-4 ASTM F1459 GB/T 34542.3
	Hydrogen-Induced Cracking Stress Intensity Factor Threshold Test	Under high-pressure hydrogen environments, monotonic loading is applied to the specimen to obtain the material's hydrogen-induced cracking stress intensity factor threshold. This test can use different methods according to different standards	ISO 11114-4 ASME BPVC VII-3 KD-10 ANSI/CSA CHMC 1 ASTM F1624 GB/T 34542.2
In-situ Performance Test	Slow Strain Rate Tensile Test	Under high-pressure hydrogen environments, tensile tests are conducted on smooth round bar specimens or notched specimens to obtain the material's yield strength, tensile strength, elongation after fracture, and other mechanical properties	ANSI/CSA CHMC 1 ASTM G142 GB/T 34542.2
	Fatigue Crack Growth Rate Test	Under high-pressure hydrogen environments, cyclic loading is applied to the specimen to obtain the relationship curve between the material's fatigue crack growth rate and the range of stress intensity factors	ASME BPVC VII-3 KD-10 ANSI/CSA CHMC 1 GB/T 34542.2
	Fatigue Life Test	Under high-pressure hydrogen environments, cyclic loading is applied to smooth round bar specimens or notched specimens to obtain the material's stress/strain-life curves	ANSI/CSA CHMC 1 GB/T 34542.2

hydrogen storage containers as much as possible, and achieve the "leak before burst" failure requirement of stationary hydrogen storage containers.

2.3.2 *Applications of Hydrogen Transportation by Pipelines*

Hydrogen pipelines can be divided into long-distance transmission pipelines and short-distance distribution pipelines, as shown in Fig. 2.17. The pressure of hydrogen transmission in long-distance pipelines is higher, the pipeline diameter is larger, and they are mainly used for long-distance and large-scale transmission of high-pressure gaseous hydrogen between hydrogen production units and hydrogen

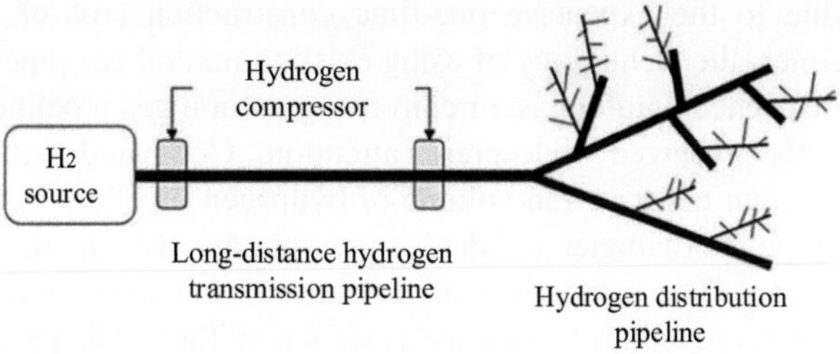

Fig. 2.17 A schematic diagram of pipeline hydrogen applications

stations; distribution pipelines have a lower hydrogen transmission pressure and a smaller pipeline diameter, and they are mainly used for the distribution of medium-pressure and low-pressure gaseous hydrogen between hydrogen stations and various users. The construction cost of hydrogen distribution pipelines is relatively low, but the construction of long-distance hydrogen pipelines is difficult and costly. Currently, the cost of long-distance hydrogen pipelines is about 630,000 $/km, while the cost of natural gas pipelines is only about 250,000 $/km, making the cost of hydrogen pipelines about 2.5 times that of natural gas pipelines [26].

Hydrogen source, hydrogen compressor, long-distance hydrogen transmission pipeline, hydrogen distribution pipeline.

The overseas development of hydrogen pipelines started earlier, and the history of pipeline transportation of hydrogen can be traced back to the late 1930s. In 1939, Germany built a pipeline about 208 km long, with a diameter of 254 mm, an operating pressure of 2 MPa, and a hydrogen transmission capacity of 9000 kg/h [27]. As of 2017, the length of hydrogen transmission pipelines in Europe is about 1598 km, with a hydrogen pressure generally between 2 and 10 MPa, mostly using seamless steel pipes, with a pipeline diameter of 0.3–1.0 m, and the pipeline material is mainly low-strength pipeline steels, such as X42, X52, X56 [28]. The longest hydrogen pipeline is owned by the French Air Liquide Group, which extends from northern France to Belgium, with a total length of about 402 km. Most of the hydrogen transmission pipelines in the United States are in Texas, Louisiana, and California, with a total length of nearly 2600 km, mostly using underground layout, operating at a hydrogen pressure of ~6.9 MPa, and the pipeline material mainly uses pipeline steel within the range of X52 to X80, with an expected service life of 15–30 years.

The construction of long-distance hydrogen pipelines in China is in its infancy, with a total mileage of about 400 km, mainly distributed in the Bohai Bay, Yangtze River Delta, etc. The hydrogen pipeline between Jiyuan and Luoyang in Henan Province is currently the longest, largest diameter, highest pressure, and largest hydrogen transmission pipeline in China, with a pipeline length of 25 km, a pipeline diameter of 508 mm, a hydrogen transmission pressure of 4 MPa, and an annual hydrogen transmission volume of 100,400 tons. According to the "China Hydrogen Energy Industry Infrastructure Development Blue Book" [29], by 2030, China's hydrogen pipelines will reach 3000 km.

However, due to the expensive one-time construction cost of long-distance hydrogen pipelines, the technology of using existing natural gas pipelines to transport hydrogen-enriched natural gas, or converting natural gas pipelines into hydrogen pipelines, has received widespread attention. Compared with natural gas pipelines, the current construction volume of hydrogen pipelines is still relatively small, and the pipeline diameter and design pressure are also smaller than those of natural gas pipelines. The comparison of the construction status of hydrogen pipelines and natural gas pipelines worldwide is shown in Table 2.8. Therefore, how to use existing natural gas pipelines for hydrogen transmission and further develop into pure hydrogen pipelines is one of the main pathways for the rapid development of hydrogen pipelines.

In 2019, the world's first hydrogen pipeline, converted from a natural gas pipeline, was put into use between Dow Benelux and Yara. In China, in 2018, the Chaoyang Renewable Energy Hydrogen Demonstration Project was launched by the State Power Investment Corporation, which uses "green hydrogen" produced by electrolyzing water with renewable energy and mixes it with natural gas for use in gas boilers. It has been safely operated for 1 year with a 10% hydrogen blending ratio. However, the domestic hydrogen-natural gas mixed transmission will be mainly experimental research and pilot demonstrations in the short term. Moreover, there is a significant difference in the physical properties of hydrogen and natural gas (Table 2.9).

1. In terms of combustion energy: hydrogen has a lower density, but its calorific value per unit mass is much higher than that of natural gas.
2. In terms of combustion properties: hydrogen is easier to ignite and its flame speed is much faster than that of natural gas.

Table 2.8 Comparison of the construction status of hydrogen pipelines and natural gas pipelines [28]

Pipeline type	Pipeline diameter/mm	Design pressure/MPa	Construction mileage/km	Common materials
Hydrogen	304–914	2–10	6000	X42,X52
Natural gas	1016–1420	6–20	1,270,000	X70,X80

Table 2.9 Comparison of the main physical properties of hydrogen and natural gas

Physical properties	Hydrogen	Natural gas
Density/(kg/m^3)	304–914	1016–1420
Specific heat at constant pressure/[kJ/(kg·K)]	2–10	6–20
Calorific value/(kJ/kg)	Approximately 142	Approximately 47.5
Flammability limit (volume fraction, %)	4–75	5–15
Detonation limit (volume fraction, %)	18.3–59	6.5–12
Minimum ignition energy/MJ	0.02	0.28
Combustion speed/(m/s)	2.65	0.4
Diffusion coefficient in iron (100 °C)/(cm^2/s)	2.5×10^{-7}	–

3. In terms of safety: hydrogen can diffuse in iron, causing hydrogen embrittlement, but this is not a concern with natural gas.

Therefore, hydrogen pipelines cannot be directly designed and constructed according to the standard specifications for long-distance natural gas pipelines. The feasibility of natural gas pipeline modification technology and the technology for transporting hydrogen-mixed natural gas still needs further assessment, and there is also a lack of relevant standards.

Overseas standards for hydrogen pipeline design mainly include ASME B31.12: 2019 "Hydrogen piping and pipelines" [30] and CGA G-5.6: 2005 "Hydrogen pipeline systems" [31], both of which are applicable to the design of long-distance hydrogen transport pipelines and short-distance hydrogen distribution pipelines. Currently, China does not have applicable standards for long-distance hydrogen transport pipelines, only GB 4962—2008 "Technical Safety Regulations for Hydrogen Use" [32] and GB 50177—2005 "Design Specifications for Hydrogen Stations" [33] are applicable to hydrogen supply stations and short-distance hydrogen distribution pipelines in workshops. The standard GB/T 34542.5 "Hydrogen Storage and Transportation System Part 5: Technical Requirements for Hydrogen Transportation Systems" is currently being compiled. During the compilation process of the standard, it is particularly necessary to consider the changes in pipeline performance caused by hydrogen embrittlement.

Pipelines are one of the keys to Hydrogen Transportation by Pipelines technology. The steel pipe design formula in hydrogen pipelines is different from that in natural gas pipelines. The hydrogen pipeline design formula adds a "material performance coefficient", which reflects the adverse effect of hydrogen on the mechanical properties of metal pipelines. After increasing the material performance coefficient, the calculated wall thickness of the pipeline will relatively increase, and the design pressure will relatively decrease, which is more conducive to ensuring the safety of long-distance hydrogen pipelines. The design formula for hydrogen pipelines is as follows [30]:

$$P = \frac{2S\delta}{D} F_f E_f T_f H_f \tag{2.5}$$

The design formula for natural gas pipelines is

$$P = \frac{2S\delta}{D} F_f E_f T_f \tag{2.6}$$

where, P is the design pressure (MPa); s is the minimum yield strength (MPa); δ is the nominal wall thickness (mm); D is the nominal diameter (mm); F_f is the design factor; E_f is the axial joint factor; T_f is the temperature reduction factor; H_f is the material performance coefficient.

The material performance coefficients of different materials are shown in Table 2.10.

Table 2.10 Material performance coefficients of hydrogen pipeline materials

Material type	X42	X52	X60	X70	X80
Yield strength/MPa	289.6	358.5	413.7	482.7	551.6
Tensile strength/MPa	413.7	455.1	517.1	565.4	620.6
Maximum allowable pressure/MPa	20.68	20.68	20.68	10.34	10.34

Note: When the design pressure is at an intermediate value, interpolation is used to determine the value

Table 2.11 Comparison of several compressors used for large-scale pipeline hydrogen transmission

Type of compression	Technical description	Advantages/main issues
Piston compression	Technically mature, suitable for various gases; hydrogen requires special design due to its extremely low molecular weight	Can output high-pressure hydrogen; requires electric drive; oil-lubricated type helps control gaseous hydrogen, but will contaminate hydrogen; oil-free type has slightly lower output pressure (within 25 MPa); large volume, many components
Centrifugal compression	Technically mature, suitable for various gases, can be used for gas compression above 30 MW; but hydrogen has an extremely low molecular weight, making it extremely difficult to compress, with many compression stages, and expensive equipment	It is difficult to maintain mechanical tolerances between multiple compression stages, and the seal between stages must be considered, resulting in low compression efficiency and high equipment prices
Membrane compression	Suitable for all gases; the three-layer membrane structure separates gaseous hydrogen and hydraulic oil, ensuring the entire compression process is pollution-free and hydrogen leak-free; can achieve high pressure (>70 MPa)	Compared to piston type, each stage has a higher compression ratio, which can reduce investment costs; lacks large-scale equipment for pipeline hydrogen transmission; difficulty and cost of size enlargement are high; maximum capacity is 2000 Nm^3/h
Metal hydride compression	Made with reversible hydrogen-absorbing and releasing metal hydrides, it is more economical and simpler to operate than traditional mechanical hydrogen compressors	Metal hydride compressors are compact, noiseless, require no dynamic seals, require minimal maintenance, and can operate unattended for long periods; however, the technology is new and it is difficult to manufacture large compressors suitable for pipeline hydrogen transmission

In addition to the design and manufacture of the pipeline itself, the hydrogen compressor is also a necessary and very critical link for pipeline hydrogen transmission. Table 2.11 introduces several large-scale high-pressure hydrogen compressors [34]. The specific choice of compressor should be considered comprehensively based on parameters, such as flow rate, suction, and exhaust pressure.

The cost of designing and constructing new hydrogen pipelines is high, which is a major obstacle to the construction of new hydrogen pipelines. In the future, with the development of the hydrogen industry and the increasing demand for hydrogen, the further research on hydrogen pipeline technology, hydrogen-natural gas mixed transmission technology, natural gas pipeline transformation technology, and large-scale hydrogen compressor, as well as the long-distance hydrogen pipeline and hydrogen-natural gas pipeline standards in China, will be significantly improved. Pipeline hydrogen transmission will become one of the main ways of long-distance hydrogen transportation due to its economic effectiveness. When the mixed hydrogen concentration is <10%, it can be analyzed according to the American Compressed Gas Association standard CGA-5.6: 2005 "Hydrogen Pipeline System", and when the hydrogen concentration is less than 10%, it can be analyzed according to the American Society of Mechanical Engineers standard ASME B31.12: 2019. The specific analysis process is shown in Fig. 2.18.

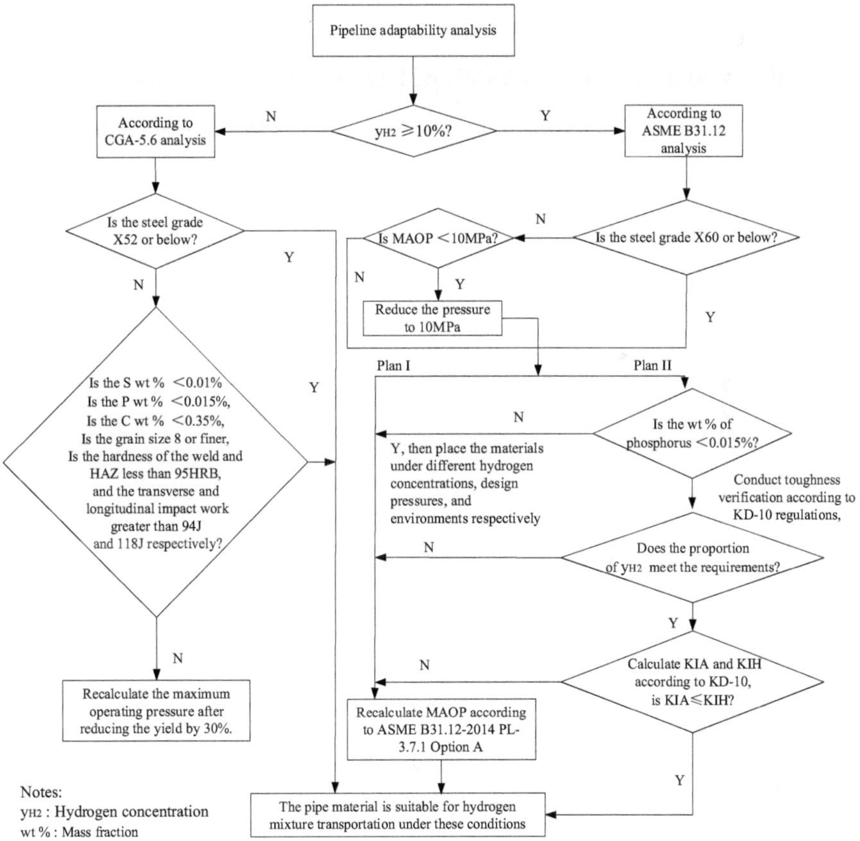

Fig. 2.18 A pipeline applicability analysis flowchart

Pipeline adaptability analysis, hydrogen concentration ≥10%?

According to CGA-5.6 analysis, is the steel grade X52 or below?

Is the mass fraction of sulfur <0.01%, the mass fraction of phosphorus <0.015%, the mass fraction of carbon <0.35%, the grain size 8 or finer, the hardness of the weld and HAZ less than 95HRB, and the transverse and longitudinal impact work greater than 94 and 118 J respectively?

Recalculate the maximum operating pressure after reducing the yield by 30%.

According to ASME B31.12 analysis, is the steel grade X60 or below? Is MAOP <10 MPa?

Reduce the pressure to 10 MPa, Plan I, Plan II.

Is the mass fraction of phosphorus <0.015%?

Y. Then place the materials under different hydrogen concentrations, design pressures, and environments respectively.

Conduct toughness verification according to KD-10 regulations, does the proportion of hydrogen concentration meet the requirements?

Calculate KIA and KIH according to KD-10, is KIA ≤ KIH?

Recalculate MAOP according to ASME B31.12-2014 PL-3.7.1 Option A

The pipe material is suitable for hydrogen mixture transportation under these conditions.

Exercises

1. Single atom hydrogen forms atomic bonds at ( ) defect positions, leading to an increase in internal pressure, resulting in bubbles or cracking.

   A. Vacancies
   B. Doping
   C. Grain boundaries
   D. Dislocations

2. The main advantages of diaphragm compressors are ( ).

   A. High compression ratio
   B. Good sealing
   C. Large exhaust volume
   D. High thermal efficiency

3. The main advantages of steel tape miswound high-pressure hydrogen storage container (MSLV) are ( ).

   A. Explosion suppression and explosion resistance
   B. Defect dispersion

C. Reliable structure
D. Low cost

4. The material composition of Type III tanks is ( ).

A. All metal
B. Metal liner + GFRP layer
C. Metal liner + CFRP layer
D. Plastic liner + CFRP layer

5. High-pressure hydrogen transportation in pipelines is carried out by burying or overhead seamless steel pipe systems. The correct statements among the following are ( ).

A. Suitable for short-distance transportation
B. High transportation efficiency
C. Suitable for high-pressure transportation
D. High energy consumption

6. Hydrogen embrittlement is caused by the presence of hydrogen in material defects, usually manifested as a decrease in mechanical properties such as ____ and ____.
7. Hydrogen embrittlement is mainly divided into ____, ____, and ____, a total of three types.
8. According to the Joule-Thomson effect, the temperature of a few gases such as hydrogen and helium ____ after throttling expansion.
9. According to Sievert's law, the degree of hydrogen embrittlement is directly proportional to ____.
10. The design of lightweight high-pressure hydrogen storage containers for vehicles usually requires ____, ____, ____, ____, and ____.
11. The hydrogen pressure of the mainstream long-tube trailers on the market is ____, and the main forms of long-tube trailers are ____ and ____.
12. For a certain model of fuel cell vehicle, the maximum storage pressure of the hydrogen storage system is 70 MPa, the total water volume of the hydrogen cylinder is 141 L, the limit pressure of hydrogen used in the hydrogen cylinder is 2 MPa, assuming that the hydrogen consumption of this fuel cell vehicle is 0.6 kg H_2/100 km, try to estimate the maximum driving range of this fuel cell vehicle under the condition of 15 °C? ($\alpha = 1.9155 \times 10^{-6}$ K/Pa, keep the result to two decimal precision, the pressure conditions are all absolute pressure).

References

1. Del-Pozo A, Villalobos JC, Serna S (2020) A general overview of hydrogen embrittlement (Chapter 6). In: Current trends and future developments on (bio-)membranes: recent advances in metallic membranes, pp 139–168

2. Huang F, Liu J, Deng ZJ et al (2010) Effect of microstructure and inclusions on hydrogen induced cracking susceptibility and hydrogen trapping efficiency of X120 pipeline steel. Mater Sci Eng A 527:6997–7001
3. Dong C, Liu Z, Li X et al (2009) Effects of hydrogen-charging on the susceptibility of X100 pipeline steel to hydrogen-induced cracking. Int J Hydrog Energy 34:9879–9884
4. Yen SK, Huang IB (2003) Critical hydrogen concentration for hydrogen-induced blistering on AISI 430 stainless steel. Mater Chem Phys 80:662–666
5. Mohtadi-Bonab MA, Szpunar JA, Basu R et al (2015) The mechanism of failure by hydrogen induced cracking in an acidic environment for API 5L X70 pipeline steel. Int J Hydrog Energy 40:1096–1107
6. Kittel J, Smanio V, Fregonese M et al (2010) Hydrogen induced cracking (HIC) testing of low alloy steel in sour environment: impact of time of exposure on the extent of damage. Corros Sci 52:1386–1392
7. Carneiro RA, Ratnapuli RC, De Freitas Cunha Lins V (2003) The influence of chemical composition and microstructure of API linepipe steels on hydrogen induced cracking and sulfide stress corrosion cracking. Mater Sci Eng A 357:104–110
8. Sugimoto H, Fukai Y (1992) Solubility of hydrogen in metals under high hydrogen pressures: thermodynamical calculations. Acta Metall Mater 40(9):2327–2336
9. Mori D, Hirose K (2009) Recent challenges of hydrogen storage technologies for fuel cell vehicles. Int J Hydrog Energy 34(10):4569–4574
10. Barthelemy H, Weber M, Barbier F (2017) Hydrogen storage: recent improvements and industrial perspectives. Int J Hydrog Energy 42(11):7254–7262
11. San Marchi C, Somerday BP (2008) Technical reference on hydrogen compatibility of materials. Sandia National Laboratories, California
12. Iijima T, Abe T, Itoga H (2015) Development of material testing facilities in high pressure gaseous hydrogen and international collaborative work of a testing method for a hydrogen society—toward contribution to international standardization. Synthesiology 8(2):62–69
13. Zheng J, Zhou C, Xu P et al (2013) Research progress on durability testing equipment for materials in high pressure hydrogen environment. J Solar Energy 34(8):1477–1483
14. Huang S, Hui H (2021) A jacketed high-pressure hydrogen storage device for hydrogen refueling stations: CN112178446A
15. Xu H (2008) Research on multi-functional full-layer high-pressure hydrogen storage container for hydrogen refueling stations. Zhejiang University, Hangzhou
16. Ye S, Zheng J, Yu T et al (2019) Light weight design of multi-layered steel vessels for high-pressure hydrogen storage. In: Proceedings of the pressure vessels and piping conference. American Society of Mechanical Engineers [s. n.: S. l.]
17. Wang H, Zhang Q, Zhang H et al (2016) A large-capacity steel liner fully wrapped high-pressure hydrogen storage container: CN205101849U
18. National Boiler and Pressure Vessel Standardization Technical Committee (2011) Fixed high-pressure hydrogen storage steel tape miswinding container: GB/T 26466—2011. China Standards Publishing House, Beijing
19. Hydrogen Energy Association (2009) Hydrogen energy technology. Translated by Song Yongchen, Ning Yadong, Jin Dongxu. Science Press, Beijing
20. State Administration for Market Regulation (2021) Safety technical regulations for gas cylinders: TSG 23—2021. Xinhua Publishing House, Beijing
21. General Administration of Quality Supervision, Inspection and Quarantine of the People's Republic of China, Standardization Administration of China (2016) Large volume seamless steel gas cylinders: GB/T 33145—2016. China Standards Publishing House, Beijing
22. Luo H, Bo K, Li B et al (2016) Structural characteristics and periodic inspection issues of high purity gas long tube trailers. China Spec Equipment Saf 32(3):35–39
23. Xu S, Gai X, Wang N (2015) Application of container tube bundle transport vehicle in hydrogen transport. Shandong Chem Ind 44(2):88–89

24. General Administration of Quality Supervision, Inspection and Quarantine of the People's Republic of China (2017) Aluminum inner liner carbon fiber fully wrapped gas cylinder for compressed hydrogen vehicles: GB/T 35544—2017. China Standards Publishing House, Beijing
25. Zheng J, Ma K, Zhou W et al (2018) High pressure hydrogen storage container for hydrogen refueling station. Pressure Vessel 35(9):35–42
26. Drive U (2013) Hydrogen delivery technical team roadmap. Hydrogen Delivery Technical Team, California
27. Gillette JL, Kolpa RL (2008) Overview of interstate hydrogen pipeline systems. Argonne National Lab, Lemont
28. Liu Z, Xiong S, Zheng J et al (2020) Comparative analysis of hydrogen pipelines and natural gas pipelines. Pressure Vessel 37(2):56–63
29. National Hydrogen Energy Standardization Technical Committee (2016) China hydrogen energy industry infrastructure development blue book. China Standards Publishing House, Beijing
30. Hydrogen piping and pipelines: ASME B31. 12: 2019
31. Hydrogen pipeline systems: CGA G-5. 6: 2005
32. General Administration of Quality Supervision, Inspection and Quarantine of the People's Republic of China, Standardization Administration of China (2008) Safety technical regulations for hydrogen use: GB 4962—2008. China Standards Publishing House, Beijing
33. Ministry of Construction of the People's Republic of China, General Administration of Quality Supervision, Inspection and Quarantine of the People's Republic of China. Design code for hydrogen stations: GB 50177—2005. Beijing: China Planning Publishing House, 2005
34. Wu C, Li Y, Li Y (2020) Hydrogen storage and transportation. Chemical Industry Press, Beijing

Chapter 3
Low-Temperature Liquid Hydrogen Storage and Transportation

With the purchase of this book, you can use our “SN Flashcards” app to access questions using ▶ www.sn.pub/5lls30 free of charge in order to test your learning and check your understanding of the contents of the book. To use the app, please follow the instructions in Chap. 1.

Liquid hydrogen storage technology refers to the technology of condensing gaseous hydrogen into a liquid state and storing it. Under a lower pressure, liquid hydrogen is much denser than gaseous hydrogen. Liquid hydrogen is a colorless, odorless, and non-toxic liquid. Under standard atmospheric pressure, the temperature of saturated liquid hydrogen is 20.37 K (−252.78 °C), with a density of about 70.85 kg/m^3, which is about 790 times the density of gaseous hydrogen under standard conditions. Due to the low critical temperature of liquid hydrogen, 33.19 K (−239.97 °C), the liquefaction and storage of hydrogen gas is much more difficult than that of common gases.

3.1 Principle of Hydrogen Liquefaction

Liquefaction of hydrogen requires precooling to below the conversion temperature, and then further cooling to below the critical temperature through isentropic expansion or isenthalpic throttling to liquefy the hydrogen. Ortho-hydrogen and para-hydrogen are two spin isomers of molecular hydrogen. Ordinary hydrogen at room temperature contains 75% ortho-hydrogen and 25% para-hydrogen. At low temperatures, ortho-hydrogen gradually converts to para-hydrogen and releases heat. To avoid the loss of liquid hydrogen vaporization caused by the conversion heat

J. Zou, *Hydrogen Storage and Transportation*,
https://doi.org/10.1007/978-981-96-2875-9_3

during the storage of liquid hydrogen, most of the ortho-para hydrogen conversion process must be completed during the production process [1].

3.1.1 Ortho-Para Hydrogen Conversion

Based on the different spin directions of the two atomic nuclei in a hydrogen molecule, the hydrogen molecules can be divided into ortho-hydrogen and para-hydrogen, two spin isomers. Ortho-hydrogen refers to hydrogen molecules where the atomic nuclei spin in parallel directions, while the two atomic nuclei of para-hydrogen spin in opposite directions. Ortho-hydrogen and para-hydrogen are chemically identical, but their physical properties are slightly different, with ortho-hydrogen having a higher energy state, and its specific heat and latent heat are slightly higher than those of para-hydrogen.

In thermal equilibrium, there is a stable ratio between ortho-hydrogen and para-hydrogen, which changes with temperature, as shown in Fig. 3.1. At room temperatures and above, the proportion of para-hydrogen remains around 25%; when the temperature drops to about 120 K, ortho-hydrogen begins to convert to para-hydrogen, and the proportion of para-hydrogen rapidly increases; when the temperature drops to about 20 K, the proportion of para-hydrogen in thermal equilibrium reaches 99.8%, at which point hydrogen molecules almost only have the spin form of para-hydrogen. However, without human intervention, the conversion of ortho-para hydrogen is extremely slow. If room-temperature hydrogen gas is rapidly liquefied, the proportion of ortho-para hydrogen in liquid hydrogen is far from thermal equilibrium.

Para-hydrogen ratio, Temperature.

Spin direction, Ortho-hydrogen, Para-hydrogen, Hydrogen atom, Covalent bond.

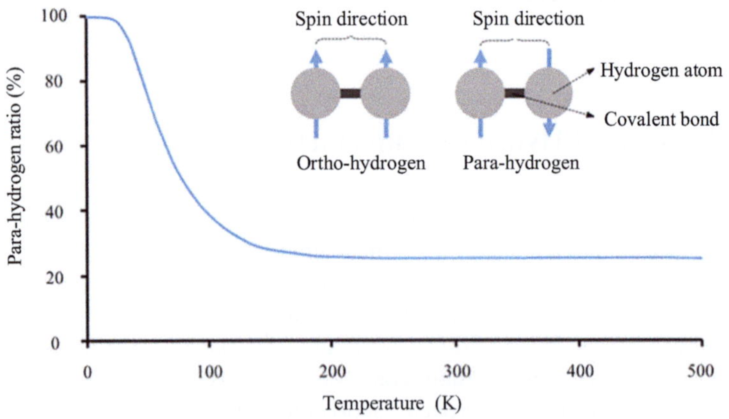

Fig. 3.1 Spin directions of ortho-hydrogen and para-hydrogen and the ratio of ortho and para hydrogen with the change of temperature

The conversion of ortho-hydrogen to para-hydrogen releases about 1.42 kJ/mol of energy, while the latent heat of vaporization of liquid hydrogen at 20 K is only 0.89 kJ/mol. This means that the energy released by the conversion of saturated liquid ortho-hydrogen to para-hydrogen can completely cause it to evaporate. Experiments show that the heat released due to the change in the spin form of hydrogen within 1h is enough to cause 1% of liquid hydrogen to evaporate. Liquid hydrogen that has not reached ortho-para hydrogen equilibrium will lose more than half in a few days and nights.

Therefore, the para-hydrogen content in the liquid hydrogen produced by the factory must be at least 95%, while in the liquid hydrogen that needs to be stored and transported for a long time, the content of para-hydrogen must reach 98%. It is necessary to carry out ortho-para hydrogen conversion during the liquefaction processes of hydrogen to reduce the evaporation of liquid hydrogen and extend the storage time of liquid hydrogen. An ortho-para hydrogen converter is set up in the hydrogen liquefaction device and immersed in liquid hydrogen, so that the conversion heat is absorbed by the surrounding liquid hydrogen, and a catalyst is used to increase the conversion. Efficient catalysts are mainly chromium nickel catalysts and iron hydroxides, including Cr_2O_3 + NiO, $Cr(OH)_3$, $Fe(OH)_3$, etc.

The catalyst must be activated before use. Among them, the activation of the chromium nickel catalyst is to heat the reactor and the catalyst together to 150 °C and blow it off with hydrogen. The activation of the iron hydroxide catalyst is to heat it in the reactor to 130 °C while pumping it to vacuum, after 24 h, then replace its vacuum with room-temperature hydrogen. However, the activated chromium nickel catalyst is prone to spontaneous combustion, and once burned, it will cause irreversibly poisoning. Therefore, in production, the iron hydroxide catalyst, which is less efficient but not easy to poison, is usually chosen.

For large-scale hydrogen liquefaction projects, to improve the efficiency of ortho-para hydrogen conversion, the conversion is usually divided into two or more stages. The first stage is to achieve conversion in the 80 K temperature zone, where the conversion heat of ortho-hydrogen is absorbed by pre-cooled liquid nitrogen or cold hydrogen gas, and this process can produce about 50% of para-hydrogen; the second stage is carried out in the 20 K temperature zone, where ortho-hydrogen is almost completely converted into para-hydrogen. During the conversion process, hydrogen molecules do not directly split into atoms and recombine, but they are reoriented within a molecular range through nuclear spin.

3.1.2 *Joule-Thompson Effect*

The Joule-Thompson effect refers to the phenomenon where, under isenthalpic conditions, when a gas flow is forced through a porous plug, or a small gap, or a small tube mouth, the pressure decreases due to volume expansion, causing a change in temperature. Most gases at normal temperature and pressure, after throttling expansion, the temperature drops, producing a refrigeration effect, while a few gases such as hydrogen and helium, after throttling expansion, the temperature rises, producing a heating effect. The enthalpy of the gas before and after adiabatic throttling does not change, and the change in the state properties of the gas in this process is shown in the following formula [2]:

$$\mathrm{H} = \mathrm{U}_1 + \mathrm{P}_1\mathrm{V}_1 = \mathrm{U}_2 + \mathrm{P}_2\mathrm{V}_2 \tag{3.1}$$

where, H is the enthalpy of the gas (J); U is the internal energy of the gas (J); P is the gas pressure (Pa); V is the gas volume (m^3); the subscript 1 is the state before throttling; the subscript 2 is the state after throttling.

The Joule-Thompson coefficient "μ" is commonly used to characterize the Joule-Thompson effect, which is defined as the change in temperature with pressure under isenthalpic conditions:

$$\mu = \left(\delta T / \delta P\right)_{\mathrm{H}} \tag{3.2}$$

For different gases, at different pressures and temperatures, the "μ" values are different. For any real gases, on the pressure-temperature curve, when the decrease in pressure cannot change the temperature, the curve formed by these points is called the transformation curve of the gas. For helium and hydrogen at 1 atm (1.01×10^5 Pa), the transformation temperature is very low, so the temperature of helium and hydrogen will rise when they expand at room temperature. Real gases will change temperature (either rise or fall depending on the initial temperature) when they expand freely under isenthalpic conditions. For a real gas, there will be a Joule-Thomson transition temperature under given pressure conditions. Above this temperature, the gas temperature will rise, below this temperature, the gas temperature will fall, and at this temperature, the gas temperature will remain unchanged. The reasons for the temperature change are as follows:

Temperature rises: When molecules collide, potential energy is temporarily converted into kinetic energy. As the average distance between molecules increases, the average number of collisions per unit time decreases and the potential energy decreases, so the kinetic energy increases, and the temperature rises accordingly.

Temperature drops: When the gas expands, the average distance between molecules increases. Because there is an attraction between molecules, the potential energy of the gas increases. This process is an isentropic process, the total energy of the system is conserved, so the increase in potential energy will cause a decrease in the kinetic energy, and the gas temperature drops accordingly.

The transformation temperature of hydrogen under standard atmospheric pressure is only 204.6 K (−68.55 °C), which is an important reason for the sharp rise in temperature during high-pressure hydrogen refueling processes at room temperature. This also means that the high-pressure hydrogen must be sufficiently pre-cooled to achieve throttling cooling and liquefaction.

3.1.3 Hydrogen Liquefaction Processes

The liquefaction process of hydrogen requires energy inputs. Cooling consumes most of the energy required for liquefaction, including the cooling of hydrogen and the transformation of ortho-hydrogen to para-hydrogen, while other energy consumption is mainly used for the work of compressing the fluid. Although some studies have pointed out that the minimum theoretical energy consumption for liquefying hydrogen is 3.92 kW h/kg H_2, unavoidable heat transfer losses make the energy consumption of hydrogen liquefaction in actual engineering ranges from 6.5 to 15 kW h/kg H_2. The amount of the energy consumption is related to the scale capacity and insulation efficiency of the hydrogen liquefaction system.

The process flow of liquefied hydrogen has the following characteristics:

1. To produce liquid hydrogen through expansion or throttling, hydrogen needs to be pre-cooled to below the critical temperature.
2. The entire system needs efficient insulation.
3. In the 20 K liquid hydrogen temperature zone, most gases will solidify, so other gas impurities except helium need to be removed before hydrogen liquefaction.
4. The system materials need to have ultra-low temperature resistance and high hydrogen embrittlement resistance.
5. It is indispensable to reduce the possibilities of heat transfer with the outside world and improve the sealing of the system.
6. It must have ortho-para hydrogen conversion capability [2].

Based on the above characteristics, the industrial preparation of liquid hydrogen mainly includes the following steps: (1) purification and drying of hydrogen; (2) compression of hydrogen; (3) cooling of hydrogen; (4) expansion/throttling liquefaction of hydrogen; (5) ortho-para hydrogen conversion. The following introduces two common hydrogen liquefaction cycle processes: the Linde-Hampson cycle and the Claude cycle.

In 1895, German low-temperature engineer Linde and British engineer Hampson independently applied for patents for gas liquefaction, which are later known as the Linde-Hampson cycle. The core of this process is that hydrogen gas below the transition temperature is liquefied by reducing its temperature after passing through a throttling expansion valve, as shown in Fig. 3.2. Hydrogen gas is pressurized by a

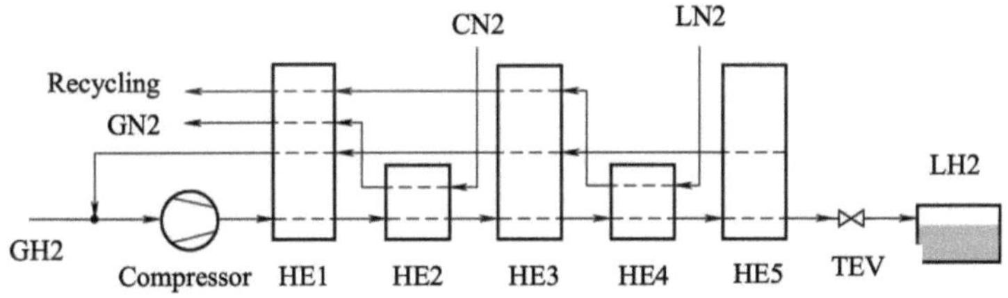

Fig. 3.2 A typical Linde-Hampson cycle hydrogen liquefaction process flow diagram

compressor and initially cooled in the subsequent heat exchanger 1 (HE1) with supercooled nitrogen gas, then further cooled in the remaining four heat exchangers (HE2 to HE5) by compressed nitrogen gas and liquid nitrogen. Once the temperature of the hydrogen gas drops below its transition temperature, the temperature drop is completed by the throttling expansion valve through the Joule-Thomson effect, which successfully cools a portion of the hydrogen gas into liquid hydrogen. The liquid hydrogen is then separated by a separator, and the remaining hydrogen gas is recycled and mixed with the incoming hydrogen gas. The equipment required for this cycle is simple, reliable, and easy to handle; the drawback is its low efficiency. Only when the pressure reaches 10–15 MPa and the temperature drops to 50–70 K can the relatively ideal liquefaction rate (24–25%) be achieved. This cycle is mainly used for small-scale hydrogen liquefaction in scientific research and experimental setups.

Room temperature hydrogen gas, compressor, throttling expansion valve, liquid hydrogen.

Recycling, nitrogen gas, compressed nitrogen gas, liquid nitrogen.

In 1902, French engineer Claude invented an air liquefaction process with a piston-type expander, which is named as the Claude cycle referring to the gas liquefaction cycle with an expander. The role of the expander is to use the adiabatic expansion effect to allow the hydrogen gas to perform works externally on the environment using its internal energy, thereby reducing the temperature of the hydrogen gas. Compared with the Linde-Hampson cycle, the Claude cycle uses the adiabatic expansion effect of the expander to reduce the temperature of part of the hydrogen gas and uses its own cooling capacity as a pre-cooling source to cool the hydrogen gas, thus it is more efficient. Hydrogen liquefaction systems usually use a turbine expander.

Figure 3.3 shows a typical Claude cycle hydrogen liquefaction process. Similar to the Linde-Hampson cycle, the hydrogen gas still needs to be cooled through several heat exchangers (HE1 to HE3) after compression, but an expander is installed between the heat exchangers, and the liquid nitrogen and high-pressure nitrogen gas are no longer required as refrigerants. After the cooled high-pressure hydrogen gas exchanges heat with the supercooled hydrogen gas in heat exchanger 1 (HE1), one part of the hydrogen gas enters the expander to cool the remaining gas; the other

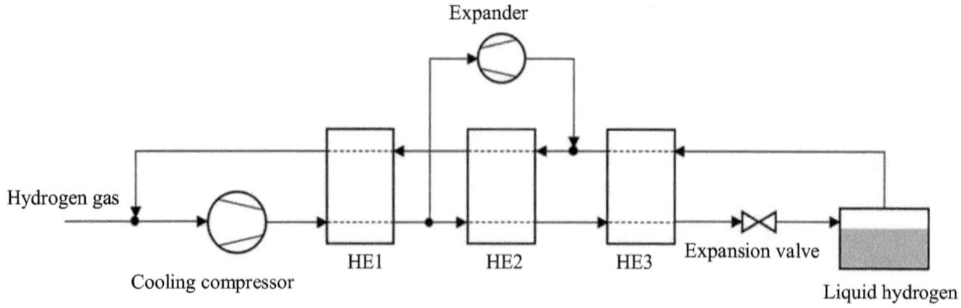

Fig. 3.3 A typical Claude cycle hydrogen liquefaction process flow diagram

part enters the throttling expansion valve after multi-stage cooling to undergo the Joule-Thomson effect, finally obtaining liquid hydrogen. The part of the hydrogen gas that is not successfully condensed will be used as a refrigerant in the heat exchanger together with the hydrogen gas cooled by the expander, and then mixed with the incoming hydrogen gas. Since there is no need to use high-pressure nitrogen and liquid nitrogen as pre-cooling sources for hydrogen gas, the Claude cycle has lower unit energy consumption than the Linde-Hampson cycle and is suitable for medium and large-scale hydrogen liquefaction. It is widely used in commercial liquid hydrogen factories around the world.

Hydrogen gas, cooling compressor, expander, expansion valve, liquid hydrogen.

Interestingly, after inventing the corresponding gas liquefaction methods, both Linde and Claude used their respective technologies to establish companies to provide industrial gases for society. Today, they have developed into the German Linde Group and the French Air Liquide Group, both of which have become the leading multinational engineering and industrial gas supply companies.

3.2 Key Materials and Equipment of Liquid Hydrogen Storage Tanks

3.2.1 Liquid Hydrogen Storage Tank

As a container for liquid hydrogen, the application scenarios of liquid hydrogen storage tanks are diverse, ranging from giant spherical tanks used in liquid hydrogen factories and offshore receiving stations, to liquid hydrogen tank containers and tank trucks for transportation, and quasi-cylindrical liquid hydrogen fuel tanks for hydrogen refueling stations and vehicles, and even irregular miniature liquid hydrogen tanks for laboratories and orbit change of satellites. Although there are huge differences in volume and shape, they are basically double-layer containers with vacuum jackets in structure. This is because, under standard atmospheric pressure, the temperature of saturated liquid hydrogen is as low as −253 °C and the latent heat of vaporization is very small. To maintain the liquid hydrogen state and minimize evaporation, it is necessary to isolate heat leakage as much as possible, cutting off the pathways of heat transfer from the aspects of heat conduction, heat convection, and heat radiation. This is also the key technology in the design of liquid hydrogen storage containers and tanks [3].

The use of a double-layer container structure with a vacuum jacket between the inner liner and the outer tank, especially under high vacuum conditions of the order of 1×10^{-3} Pa, the gas molecules in the interlayer space are extremely rare making it difficult to achieve heat transfer through contact between materials and convection of gas, and its heat leakage impact can basically be ignored. The support structure connecting the inner liner and the outer tank uses low thermal conductivity and high strength non-metallic materials. For example, using high-strength carbon fiber reinforced composites in liquid hydrogen containers can minimize the heat transfer by thermal conduction. Liquid hydrogen is the least dense cryogenic medium, so the cross-sectional area of the liquid hydrogen tank support structure can be made smaller. However, heat radiation does not require a medium, and according to the Stefan-Boltzmann law, the heat transfer by radiation is related to the fourth power of the temperature difference, not a linear relationship with the temperature difference for heat conduction and heat convection. Therefore, for liquid hydrogen tanks, heat radiation is the most important factor affecting heat transfer and requires more cautious measures than ordinary cryogenic tanks [4].

Figure 3.4 shows a cross-sectional view of a typical structure of a liquid hydrogen tank. Structurally, the liquid hydrogen tank consists of an inner liner, an outer tank, a support structure, and interlayer insulation material. The inner liner is used to hold liquid hydrogen and its evaporated gas, the outer tank provides a closed vacuum environment for the inner liner, and the support structure between the inner liner and the outer tank is used to maintain the relative stability of the inner liner installation position. In the vacuum layer between the inner liner and the outer tank, insulation material needs to be filled, and the high vacuum multilayer insulation material used in Fig. 3.4 is one of them.

Outer tank, inner liner, support structure, vacuum layer, high vacuum multilayer insulation material, liquid hydrogen.

To meet the functional requirements of liquid hydrogen storage tanks in application scenarios, pipelines and other accessories are also needed. Figure 3.5 provides a flowchart of a typical liquid hydrogen storage and transportation container, showing that the commonly used pipelines and functions of the liquid hydrogen tank are mainly divided into the following parts: top/bottom filling pipelines, liquid discharge pipelines, pump suction pipelines, liquid level gauges, gas-liquid phase pipelines, full-capacity measurement pipelines, overpressure discharge pipelines, pressurization pipelines, and pump return gas pipelines. Among them, the filling pipelines, pump suction, and return gas pipelines are in direct contact with liquid hydrogen and are extremely cold, so the pipelines and valves are designed with a vacuum jacket structure.

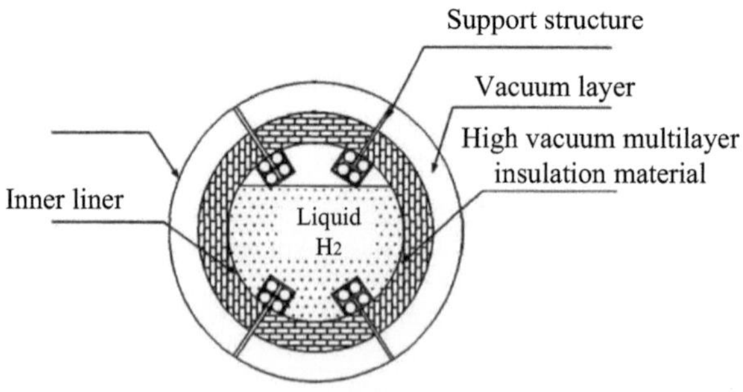

Fig. 3.4 A cross-sectional view of a typical high vacuum insulated liquid hydrogen tank structure

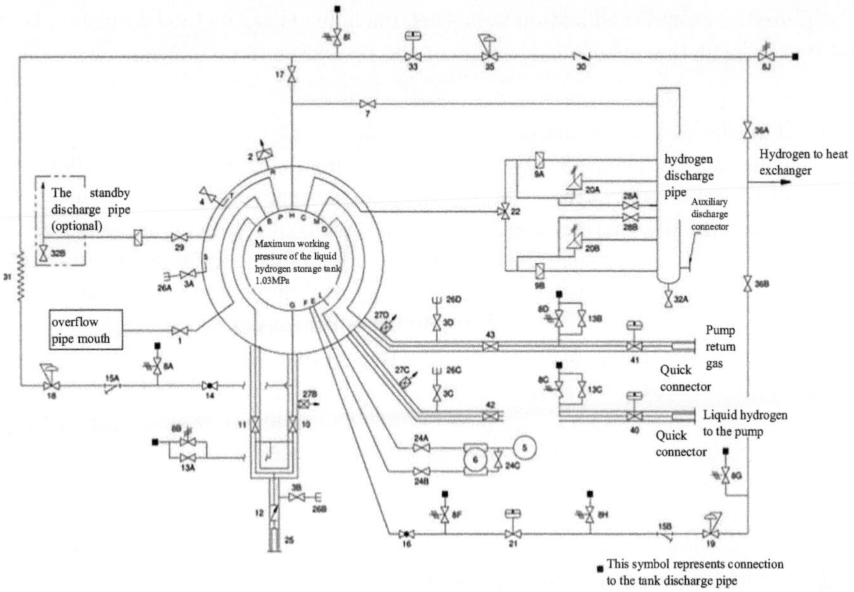

Fig. 3.5 A flowchart of a typical liquid hydrogen storage and transportation container

The standby discharge pipe (optional), overflow pipe mouth, maximum working pressure of the liquid hydrogen storage tank, hydrogen discharge pipe, auxiliary discharge connector, hydrogen gas to the heater, quick connector, pump return gas, liquid hydrogen to the pump, this symbol represents connection to the tank discharge pipe.

It should be pointed out that the hydrogen molecules are extremely prone to leakage, thereby reducing the vacuum degree of the interlayer. Therefore, the inner liner and pipelines in contact with liquid hydrogen not only need to use low-temperature resistant and high hydrogen embrittlement resistant materials (such as austenitic stainless steel), but also need to consider the tightness of the welding connection, which has a lower tolerance for defects than ordinary low-temperature containers. In addition, considering the expansion coefficient of the material, the inner liner and interlayer pipelines are manufactured at room temperature, but the size shrinkage caused by the sharp drop in temperature under service conditions needs to be fully considered in the design of the support structure and interlayer pipelines.

3.2.2 Insulation Methods of Liquid Hydrogen Storage Tanks

Insulation methods of low-temperature liquid hydrogen storage tanks include stacked insulation, low-vacuum insulation, high-vacuum multilayer insulation, and high-vacuum multi-screen insulation, which are suitable for liquid hydrogen tanks of different sizes and -application scenarios. Each insulation method corresponds to a different insulation structure component and insulation material [5].

1. Stacked insulation

Stacked insulation is an insulation solution of the lowest cost. The interlayer of the container does not consider the vacuum environment but directly places the container in perlite (pearl sand), foam plastic, polystyrene, and other materials with small density, low thermal conductivity, and low cost. Although these materials can effectively reduce heat conduction, they do not form a vacuum environment, and the insulation material on the periphery of the liquid hydrogen container will affect the insulation performance due to the condensation or even solidification of air and water. The performance of this type of insulation material is not satisfactory, but for liquid hydrogen tanks with extremely large volumes, especially for super projects with a single tank volume exceeding 10,000 m^3, considering construction, operation, and maintenance costs, stacked insulation is still the preferred solution.

2. Low-vacuum insulation

Based on the improvement of the stacked insulation scheme, the insulating interlayer filled with powder or fiber is pumped and maintained to a low vacuum environment, which can effectively reduce gas heat conduction and eliminate convective heat transfer, and the final actual thermal conductivity is only a fraction of that of stacked insulation [6]. To isolate radiation heat exchange, metal powder can be added to the insulation material, or multiple layers of reflective materials can be added to the low-temperature side close to the outer wall of the inner liner.

In recent years, an insulation technology scheme using vacuum glass microspheres has received widespread attention. As shown in Fig. 3.6, vacuum glass microspheres are a series of glass bubbles with diameters less than 0.001 mm. They are lightweight and high-strength. Due to the vacuum environment inside, their insulation effect can approach the thermal conductivity under low vacuum conditions. In 1998, the Cryogenic Test Laboratory at the Kennedy Space Center began testing and researching glass sphere products, and in 2003 submitted a proposal to NASA to replace perlite powder with vacuum glass microspheres for the modification of cryogenic storage tanks.

Between 2008 and 2016, NASA conducted a comparative experiment on the insulation performance of a 50,000 US gallon (approximately 189 m^3) spherical liquid hydrogen storage tank. The liquid hydrogen container using vacuum glass microspheres as insulation material underwent three complete thermal cycle tests over 9 years. The comparative experimental results showed that the insulation performance had been significantly improved compared to that when perlite particles were used, with the average evaporation rate of the spherical liquid hydrogen tank decreasing by 46%, reaching 0.10%/day. In 2016, NASA launched the construction plan for the world's largest spherical liquid hydrogen storage tank, with vacuum glass microsphere insulation as the preferred solution. The

Fig. 3.6 A 200-time magnified image of K1 vacuum glass microspheres by 3 M Company, USA

Fig. 3.7 A spherical liquid hydrogen storage tank insulated with vacuum glass microspheres by NASA, USA

hydrogen storage tank has a capacity of 1.25 million US gallons (4730 m^3) and was completed and put into use in 2022, as shown in Fig. 3.7.

3. High-vacuum multilayer insulation

To reduce the impact of thermal radiation on the insulation performance of the hydrogen storage tank, a multilayer insulation structure composed of many layers of reflective screens and spacer materials is usually installed in the insulation space. This structure causes the heat from thermal radiation to attenuate layer by layer during transmission, with only a very small part reaching the outer wall of the liquid hydrogen tank inner liner [7]. Relevant technical requirements can be referred to the GB/T31480—2015 "High-Vacuum Multilayer Insulation Material for Cryogenic Containers".

The most common material for reflective screens is aluminum foil or double-sided aluminum-coated film, which has a high emissivity, low cost, and light weight. Compared with aluminum foil with higher emissivity, the double-sided aluminum-coated film with lower emissivity can establish a larger temperature gradient on the cold side, while aluminum foil with higher emissivity can establish a larger temperature gradient on the hot side. This difference decreases as the number of reflective screens increases. For liquid hydrogen containers, since the cold side temperature is lower, it is desirable for the insulation material to establish a larger temperature gradient, so double-sided aluminum-coated film is suitable as the radiation screen material. When the number of radiation screens is less than 30, choosing different radiation screen materials for both the cold side and hot side can achieve better insulation performance. When the number of radiation screens is more than 30, the performance difference between the multilayer insulation structure of double-sided aluminum-coated film and the aluminum foil is not significant, and the choice of reflective screen material is more about considering the needs of the process. For example, if the strength requirement for the radiation screen is high, double-sided aluminum-coated film should be chosen, while if the heating temperature requirement during the vacuuming process is high, aluminum foil with better temperature resistance should be chosen.

The function of the spacer materials is to increase thermal resistance and reduce heat conduction, so it is better to choose materials with a short fiber length and a low thermal conductivity as much as possible. Commonly used spacer materials include glass fiber paper, chemical fiber paper, and plant fiber paper. The characteristics of chemical fiber paper and plant fiber paper are low unit area weight, short fibers, good strength, but poor high-temperature resistance. When the baking temperature in a vacuum reaches 150 °C and lasts for more than 4 h, there will be phenomena of carbonization, yellowing, and brittleness, so the highest heating temperature during the vacuuming process generally does not exceed 120 °C. The thermal insulation performance of glass fiber paper has a great relationship with the fiber length. Although ultrafine glass fiber paper made of short fibers has good insulation effects, its strength is very poor and it is very prone to brittleness. The commonly used glass fiber paper is produced by a special processing technique using ultrafine glass fiber cotton raw materials produced by the all-electric kiln flame method. It does not burn when exposed to open flames, and has low weight, good uniformity, low thermal conductivity, and large contact thermal resistance. The disadvantages are low tensile strength, brittleness, generating dust harmful to the human respiratory system, and itching when in contact with the skin.

To improve the production efficiency and structural performance consistency of low-temperature tank products, high-vacuum multilayer insulation materials can be compounded into the form of a quilt. The most common combination is to sew every ten reflection layers and ten layers of insulation material alternately into an insulation quilt, and then cut and design the insulation structure according to the size of the container. Due to the edge effect, although the performance of the multilayer insulation structure will decrease, it can also be exchanged for structural stability and layer density assurance, so that the multilayer insulation structure can obtain stable insulation performance in large-scale applications. At the same time, the use of insulation quilts can also ensure that the multilayer insulation structure will not peel off under a vibration impact environment. To speed up the vacuuming of the container and ensure the vacuum degree between the layers of the insulation structure, it is necessary to open air holes of a certain size in the multilayer insulation structure. Compared with single-layer materials with relatively low strength, the insulation quilts will not tear the material due to the opening of holes.

High-vacuum multilayer insulation structures are widely used in liquid hydrogen storage technology and can meet the insulation requirements of most liquid hydrogen tanks, achieving a non-emission maintenance time of 5–15 days or even longer. From vehicle-use fuel tanks, to liquid hydrogen storage tank trucks and tank containers, and to storage tanks used in liquid hydrogen refueling stations and liquid hydrogen factories, the volume of liquid hydrogen storage tanks manufactured by factories ranges from several hundred liters to several hundred cubic meters, including on-site construction of spherical liquid hydrogen tanks. The largest high-vacuum multilayer insulated liquid hydrogen spherical tank in Russia can reach a volume of 1400 m^3.

4. High vacuum multi-screen insulation

If further improvement of the insulation performance and extension of the non-emission maintenance time of liquid hydrogen storage tanks, metal screens and liquid nitrogen protection screens based on the high-vacuum multilayer insulation structure to form a multi-screen insulation structure can be adopted. This composite insulation structure is very complex and costly, and is more used in liquid helium storage tank containers for long-distance maritime transport, which can achieve a maintenance time of more than 35 days.

5. Adsorbents for vacuum maintenance

Experiments show that within the vacuum range of 10^{-2}–10 Pa, the effective thermal conductivity of high-vacuum multilayer insulation increases exponentially, while the change in vacuum under other conditions has little effect on the effective thermal conductivity. Therefore, the vacuum in the insulation layer should not exceed 10^{-2} Pa. However, the vacuum will inevitably decrease gradually due to leakage of the inner liner, outer tank, and gas release from the insulation material. The gas source leaking into the insulation layer from the outer tank is air, mainly composed of N_2 and O_2, the gas source leaking into the insulation layer from the inner liner is H_2, and the main component of the gas released by the material is also H_2. In order to achieve and maintain the high vacuum required for liquid hydrogen tank insulation, adsorbents need to be used in the insulation space to adsorb residual gases. For liquid hydrogen containers, the main component of the residual gas is H_2.

According to the distribution of residual gases in high-vacuum multilayer insulation containers and different gas absorption principles, liquid hydrogen containers need to use low-temperature adsorbents and room-temperature adsorbents at the same time, and they are placed on the inside and outside of the multilayer insulation structure, respectively. Common low-temperature adsorbents include activated carbon and 5A molecular sieves. Table 3.1 shows the adsorption capacity of two common adsorbents under different adsorption temperature conditions. It can be seen that: as the temperature decreases, the adsorption capacity of the adsorbent will significantly increase; activated carbon has a better ability to adsorb hydrogen. The adsorbent used on the room temperature side is palladium oxide (PdO).

Activated carbon adsorbents include microspherical activated carbon, and carbonized and activated coconut shells, etc. High-carbon organic substances first remove internal water contents under high-temperature isolation air conditions, and undergo carbonization treatment with tar, and ultimately form activated carbon after introducing carbon dioxide, steam, and oxygen treatment. Activated carbon has a large specific surface area and abundant micropores, and its chemical properties are stable, cost-effective, and easy to regenerate.

The main component of the molecular sieve is an aluminosilicate with a skeletal structure formed by tetrahedral connections, and after high-temperature baking, it becomes a porous crystal with many crystal cavities (diameter of 0.25–0.8 nm). Therefore, the molecular sieve not only has a large internal surface area that can adsorb gas molecules smaller than the crystal cavity size, but also has an excellent adsorption capacity for polar molecules (such as water). Although the size of hydrogen molecules is far smaller than the crystal cavity size, the molecular sieve has poor adsorption capacity for it, but it can achieve adsorption of hydrogen under low temperatures, especially in the liquid hydrogen temperature zone.

The principle of palladium oxide (PdO) adsorbing hydrogen is more of a chemically catalytic reaction rather than physical adsorption [8]. PdO can react with H_2 to produce Pd and H_2O, and it can also catalyze the reaction of O_2 in the air with H_2 to directly produce H_2O, while water can be easily adsorbed by the molecular sieve.

Table 3.1 Adsorption of molecular sieves and activated carbon under the same condition

Adsorption conditions	Adsorption capacity (mL/g)			Type of adsorbent
	N_2	O_2	H_2	
288 K (15 °C)	8	>8	1×10^{-3}	5A Molecular Sieve
	>8	>8	5	GH-0 Coconut Shell Activated Carbon
77 K	85	>85	1×10^{-3}	5A Molecular Sieve
	>85	>85	55	GH-0 Coconut Shell Activated Carbon
20 K			160	5A Molecular Sieve
			200	GH-0 Coconut Shell Activated Carbon

Therefore, palladium oxide combined with the molecular sieve can adsorb large amounts of room temperature hydrogen.

3.3 Applications of Liquid Hydrogen Storage and Transportation Technology

Liquid-state hydrogen storage and transportation technology refers to the technology of storing and transporting hydrogen gas using liquid hydrogen or hydrogen-rich liquid compounds. Liquid hydrogen storage and transportation technology involves cooling hydrogen gas to below its liquefaction temperature and storing and transporting it in the form of liquid hydrogen. The greatest advantage of liquid hydrogen transportation is its high gravimetric storage density. However, it has issues, such as high storage container requirements, easy evaporation (evaporating 0.3–2% daily), high liquefaction energy consumption (>20 kW h/kg), and significant safety risks. Currently, it is only used in fields like aerospace in China. When the centralized production base of liquid hydrogen is far away from the users, it is necessary to store the liquid hydrogen in specialized low-temperature insulated hydrogen tanks and transport it by heavy trucks, trains, or ships. Liquid hydrogen transportation is an economic method of hydrogen transportation with high density, high speed, and long-distance capabilities. At present, the relevant technology for the preparation, transportation, and storage of liquid hydrogen has been put into practice, but the scale is relatively small, and the core patents are mostly held by the gas companies such as Linde Group and Air Liquide Group.

A renewable energy electricity-hydrogen system of liquid hydrogen storage and transportation is shown in Fig. 3.8. The energy consumption of hydrogen liquefaction is high, but in the liquid hydrogen technology route, the liquid hydrogen factory can be built in areas with cheap electricity next to wind and solar power plants. In big cities where large-scale users are located, the unit energy consumption of liquid

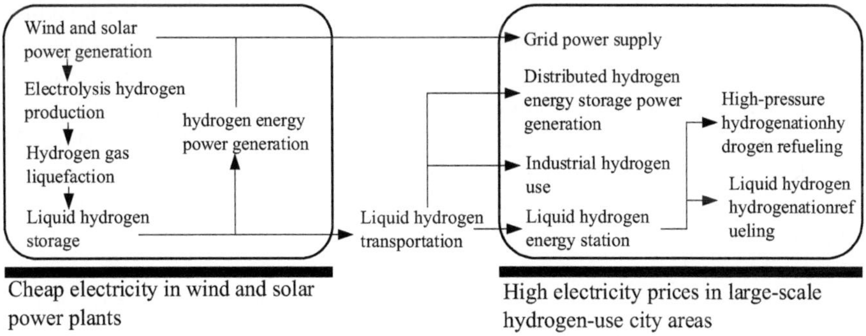

Fig. 3.8 A renewable energy electricity-hydrogen system based on liquid hydrogen storage and transportation

hydrogen at the refueling terminal is very low, only 1 kW h/kg, which is 1/6 to 1/4 of high-pressure hydrogen storage and refueling; coupled with the advantages of transportation and hydrogen refueling station operation, the comprehensive hydrogen use cost is even lower than that of high-pressure gaseous hydrogen route. Compared with the electrolysis energy consumption of 45–55 kW h/kg for green hydrogen production, the energy consumption of hydrogen liquefaction accounts for 1/9 to 1/5 of the hydrogen production energy consumption, and its impact on energy consumption is relatively weak. Of course, in the early stages of hydrogen energy demonstration and development, when the scale of hydrogen energy application is not large, the distance of hydrogen gas transportation is short, and under seasonal energy storage regulation and other situations, the advantages of the liquid hydrogen storage and transportation technology route are not obvious.

Wind and solar power generation, electrolysis hydrogen production, hydrogen gas liquefaction, liquid hydrogen storage, hydrogen energy power generation, cheap electricity in wind and solar power plants.

Liquid hydrogen transportation.

Grid power supply, distributed hydrogen energy storage power generation, industrial hydrogen use, liquid hydrogen energy station, high-pressure hydrogen refueling, liquid hydrogen refueling, high electricity prices in large-scale hydrogen-use city areas.

Generally speaking, the larger the scale of the hydrogen liquefaction station, the lower the relative liquefaction energy consumption. The domestic Aerospace Information Research Institute (AIR) NO.101 of Chinese Academy of Sciences (CAS), Beijing Sinoscience Fullcryo Technology Co., Ltd., Jiangsu Guofu Hydrogen Energy Equipment Co., Ltd., and other related companies have been working on the localization of hydrogen liquefaction equipment. In September 9, 2021, Aerospace Information Research Institute (AIR) NO.101 developed China's first hydrogen liquefaction system based on a helium expansion refrigeration cycle with independent intellectual property rights, which was successfully debugged and produced liquid hydrogen, with a hydrogen content of 97.4%. The efficiency of the turbo expander under rated conditions reached 80%, the design output of liquid hydrogen was 1.7 t/day (1 m^3/h), and the full load output reached 2 t/day. Nowadays, the largest liquid hydrogen storage container currently made in China has a volume of 300 m^3, used for rocket launches in Wenchang, and the largest liquid hydrogen transport semi-trailer has a volume of 25 m^3, which operates all year round in the Beijing Aerospace Test Technology Research Institute, supplying liquid hydrogen for various tests. In terms of standards, China passed the GB/T 40060—2021 "Technical Requirements for Liquid Hydrogen Storage and Transportation" [9] and GB/T 40061—2021 "Technical Specifications for Liquid Hydrogen Production Systems" [10] in 2021, providing standards for the storage, transportation, and production of liquid hydrogen.

In the field of on-board liquid hydrogen storage systems, China has made a breakthrough in 2021 with the application of on-board liquid hydrogen storage systems in heavy-duty trucks. The world's first 35-ton and 49-ton scale distributed

Fig. 3.9 A liquid hydrogen fuel cell powered heavy-duty truck head

liquid hydrogen fuel cell powered heavy-duty commercial vehicles (Fig. 3.9) were developed. In September 11, 2021, they successfully passed the first comprehensive test of liquid hydrogen fuel cell vehicles in China. They completed the insulation, refueling, and evaporation rate tests of the on-board liquid hydrogen fuel cell system, as well as the overall vehicle performance evaluation. This marked a key step in the transition of on-board liquid hydrogen fuel cell technology from research exploration to industrial application.

In terms of liquid hydrogen refueling stations, there are currently more than 120 liquid hydrogen storage and refueling stations worldwide, accounting for over one-fifth of the total number of the world's refueling stations. The longest operation one has been in operation for more than 10 years. The previous GB 50516—2010 "Hydrogen Refueling Station Technical Specifications" in China did not mention any applications of liquid hydrogen at refueling stations. However, the revised GB 50516—2010 "Hydrogen Refueling Station Technical Specifications (2021 Edition)" clearly states that refueling stations should be designed in conjunction with the hydrogen supply method, which can include a hydrogen tube trailer, a hydrogen tube bundle container, a liquid hydrogen tank truck, liquid hydrogen tank container transportation or pipeline transportation. In November 2021, China's first liquid hydrogen refueling station—Zhejiang Petroleum Hongguang (Sakura) integrated energy service station was built in Pinghu, Zhejiang, with the design, construction, installation, and debugging tasks undertaken by the Beijing Institute of Aerospace Testing Technology. The station has a 14 m^3 liquid hydrogen storage tank, two 90 MPa high-pressure hydrogen storage cylinders, and a 35 MPa hydrogen refueling machine for hydrogen fuel cell powered vehicles. It also includes a 120-kW charging pile rectifier cabinet and two charging parking spaces (Fig. 3.10). This refueling station can meet a maximum daily hydrogen refueling capacity of 1000 kg for fuel cell vehicles.

Fig. 3.10 China's first liquid hydrogen refueling station

In terms of liquid hydrogen transportation, it can be divided into land transportation, waterway transportation, and multi-modal transportation of liquid hydrogen storage tank containers.

1. Liquid Hydrogen Land Transportation

Similar to liquefied natural gas (LNG), liquid hydrogen as a commodity can be transported from liquid hydrogen factories or receiving stations to refueling stations and industrial vaporization stations for end users. It can also be transported from renewable energy bases that produce green hydrogen or transfer and receiving stations for liquid hydrogen trade to cities and regions with high hydrogen demand for energy or feedstocks. Land transportation equipment for liquid hydrogen includes liquid hydrogen tank containers, liquid hydrogen road tankers, and railway tankers.

Liquid hydrogen is classified as a dangerous commodity according to the national standards of China. According to the China's "List of Dangerous Chemicals (2015 Edition)" and the GB 12268—2012 "List of Dangerous Goods", the domestic code for liquid hydrogen is 21002, and the United Nations code is UN1966. When transported on roads and railways, it should comply with the relevant regulations on dangerous goods transportation issued by the Ministry of Transport. A liquid hydrogen tanker is shown in Fig. 3.11. It should be noted that the road transport regulations and management requirements are various in different countries and regions, and the maximum size and volume allowed for tankers in land transportation also differ. In China, the maximum volume of a cryogenic liquid vehicle tanker does not exceed 52.6 m^3.

In the tank truck transportation, the liquid hydrogen storage tank is fixedly connected to the vehicle's power structure, so the liquid hydrogen storage tank truck can only be used on land. However, a tank container is a tank body with a standardized size frame structure and corner piece, which can be conveniently loaded, unloaded, and transported, as shown in Fig. 3.12. When transported on the highway, the tank container can be directly placed on a general flatbed transport vehicle. After reaching the destination, the tank container can be unloaded,

a. Praxair Company, USA

b. Air Liquide Company, France

Fig. 3.11 Application of liquid hydrogen storage tankers in the USA and France. (**a**) Praxair Company, USA. (**b**) Air Liquide Company, France

a. Kawasaki Heavy Industries, Japan

b. Linde Company, Germany

Fig. 3.12 Applications of liquid hydrogen tank containers in Japan and Germany. (**a**) Kawasaki Heavy Industries, Japan. (**b**) Linde Company, Germany

and multiple standard-sized tank containers can be stacked and layered. The external dimensions of the International Standardization (ISO) tank container are consistent, which can be standardized for mass production and used globally. The most commonly used liquid hydrogen tank container specification is a 40 ft (1 ft = 30.48 cm) ISO tank container.

The transportation and application methods of tank containers are flexible, and the steel frame structure is not only suitable for stacking, but also provides good protection for the tank body, which is safer than road tank trucks. The disadvantage is that the effective volume and load are less than that of tank trucks. The most commonly used 40 ft ISO liquid hydrogen tank container has a volume of about 40 m^3.

2. Waterway transportation of liquid hydrogen

In terms of liquid hydrogen transport ships, the United States, which first used ships to transport liquid hydrogen, carried out the “Apollo Space Project” in the 1960s and 1970s. This project once used a 947 m^3 storage tank to transport liquid hydrogen by barge, transporting liquid hydrogen from Louisiana to the Kennedy Space Launch Center in Florida via seaway routes, which is more efficient, safe, and economical than land transportation. In addition, Germany also tried to build a SWATH (Small Waterplane Area Twin Hull) ship in 2004, with a hull length of over 300 m and equipped with five spherical storage tanks.

Due to the demand for large-scale hydrogen liquefaction and import and export of liquid hydrogen, the technology of liquid hydrogen ships for long-distance maritime transportation is also constantly developing. Since 2014, Kawasaki Heavy Industries has been developing the world's first liquid hydrogen transport ship dedicated to maritime trade. In December 11, 2019, this liquid hydrogen transport ship named "Suiso Frontier" was launched from Kawasaki Heavy Industries' shipyard in Kobe Port, Japan, as shown in Fig. 3.13. The ship is 116 m in length, 19 m in width, 8000 ton in weight, and is equipped with two 25 m length and 16 m height vacuum insulated liquid hydrogen storage tanks, each of which can store 1250 m^3 of liquid hydrogen.

Kawasaki Heavy Industries Group of Japan has a long history of planning for liquid hydrogen maritime transportation and building a liquid hydrogen supply chain. In early 2018, it cooperated with the Australian government to build a 770 ton/day brown coal hydrogen production and hydrogen liquefaction project in several stages, which would be fully operational in 2030. At that time, two more 160,000 m^3 liquid hydrogen transport ships will be built to achieve large-scale seaway transportation. The Osaka Kobe airport liquid hydrogen receiving station was also completed and put into use in 2021. The liquid hydrogen transported by ship is first transferred to a 19 m in diameter and 2500 m^3 liquid hydrogen spherical tank, and then unloaded to the liquid hydrogen transport vehicle through the raw materials input equipment, and distributed to various liquid hydrogen end-user points by road. The launch of Japan's liquid hydrogen transport ship and the opening of the liquid hydrogen seaway receiving station will open a new chapter in global hydrogen energy trade. In the future, liquid hydrogen will be like liquefied natural gas, transported thousands or even tens of thousands of miles by sea through transport ships of tens of thousands or even hundreds of thousands of cubic meters, becoming the most efficient hydrogen energy storage and transportation method.

Fig. 3.13 The Japanese "Suiso Frontier" liquid hydrogen transport ship and the Osaka Kobe airport liquid hydrogen receiving station

3. Multimodal transport of liquid hydrogen storage tank containers

The transportation method of liquid hydrogen ships to various places on land involves multiple liquid hydrogen transfer processes: from the liquid hydrogen ship to the liquid hydrogen receiving station, from the liquid hydrogen receiving station to the liquid hydrogen tank truck, and from the liquid hydrogen tank truck to the hydrogen refueling station or vaporization station. Since the temperature of liquid hydrogen is lower than the freezing temperature of air, each additional transfer increases the risk of air entering the liquid hydrogen system to form solid particles, and each transfer requires a certain amount of hydrogen to be wasted for purging. This mode of transportation will only be economical if liquid hydrogen enters large-scale applications. However, judging from the current global application of liquid hydrogen, this technological route still has a long way to go.

The multimodal transport of liquid hydrogen tank storage containers can not only be transported by sea and road but also be directly placed at hydrogen refueling stations and vaporization stations as fixed containers, realizing the direct delivery of liquid hydrogen from the production factory to the end user, reducing to three transfer processes. When offshore liquid hydrogen production does not yet form a large scale, it can quickly realize the industrial promotion of the liquid hydrogen industry chain, which is more economical than liquid hydrogen ships. The multimodal transport of liquid hydrogen tank containers is shown in Fig. 3.14.

From the perspective of the safety of liquid hydrogen land transport, the safety of liquid hydrogen tank container transport is much higher than that of liquid hydrogen tank trucks. Due to the requirements of container stacking, liquid hydrogen tank containers must be protected by a frame, so even in the event of a rear-end collision, collision rollover and other traffic accidents during road transport, the liquid hydrogen tank container can still ensure no damage and leakage under the protection of

a. Land transport

b. Waterway transport

Fig. 3.14 Stacking of containers and application of multimodal transport by road and water. (**a**) Land transport. (**b**) Waterway transport the frame, thereby greatly improving its safety. Tank trucks, because they lack rigid frame protection, can easily leak and cause secondary combustion explosions in the event of a rear-end collision, resulting in disastrous accidents in densely populated areas or on highways, such as the Wenling tank truck explosion incident in 2020. Enhancing the safety of equipment and reducing the probability of accidents are of great significance for the healthy and rapid development of the hydrogen industry. From this perspective, vigorously developing multimodal transport of liquid hydrogen tank containers can more easily and quickly promote the industrialization process.

Overall speaking, domestic liquid hydrogen liquefaction and storage and transportation technology is falling far behind foreign countries. Among them, the development of medium-scale and large-scale hydrogen liquefaction devices, large-scale liquid hydrogen spherical tanks, liquid hydrogen high-pressure pumps, liquid hydrogen tank containers, liquid hydrogen refueling guns, liquid hydrogen delivery pumps, and other key process equipment all need to make independent breakthroughs.

Exercises

1. For a high-vacuum liquid hydrogen tank with good vacuum performance, the heat transfer mode that has the greatest impact on the thermal insulation performance is ( ).

    A. Heat conduction
    B. Heat convection
    C. Heat radiation
    D. All of the above

2. At room temperature, the proportion of para-hydrogen molecules to the total number of hydrogen molecules is approximately ( ).

    A. 99%
    B. 75%
    C. 50%
    D. 25%

3. In the vacuum interlayer of the liquid hydrogen tank, the main gas sources leaking into the interlayer from the outer tank are ( ) and ( ), and the main gas sources leaking into the interlayer from the materialand to the inner liner are ( ).

    A. He
    B. N_2
    C. O_2
    D. CO_2
    E. H_2

4. The common specifications for liquid hydrogen tanks are ( ).

   A. 40 ft ISO tank container
   B. 40 m GB tank container
   C. 20 ft ISO tank container
   D. 20 m GB tank container

5. The commonly used catalysts for the conversion of ortho-hydrogen to para-hydrogen include two types, namely ____ and ____.
6. The effect of both pressure and temperature dropping after gas throttling is called ____, and the temperature that remains constant is referred to as the gas's ____. The conversion temperature of hydrogen under standard atmospheric pressure is ____.
7. The typical structure of a liquid hydrogen tank includes ____, ____, ____, ____, ____, ___, and ____.
8. There are four types of insulation methods that can be used for liquid hydrogen tanks, namely ____, ____, ____, and ____.
9. In the high vacuum interlayer of the liquid hydrogen tank, the commonly used low-temperature adsorbents are ____ and ____, while the adsorbent used on the normal temperature side is ____.
10. The equipment for land transportation of liquid hydrogen includes ____, ____ and ____, etc.
11. The maximum volume of cryogenic liquid tank trucks in China does not exceed ____.
12. Please briefly explain the reason for the conversion of ortho-hydrogen to para-hydrogen in the hydrogen liquefaction process.
13. Please briefly describe the characteristics of the hydrogen liquefaction process.
14. Please analyze the similarities and the differences between the Linde-Hampson cycle and the Claude cycle in the hydrogen liquefaction process.
15. High vacuum multilayer insulation is also known as "super insulation". Please briefly explain the principle of this insulation method and understand the reasons for its excellent insulation performance.

References

1. Chen B, Lin L (1989) Cryogenic insulation and heat transfer. Zhejiang University Press, Hangzhou
2. Xu C, Zhang S, Xie Y (2007) Vacuum cryogenic technology and equipment. Metallurgical Industry Press, Beijing
3. Xu L (1999) Cryogenic insulation and storage and transportation technology. Mechanical Industry Press, Beijing
4. Xu WQ, Yang GD, Lou P (2015) Design and simulation of airborne liquid-hydrogen tank for unmanned aerial vehicle. Chin J Vacuum Sci 35(3):266–270
5. Barron RF, Nellis GF (2007) Cryogenic heat transfer. CRC Press, Florida

6. Chen S, Tan Y, Yang S et al (2012) Research on the gas release performance of cryogenic insulated gas cylinders. J Vacuum Sci Technol 32(5):5
7. Wei W (2013) Research on the heat transfer mechanism of perforated high vacuum multilayer insulation and the development of its composite structure. Shanghai Jiaotong University, Shanghai
8. Zeng Y (2009) Research on the hydrogen absorption characteristics of palladium monoxide in high vacuum multilayer insulation structure. Shanghai Jiaotong University, Shanghai
9. National Hydrogen Energy Standardization Technical Committee (2021) Technical requirements for liquid hydrogen storage and transportation: GB/T 40060—2021. China Standard Publishing House, Beijing
10. National Hydrogen Energy Standardization Technical Committee (2021) Technical specifications for liquid hydrogen production systems: GB/T 40061—2021. China Standards Publishing House, Beijing

Chapter 4
Hydrogen Storage and Transportation Technologies Using Hydrogen-Rich Liquid Compounds

In addition to high-pressure gaseous hydrogen storage and low-temperature liquid hydrogen storage, hydrogen-rich liquid compounds can also serve as a type of chemical hydrogen storage carrier. Among them, the gravimetric hydrogen storage density of reversible hydrogen storage molecules, such as aromatic compounds containing benzene rings, is between 5–7.6 wt%, and the volumetric density is about 45–85 g/L (Fig. 4.1). Liquid ammonia (NH_3) and methanol (CH_3OH) have significantly higher gravimetric hydrogen storage densities than aromatic compounds because they can completely release the hydrogen through catalytic cracking. For example, NH_3 can undergo direct dehydrogenation, and its theoretical gravimetric hydrogen storage density can reach 17.6 wt%, while the theoretical gravimetric hydrogen storage density of methanol through direct dehydrogenation is about 12.5 wt%. Since the raw materials of N_2 and CO_2 come from the air and are released back into the air after use, liquid ammonia and methanol are commonly referred as circular hydrogen carriers (CHCs). It should be noted that methanol can release additional hydrogen molecules from water by steam reforming, thereby breaking the theoretical limit and reaching a gravimetric hydrogen storage density of 18.75 wt% (Fig. 4.1). Based on the steam reforming reaction, if dual-carbon molecules, such as ethanol and dimethyl ether, can be fully reformed to produce hydrogen, their gravimetric hydrogen storage densities can be further increased to 26 wt%.

In the practical process of hydrogen storage and transportation, despite having a relatively higher theoretical gravimetric hydrogen storage density, the recyclable

J. Zou, *Hydrogen Storage and Transportation*,
https://doi.org/10.1007/978-981-96-2875-9_4

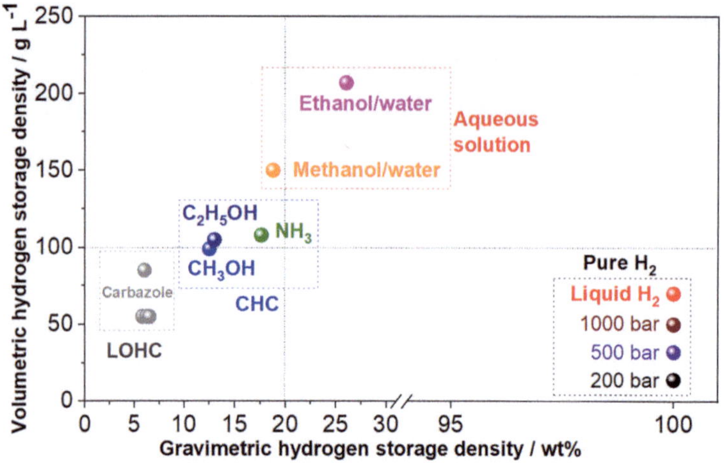

Fig. 4.1 Comparison of gravimetric hydrogen storage and volumetric hydrogen storage densities of hydrogen-rich liquid compound hydrogen storage technologies

hydrogen density needs to be realized through efficient catalytic hydrogenation and dehydrogenation reactions. For example, the C–C bond in ethanol molecules needs to be converted at high temperatures to achieve complete steam reforming to produce hydrogen, but with a high selectivity of the by-products, such as CO, CH_4, etc., leading to high costs in downstream separation processes and difficulties in fuel cell applications. Dimethyl ether and ethanol molecules are isomers with the same gravimetric hydrogen storage density. Compared to ethanol, the reforming conditions for dimethyl ether are mild. However, compared to methanol, which is a liquid at room temperature, dimethyl ether is a gas at room temperature and atmospheric pressure, which presents transportation difficulties. Among all circular hydrogen carriers, the hydrogen production technology based on the catalytic reforming reactions of organic liquids, liquid ammonia, and methanol is relatively mature and reliable, under mild conditions, with fewer by-products, and high efficiency, and has thus received widespread attention. Therefore, this chapter will focus on introducing the hydrogen storage technologies based on organic liquids, liquid ammonia, and methanol.

4.1 Liquid Organic Hydrogen Carriers (LOHCs)

Organic liquid hydrogen storage materials, also known as liquid organic hydrogen carriers (LOHCs), refer to a class of organic molecules that can absorb and release hydrogen through chemical reactions under a catalyst [1]. In principle, every compound containing unsaturated bonds (C = C double bonds or C ≡ C triple bonds) can store hydrogen through hydrogenation (an exothermic reaction) and release

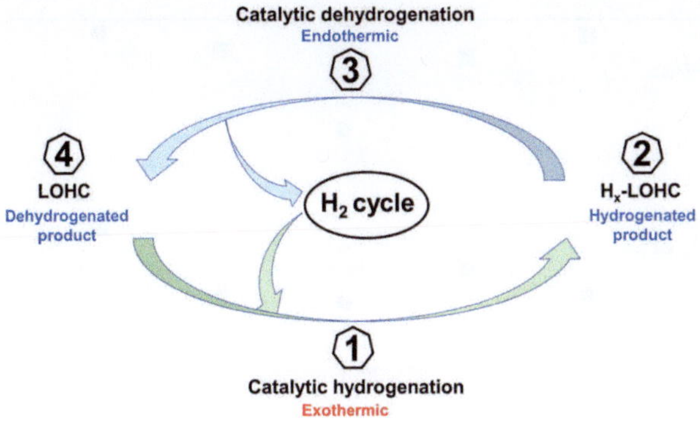

Fig. 4.2 Basic principles of liquid organic hydrogen carriers (LOHCs)

hydrogen through dehydrogenation (an endothermic reaction) (Fig. 4.2) [2]. Common unsaturated liquid organic compounds mainly include cycloalkanes, N-ethylcarbazole, toluene, 1,2-dihydro-1,2-azaborine, etc. LOHC hydrogen storage technology has a high hydrogen storage density, can store hydrogen under environmental conditions, has high safety, and is convenient for transportation. Its main disadvantage is that the absorption and release process of hydrogen is more difficult than physical adsorption-based hydrogen storage, requiring additional reaction equipment and hydrogen purification devices. The hydrogen release process requires heating, which leads to a high energy consumption and high cost [3]. In 2020, Japan used toluene-based hydrogen storage technology and successfully established the world's first international hydrogen energy supply chain between Brunei and Kawasaki [4]. Hydrogenious LOHC, a company based in Erlangen, Germany, has also been developing dibenzyltoluene-based LOHC hydrogen storage and transportation technology. Overall speaking, LOHC hydrogen storage technology is developing rapidly in Japan and Europe and is still in the demonstration stage in China.

4.1.1 Hydrogen Storage Principle of LOHCs

Organic liquid hydrogen storage technology is realized by the reversible hydrogen hydrogenation-dehydrogenation reaction of unsaturated liquid organic compounds. Theoretically, unsaturated aromatic hydrocarbons, such as olefins, acetylenes, cyclohydrocarbons, and their corresponding hydrogenated compounds, such as benzene-cyclohexane, methylbenzene-methylcyclohexane, etc., can undergo reversible hydrogen hydrogenation and dehydrogenation reactions without destroying the main structure of the carbon ring. An ideal organic liquid hydrogen storage carrier not only needs to have a high hydrogen storage density, but also needs to meet a series of other requirements, such as hydrogen storage stability, dehydrogenation rate or efficiency, cost, storage and transportation safety, and it needs to retrofit with the current existing hydrogen storage and transportation facilities. At present,

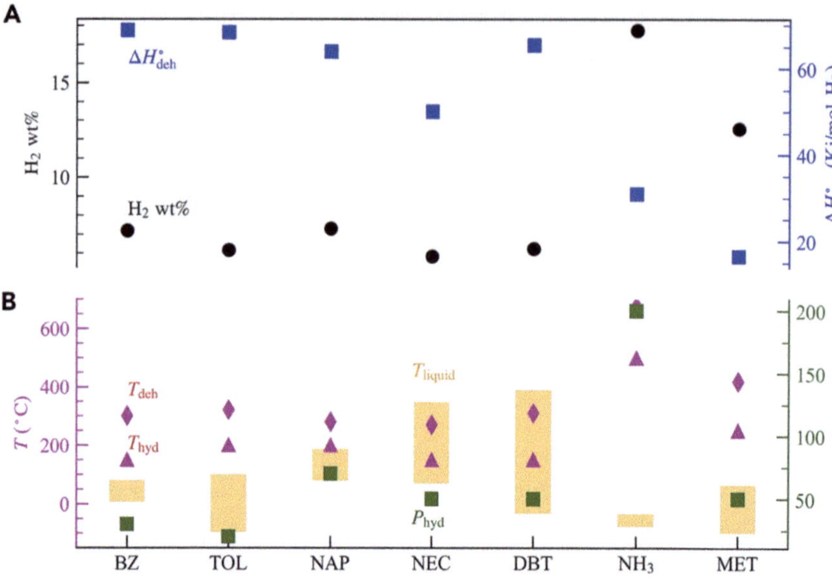

Fig. 4.3 Common LOHC hydrogen storage densities and dehydrogenation reaction enthalpy change (**a**) and dehydrogenation reaction temperatures and pressures (**b**) [5]

commonly used LOHC substances include (Fig. 4.3, Table 4.1) benzene, toluene, naphthalene, carbazole, N-ethylcarbazole (NEC), dibenzyltoluene (DBT) [5].

Benzene/cyclohexane (BZ/CHE) hydrogen storage system uses the reversible chemical reaction of benzene-hydrogen-cyclohexane to achieve hydrogen storage cycles. Its gravimetric hydrogen storage density is 7.2 wt%, and cyclohexane is in a liquid state at normal temperature of 20–40 °C. Its dehydrogenation product, benzene, is also in a liquid state at normal temperature and pressure, which can be stored and transported using existing liquid fuel delivery methods. 1 mol of cyclohexane can carry 3 mol of hydrogen, and the heat required for the reaction is much lower than the heat released when hydrogen burns, thus providing relatively abundant hydrogen energy. However, benzene is highly toxic, and the low boiling point (high volatility) of cyclohexane makes the separation and purification of the produced hydrogen gas difficult [5].

Toluene/methylcyclohexane (TOL/MCH) hydrogen storage system has a gravimetric hydrogen storage density of 6.2 wt%. Due to the relatively low toxicity of toluene, its dehydrogenation can produce hydrogen and methylcyclohexane. Both methylcyclohexane and toluene are liquid at normal temperature and pressure, and the hydrogenation and dehydrogenation processes are reversible and relatively simple. Therefore, the TOL/MCH hydrogen storage system has been commercialized by Chiyoda Corporation in Japan [6].

Naphthalene/decahydronaphthalene (NAP/DEC) hydrogen storage system has a relatively strong hydrogen storage capability, with a gravimetric hydrogen storage density of 7.3 wt% and a volumetric hydrogen storage density of 62.9 g/L. It is

Table 4.1 Comparison of common organic liquid hydrogen storage characteristics

Indicators	Toluene	DBT	NEC	Ammonia	Methanol
Temperature (°C)	Room temperature				
Pressure (MPa)	Room pressure			1	Room pressure
System gravimetric density (wt%)	4–6.5	4–6.2	4–5.8	17.6	12.5
System volumetric density (g/L)	40–55	40–55	40–55	108	99
H_2 charging electrical consumption (kWh/kg H_2)	Exothermic (3–5)			Endothermic (~6.8) [5]	Exothermic (~4.4) [5]
H_2 refueling pressure (MPa)	1–2	1–5	1–5	20–30 (H-B)	4.9–9.8 (medium pressure)
H_2 discharging electrical consumption (kWh/kg H_2)	~9	~9	~8	~4.8 [5]	3.2–7.2 [5]
H_2 discharging temperature (°C)	~320	~310	~270	300–500	200–300
Common features	Require a large hydrogenation/dehydrogenation reactor; low H_2 purity; low cyclability			Require a catalyst and a large hydrogenation/dehydrogenation reactor; high H_2 purity; moderate cyclability	
Unique features	/	Dehydrogenation at liquid state			

liquid at room temperature, which is convenient for hydrogen storage, but there may be continuous loss of raw materials during the hydrogenation, dehydrogenation, and transportation processes. 1 mol of trans-decahydronaphthalene can carry 5 mol of hydrogen, and the heat required for the reaction is about 66.7 kJ/mol H_2, which is about 27% of the heat released by the combustion of hydrogen, providing abundant hydrogen energy.

N-ethylcarbazole/perhydro-N-ethylcarbazole (NEC/PNEC) hydrogen storage system has a gravimetric hydrogen storage density of 5.8 wt%. N-ethylcarbazole is a colorless flaky crystal, soluble in hot ethanol, ether, acetone, but insoluble in water, and it easily turns black when exposed to light and is prone to polymerize [7]. Compared with other LOHC hydrogen storage systems, the NEC/PNEC system has a lower enthalpy change of hydrogenation/dehydrogenation (about 50 kJ/mol H_2), and can achieve rapid hydrogen refueling at 130–150 °C and dehydrogenation at 150–170 °C, which matches the working temperature of fuel cells, making it an ideal hydrogen storage medium [8]. However, the weaker nitrogen-alkyl bond usually causes the dealkylation reaction to occur above 270 °C. In addition, N-ethylcarbazole has a low industrial output, a high melting point (higher than room temperature), and thus needs to be heated to achieve the purpose of liquid hydrogen storage and transportation.

Dibenzyltoluene/perhydro-dibenzyltoluene (DBT/H18-DBT) hydrogen storage system has a gravimetric hydrogen storage density of 6.2 wt%, with low toxicity and high thermal stability. The DBT/H18-DBT system can achieve rapid hydrogen refueling at about 140 °C and dehydrogenation at 270–320 °C, and can use hydrogen gas mixtures for hydrogen refueling, which is compatible with the industrial hydrogen production process, without the need for hydrogen separation and purification processes [7]. However, its hydrogen storage density is greatly affected by the operation process. After several hydrogenation/dehydrogenation cycles, due to incomplete dehydrogenation, its hydrogen storage density decreases by about 77%.

LOHC hydrogen storage technology shows promising applications. It has a high gravimetric hydrogen storage density, is convenient and safe for transportation, and can be matched with the existing oil transportation facilities. It can achieve large-scale, long-distance, and long-term hydrogen energy storage and transportation, providing a hydrogen source for vehicle fuel cells. Compared with high-pressure gaseous hydrogen, low-temperature liquid hydrogen storage, and solid-state metal hydride hydrogen storage, LOHC-based hydrogen storage has many obvious advantages: LOHC has higher gravimetric and volumetric hydrogen storage densities, and commonly used materials (such as cyclohexane, methylcyclohexane, decahydronaphthalene, etc.) can all meet the specified standards; cyclohexane and methylcyclohexane are liquid at normal temperature and pressure, similar to gasoline, and can be stored and transported using the existing oil transportation pipelines, which is safe and convenient and can be transported over long distances, solving the problem of uneven energy distribution between the eastern and western regions of China; the catalytic hydrogenation and dehydrogenation reactions are reversible; they can be stored for a long time; the hydrogen storage medium can be recycled; if raw materials can be obtained at large scale, the cost of LOHC hydrogen storage can be reduced [7]. LOHC hydrogen storage technology also has shortcomings: the technical operating conditions are quite stringent, and the requirements for catalytic hydrogenation and dehydrogenation equipment are harsh; the dehydrogenation reaction needs to be carried out under low pressure, high temperature, and heterogeneous conditions, significantly limited by the heat and mass transfer effects and reaction equilibrium; the dehydrogenation reaction efficiency is low and side reactions are prone to occur, making the released hydrogen of low purity; the pore structure of the dehydrogenation catalysts can be easily destroyed under high temperature conditions, leading to the coking and deactivation.

4.1.2 Synthesis and Cracking Processes of LOHCs

A common LOHC hydrogenation and dehydrogenation process diagram is shown in Fig. 4.4. Nickel metal catalyst (Ni/Al_2O_3) or platinum metal catalyst (Pt/Al_2O_3) are commonly used for LOHC hydrogenation and dehydrogenation reactions. For the benzene/cyclohexane (BZ/CHE) hydrogen storage system, cyclohexane is the hydrogen storage carrier after benzene hydrogenation, it is a liquid at room

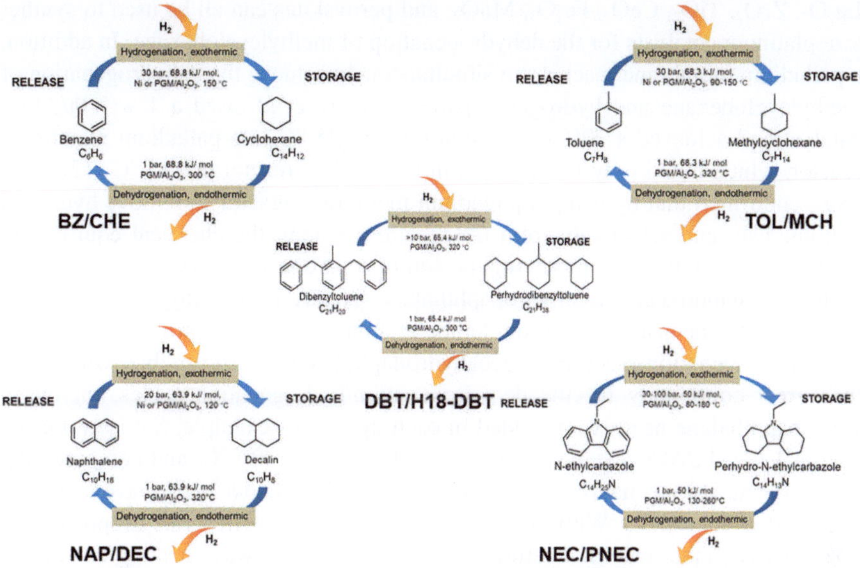

Fig. 4.4 A schematic diagram of common LOHC hydrogenation and dehydrogenation processes [3]

temperature, and the reaction is exothermic (about 68.8 kJ/mol). Biniwale et al. developed an activated carbon-supported nickel-platinum bimetallic catalyst and achieved efficient dehydrogenation of cyclohexane at 300 °C in a spray-pulsed reactor [9]. Among them, the dehydrogenation activity of the 20 wt%Ni + 0.5 wt%Pt/ACC catalyst is 60 times and 1.5 times higher than that of 0.5 wt%Pt/ACC catalyst and 20 wt%Ni/ACC catalyst, respectively, and the hydrogen production selectivity increased from 98.8% to 99.7%. They also confirmed that the hydrogen production efficiency of the 10 wt%Ag + 1 wt%Pt/ACC bimetallic catalyst is twice that of the single 10 wt%Ag/ACC catalyst. Chen et al. developed a non-precious metal bimetallic catalyst of Ni-Cu/SiO_2 (17.3 mol% Ni, 3.6 mol% Cu) for the dehydrogenation reaction of cyclohexane [10], achieving a conversion rate of 95.0% and a benzene selectivity of 99.4% at 250 °C in a plug flow reactor.

For the toluene/methylcyclohexane (TOL/MCH) hydrogen storage system, methylcyclohexane is a liquid at room temperature with a boiling point of 100.9 °C. Thermodynamically, the lower the reaction pressure, the more favorable the conversion from methylcyclohexane to toluene. Therefore, under permissible operating conditions, the reaction system pressure should be reduced as much as possible. At high temperatures, the dehydrogenation of methylcyclohexane is a gas-phase reaction. Okada used a controlled acid-base method to control the pore size distribution of alumina and obtained a highly dispersed Pt/Al_2O_3 catalyst [11]. The conversion rate of MCH was 95%, the selectivity of toluene was 99%, and the hydrogen production efficiency was 50 Nm^3/h. Besides commercial Al_2O_3 supports,

La_2O_3, ZrO_2, TiO_2, CeO_2, Fe_2O_3, MnO_2, and perovskites can all be used to synthesize platinum catalysts for the dehydrogenation of methylcyclohexane. In addition, a palladium membrane reactor can simultaneously achieve the dehydrogenation of methylcyclohexane and hydrogen separation. Gora et al. used a 1 wt%Pt/Al_2O_3 catalyst and achieved a 70% conversion rate at 225 °C in a palladium membrane reactor, which could only be achieved in a fixed bed reactor at 245 °C [12]. This work confirmed that by using a palladium membrane reactor to remove hydrogen continuously and selectively from the reaction system, the chemical equilibrium could be broken, thereby improving the equilibrium conversion rate.

For the naphthalene/decahydronaphthalene (NAP/DEC) hydrogen storage system, naphthalene is a solid at room temperature and its melting point is 80 °C, and the dehydrogenation reaction of decahydronaphthalene is irreversible, which creates great complexity for the dehydrogenation/hydrogenation cycles, therefore, fresh naphthalene needs to be added in each cycle. For example, Suttisawat et al. used a 1 wt%Pt/ACC catalyst in a fixed bed reactor at 320 °C, and compared the dehydrogenation mechanism of decahydronaphthalene under microwave and electric heating conditions. With the same catalyst, the conversion rate dropped from 85% to 10% under electric heating and 13% under microwave heating conditions [13]. Despite having a high hydrogen storage density, these defects make the NAP/DEC hydrogen storage system unsuitable for hydrogen storage and transportation, and the research on decahydronaphthalene dehydrogenation catalysts is still at the laboratory stage.

The N-ethylcarbazole/dodecahydroethylcarbazole (NEC/PNEC) hydrogen storage system has a lower hydrogenation/dehydrogenation enthalpy change (about 50 kJ/mol H_2), and rapid hydrogen refueling can be achieved at 130–150 °C, and hydrogen release can be achieved at 150–170 °C, which matches the working temperature of fuel cells, making it an ideal hydrogen storage medium [8]. However, N-ethylcarbazole is a solid at room temperature, and dealkylation reactions because of the weak nitrogen-alkyl bond start to occur above 270 °C. Therefore, it is crucial to modify the catalyst and accelerate the mass and heat transfer processes. For example, Yang et al. studied the effect of different noble metal catalysts (Pt, Pd, Rh, Ru) supported by Al_2O_3 on the dehydrogenation reaction of dodecahydroethylcarbazole at the same loading amount (5 wt%) and 180 °C [14]. Among them, the activity order follows as Pd > Pt > Ru > Rh. Sotoodeh et al. used a 5 wt% Pd/SiO_2 catalyst for dodecahydroethylcarbazole dehydrogenation reactions and achieved 100% conversion and 60% hydrogen yield at 170 °C [15].

For the dibenzyltoluene/octadecahydro-dibenzyltoluene (DBT/H18-DBT) hydrogen storage system, dibenzyltoluene is a commercial heat transfer oil, which is a liquid at room temperature (melting point −32 °C), has a high boiling point (390 °C), and is non-toxic and non-volatile. Due to its high density (0.91 g/mL), H18-DBT has an extremely high volumetric hydrogen storage density. Generally, octadecahydro-dibenzyltoluene can decompose and release hydrogen under the condition of appropriate platinum-based metal catalysts and temperatures above 250 °C. Another advantage is that dibenzyltoluene can be hydrogenated with a hydrogen gas mixture, which is compatible with the current industrial hydrogen production process, eliminating the need for high-cost hydrogen separation processes. German company Hydrogenious Technologies GmbH adopted the DBT/H18-DBT system for commercial hydrogen storage. Jorschick and others [16] attempted to hydrogenate dibenzyltoluene with a H_2/CO_2 mixture and found that the choice of catalyst was crucial to avoid methanation and carbon dioxide reduction side reactions. Among them, Pd/Al_2O_3 and Rh/Al_2O_3 are the most suitable catalysts. At 210 and 270 °C, the hydrogenation efficiency of Pd/Al_2O_3 and Rh/Al_2O_3 reached 80%. Currently, the main problems with traditional organic liquid hydrides as hydrogen storage media include: high hydrogenation and dehydrogenation temperatures, leading to high equipment costs and short service life; high cost of precious metal catalysts, which are easily poisoned and deactivated; non-precious metal catalysts are low-cost, but have lower hydrogenation and dehydrogenation efficiency during use; if used for fuel cells, the problem of high pressure during hydrogen refueling processes still needs to be solved. Improving the dehydrogenation rate and efficiency of organic liquid hydrogen storage media at low temperatures, developing more efficient catalysts under mild reaction conditions, and reducing the dehydrogenation costs are the main research directions. Based on existing research, using polycyclic aromatic hydrocarbons, a high content in low-temperature coal tar, as raw materials can solve the problem of high-value utilization of low-temperature coal tar and obtain a large amount of hydrogen storage media, saving raw material costs; at the same time, using non-precious metals to prepare bimetallic or multimetallic catalysts, reducing the use of precious metals, thereby reducing the cost of catalysts, improving the activity and stability of catalysts through experiments, and optimizing corresponding reaction conditions, can help achieve low-cost hydrogen storage media.

4.1.3 Hydrogen Storage Processes of LOHCs

LOHC hydrogen storage technology is based on the reversible hydrogenation and dehydrogenation reactions of unsaturated organic substances with hydrogen gas to achieve hydrogen storage and release. After hydrogenation, liquid organic hydrides are stable, safe, and similar to petroleum products in storage methods, so they can use existing equipment for storage and transportation, suitable for long-distance hydrogen transportation. From the perspectives of hydrogen storage contents, energy consumption in the hydrogen storage process, and the cost of hydrogen storage, single-ring aromatic hydrocarbons such as benzene and toluene exert high hydrogen storage densities and reversible hydrogen storage processes, approaching the requirements of the U.S. Department of Energy for hydrogen storage systems. In addition, the hydrogen refueling process is exothermic, and the dehydrogenation process is endothermic, making it a hydrogen storage and release cycle system could demonstrate a high thermal efficiency. The heat released during the hydrogen refueling reaction process can be recovered as the heat required for the hydrogen release process, thereby effectively reducing the heat loss and improving the overall thermal efficiency of the entire cycle system. However, the hydrogen refueling and release reaction equipment is relatively complex, and the operating cost is high. LOHC carriers suffer from problems, such as high reaction temperatures, low dehydrogenation efficiency, and catalyst poisoning by intermediate products. Currently, LOHC hydrogen storage and transportation technology is in the demonstration stage, and worldwide companies include Japan's Chiyoda Corporation and Germany's Hydrogenious Technologies Corporation.

In July 2017, under the guidance of the New Energy and Industrial Technology Development Organization of Japan, Chiyoda Corporation, Mitsubishi Corporation, Mitsui & Co., and Nippon Yusen Kaisha jointly established the Advanced Hydrogen Energy Chain Association and developed a hydrogen storage technology using the toluene/methylcyclohexane (TOL/MCH) system (Fig. 4.5) [6]. In December 2019, the Advanced Hydrogen Energy Chain Association carried out a demonstration project of long-distance maritime transport of hydrogen stored in such organic liquids, from Brunei to Kawasaki, Japan, with an annual supply scale expected to reach 210 tons. In February 2022, Chiyoda Corporation announced the achievement of the "world's first" milestone in the maritime transport of hydrogen in the form of MCH, suggesting that long-term storage and transport of hydrogen in the form of MCH is feasible on a global scale. This is crucial for practical hydrogen transport and proves that new investments in hydrogen storage equipment are not necessarily required, as existing facilities can be used for MCH storage and transport. MCH hydrogen storage and transport is a potential global clean energy solution, which will also help the international community in the generation of decarbonized supply chains.

In March 2020, Germany proposed the GET H2 plan, committed to achieving industrial production of green hydrogen in regions rich in wind and solar energy resources, and connecting with downstream application scenarios, thereby gradually building a high-efficiency hydrogen storage and transport infrastructure

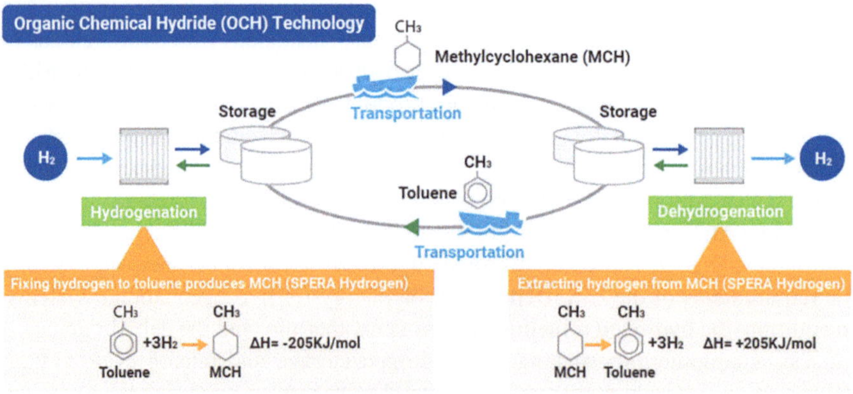

Fig. 4.5 SPERA hydrogen storage technology process developed by Chiyoda Corporation based on the TOL/MCH hydrogen storage system [6]

covering all over Germany. The plan was composed of renewable energy power generation facilities, high-temperature heat pumps for regional heating using electrolysis waste heat, hydrogen turbines, LOHC storage and transport system facilities (Hydrogenious technology), and Lingen hydrogen refueling stations. In February 2022, the automotive and industrial product supplier, Schaeffler Group, reached a cooperation agreement with Hydrogenious Technologies of Germany and the Helmholtz Erlangen-Nuremberg Renewable Energy Research Institute. The three parties would jointly develop a hydrogen fuel cell based on the LOHC technology (Fig. 4.6) [17]. The LOHC hydrogen storage technology developed by Hydrogenious uses a DBT/H18-DBT hydrogen storage system, with a hydrogen refueling pressure of 3–5 MPa, and a hydrogen release temperature of 300 °C at atmospheric pressure. H18-DBT can be transported under normal conditions. Unlike traditional fuel cell design, there are no hydrogen molecules in the LOHC fuel cells and the entire supply system. Using LOHC technology in hydrogen fuel cells allows hydrogen to be stored and transported in a liquid form, providing a more convenient and safe supply method for mobile and stationary energy end users.

China still treats hydrogen as a hazardous material, with high safety requirements for its storage, transportation, and refueling under special pressure and temperature conditions. There remain certain gaps in the policies and standards for large-scale application in many fields, creating significant obstacles for hydrogen energy utilization. One significant advantage of LOHC technologies is its safety, which can adopt mature management methods similar to those of oil and gasoline and share facilities. In 2019, the International Energy Agency (IEA) released the "Future of Hydrogen" report, in which LOHC and NH_3 were identified as the best methods for international hydrogen trade and long-distance hydrogen transportation [18]. In long-distance sea transportation, there is a significant difference in the cost among different methods of hydrogen transportation and conversion, with a LOHC cost as low as 0.6 \$/kg H_2, ammonia at 1.2 \$/kg H_2, and liquid hydrogen as high as

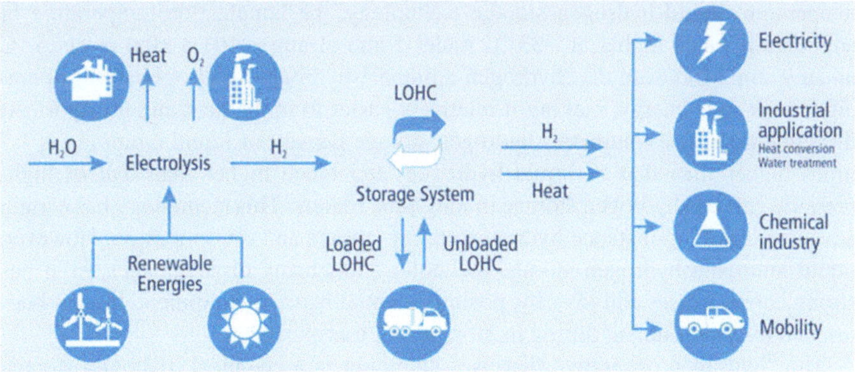

Fig. 4.6 Hydrogen storage technology process based on DBT/H18-DBT developed by Hydrogenious Company of Germany [17]

2 \$/kg H_2. In land truck transportation, the cost of transporting LOHC within a range of 500 km is only 0.8 \$/kg H_2, but considering dehydrogenation and purification processes, the cost will increase by 2.1 \$/kg H_2.

The main problems with LOHC technologies currently are [18]: (1) The cost of LOHC hydrogen storage and dehydrogenation processes is high, requiring energy inputs equivalent to 35–40% of the hydrogen energy itself; (2) LOHC hydrogen storage materials need to be returned to the factory, increasing transportation costs; (3) The cost of LOHC materials themselves is high. According to the annual output value of 50 million yuan for the first phase of the 1000-ton Yidu Project of Hynertech Co., Ltd., the price of each ton of storage oil is 50,000 yuan. Public data estimates that the hydrogen storage mass does not exceed 5.5 wt%, and the actual storage oil price after calculation is about 900 yuan/kg H_2. To reduce the cost of hydrogen storage and transportation, future LOHC technology needs to achieve breakthrough applications in three aspects [5]: (1) under the situation of the imbalance between hydrogen production companies and hydrogen-utilization companies, hydrogen storage/long-distance hydrogen transportation; (2) grid energy storage; (3) direct hydrogen oil production from ultra-high temperature conversion of waste.

4.2 Liquid Ammonia

4.2.1 Principle of Hydrogen Storage Using Ammonia

Liquid ammonia (NH_3), as a carbon-free energy source and chemical hydrogen storage vector, exerts a very high gravimetric hydrogen storage density (17.6 wt%) and volumetric hydrogen storage density (108 g/L), high energy density (22.5 MJ/kg). It is safe, reliable, and has low transportation costs, with already established storage and transportation infrastructures and operating standards [19]. Compared to the extremely low hydrogen liquefaction temperature of −253 °C required by low-temperature liquid hydrogen storage technology, the liquefaction temperature of ammonia is much higher at −33 °C under 1 atm (1 atm = 101.3 kPa) or at 25 °C under 9 atm. Moreover, the "hydrogen-ammonia-hydrogen" hydrogen storage paradigm consumes energy, making it relatively easier to implement and transport. At the same time, the volumetric hydrogen storage density of liquid ammonia is 1.7 times higher than that of liquid hydrogen, and much higher than that of high-pressure gaseous hydrogen storage in long-tube trailers. This technology has certain advantages in long-distance hydrogen energy storage and transportation. However, liquid ammonia hydrogen storage technology has many disadvantages, as it has strong corrosiveness and toxicity, posing potential harm to equipment, human bodies, and the environment during its storage and transportation.

The "hydrogen economy" that uses ammonia as a chemical hydrogen storage medium is named the "ammonia economy". Hydrogen and nitrogen are converted

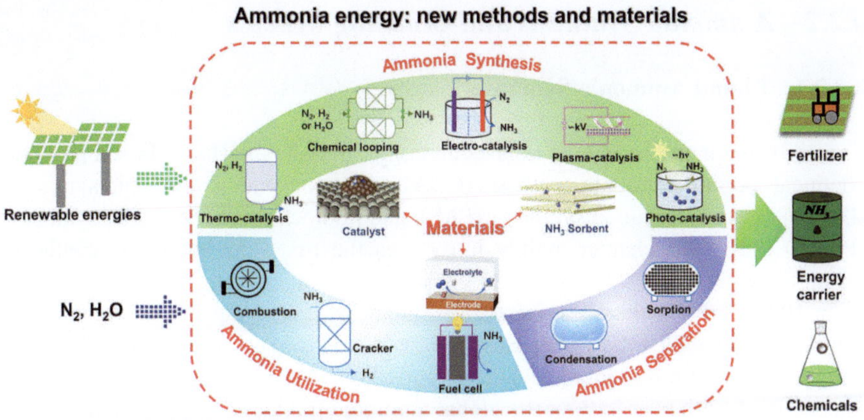

Fig. 4.7 A schematic diagram of the ammonia circular economy industry [22]

into ammonia under the action of a catalyst, stored and transported in the form of liquid ammonia, and hydrogen is released for use under the action of a catalyst (Fig. 4.7). Considering the distinct advantages of convenient ammonia fuel supply, mature liquid transport infrastructures and facilities, low costs, and zero carbon emissions, if combined with the high-power generation efficiency and high energy density of fuel cells, high-efficiency cyclic power generation applications of carbon-free ammonia fuel cells can be achieved through the electrolysis of water and nitrogen reduction processes [20]. However, since ammonia can damage the NafionTM proton exchange membranes and cannot be directly used in proton exchange membrane fuel cells (PEMFC) [21], ammonia as a hydrogen storage technology needs to achieve efficient ammonia decomposition and high-purity hydrogen separation before being used in fuel cells. To realize the entire "ammonia economy" cycle, the following issues need to be resolved:

1. Low-temperature (80–150 °C) efficient ammonia decomposition and hydrogen purification technology to integrate and apply to PEM fuel cells or direct combustion applications;
2. Efficient green ammonia synthesis technology, which combines nitrogen in the air and water to directly synthesize liquid ammonia, to avoid the energy consumption problems of hydrogen storage and ammonia synthesis from water electrolysis.

In addition, the equipment for synthesizing ammonia for hydrogen storage and decomposing ammonia for hydrogen release remains to be integrated with end-use industrial equipment.

4.2.2 *Ammonia Synthesis and Cracking Process*

4.2.2.1 Liquid Ammonia Synthesis Process

Currently, main ammonia synthesis technologies include the Haber-Bosch process, chemical looping ammonia synthesis (CLAS), electrochemical ammonia synthesis (ECAS), plasma ammonia synthesis (PAS), and photochemical ammonia synthesis (PCAS) [22]. This chapter mainly introduces the first three ammonia synthesis technologies.

In 1905, German chemists, Fritz Haber and Carl Bosch, developed an ammonia synthesis loop, called the Haber-Bosch process, and this process remains the most common method for ammonia synthesis nowadays. Ammonia synthesis is close to the end user, with over 85% of the global ammonia production, approximately 200 million tons per year. Most of them are used as agricultural fertilizer, while the rest is used in the chemical processing industry. Ammonia synthesis reaction is an exothermic process, with an equilibrium yield of only 10–20%. Traditional Haber-Bosch process is carried out under a high pressure of 20–30 MPa and a high temperature of 300–500 °C, while high temperatures are chosen to increase the reaction rate. Therefore, ammonia synthesis industry consumes a huge amount of energy, consuming 1–2% of the world's energy output each year and producing 400–500 million tons of carbon dioxide [22–24]. For ammonia synthesis, the dissociative chemisorption of nitrogen molecules (a process to break the $N \equiv N$ triple bonds into N atoms) remains the key rate-determining step and iron catalysts supported by main group oxide promoters (such as Al_2O_3, CaO, K_2O) are the most widely used catalysts [25]. Ammonia synthesis chemical plants often need to be coupled with methane reforming for hydrogen production and air separation plants.

The entire Haber-Bosch process flow diagram is shown in Fig. 4.8. Methane and water are first converted into syngas through the methane reforming reaction ($CH_4 + H_2O \rightarrow CO + 3H_2$), which is then mixed with air so that methane is further oxidized to syngas ($2CH_4 + O_2 \rightarrow 2CO + 4H_2$). Then, carbon monoxide is further fully oxidized to carbon dioxide through the water-gas shift reaction ($CO + H_2O \rightarrow CO_2 + H_2$), and CO_2 and H_2O are removed through a carbon dioxide

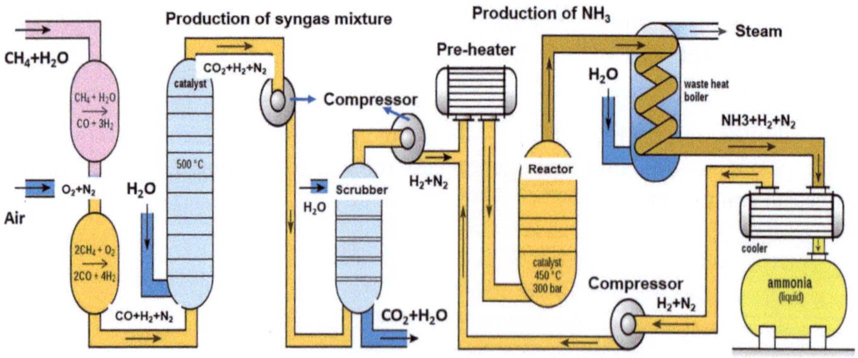

Fig. 4.8 A schematic diagram of the Haber-Bosch process for ammonia synthesis

scrubber (water scrubber), resulting in high-purity N_2 and H_2 raw materials for ammonia synthesis. Afterward, N_2 and H_2 mixture gases are mechanically compressed, preheated, and fed into the reactor at 300 atm and 450 °C, where the reaction produces ammonia gas ($N_2 + 3H_2 \rightarrow 2NH_3$), with a single process conversion of 10–20%. The outlet mixture gas ($N_2 + H_2 + NH_3$) is thus cooled to extract ammonia gas as liquid ammonia, and the remaining N_2 and H_2 mixture gases continue to be recycled into the ammonia synthesis reactor, ultimately achieving a total conversion of up to 97%. Compared to traditional ammonia synthesis technology, where the hydrogen source comes from methane steam reforming, green ammonia synthesis technology uses green hydrogen produced by water electrolysis and nitrogen separated by an air separation unit (ASU), producing 1 ton/day synthetic ammonia requiring 27 kW, while the ASU device and mechanical compression require 3.5 and 1.5 kW, respectively [26].

Chemical looping ammonia synthesis (CLAS) is often used in the process of converting carbon feedstocks into chemicals or energy, with a high energy efficiency, product selectivity, and separation performance. Integrating chemical looping technology with ammonia synthesis technology can achieve economic and efficient chemical looping ammonia synthesis (Fig. 4.9) under three specific modes [27]. The first CLAS technology is H_2O-CLAS, which uses H_2O and metal nitrides as sources of H and N, respectively, where CH_4 reduces metal oxides (M-O) to produce metals, metals react with nitrogen to produce metal nitrides (M-N), and metal nitrides undergo water hydrolysis to synthesize ammonia (Fig. 4.9a). However, its synthesis temperature is high (>1000 °C) and corresponding energy consumption for regenerating metals/metal nitrides is also significantly high. The second CLAS technology is H_2-CLAS, which uses H_2 as the reducing agent, where metals generate metal nitrides (M-N) through nitrogen fixation reactions, and metal nitrides (M-N) react with hydrogen to synthesize ammonia (Fig. 4.9b) and regenerate

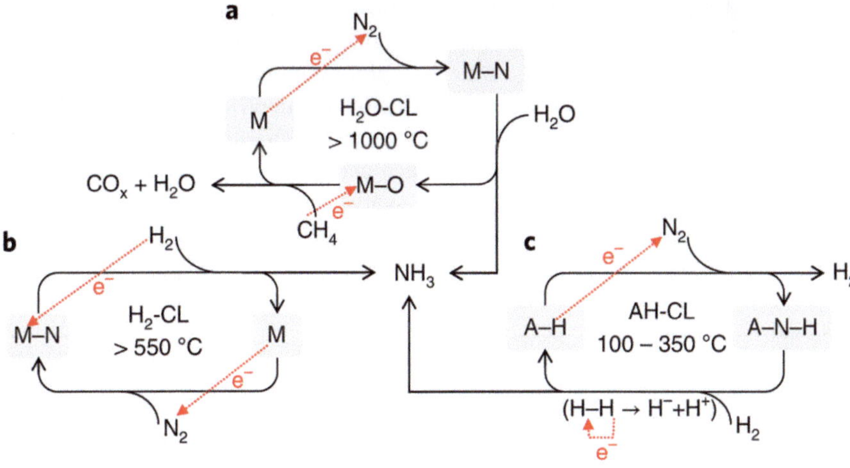

Fig. 4.9 A schematic diagram of chemical loop ammonia synthesis technology based on H_2O (**a**), H_2 (**b**) and hydrocarbon (**c**): M, M-N, and M-o represent metals, metal nitrides, and metal oxides, respectively. A-H and A-N-H represent alkali metal or alkaline earth metal hydrides and amide salts, respectively

metals. This reaction occurs between 400 and 550 °C, and the bond energy of metal nitrides and the activation energy of hydrogen significantly affect the economics of the entire process. In summary, the first two types of chemical loop ammonia synthesis technologies are based on metal nitrogen fixation reactions (metal reduces nitrogen). The third CLAS technology is AH-CLAS, which is based on alkali metal/alkaline earth metal hydrides or amide salts, where nitrogen is reduced by hydride ions (H^-), and alkali metal/alkaline earth metal amide salts synthesize ammonia through hydrogenation reactions (Fig. 4.9c). In this process, ammonia can be synthesized under conditions of 1 atm and 100 °C.

Electrochemical ammonia synthesis (ECAS) is another synthesis route for green ammonia with broad application prospects, with the following advantages [22]:

1. Compared with traditional thermochemical ammonia synthesis, electrochemical ammonia synthesis can activate and transform nitrogen under mild conditions;
2. Mild operating conditions are conducive to the dispersion of the synthesis process, suitable for small-scale synthesis and distributed ammonia synthesis based on renewable energy generation;
3. Using water as the hydrogen source for ammonia synthesis avoids the carbon emission problem of traditional syngas hydrogen production.

Electrochemical ammonia synthesis also has some problems:

1. Lack of efficient $N \equiv N$ triple bond activation and breaking electrocatalysts;
2. Low selectivity over hydrogen evolution reaction (HER) processes.

Overall speaking, considering the use of anion exchange membrane (AEM) for electrochemical ammonia synthesis and direct ammonia fuel cells, alkaline electrolytes are more suitable for electrochemical nitrogen fixation and utilization cycles. Given the ubiquity of proton exchange membrane (PEM) and high ammonia adsorption selectivity of acidic electrolytes (vs.N_2H_2), acidic electrolytes are also commonly used for electrochemical ammonia synthesis. However, hydrogen evolution reaction (HER) is more significant in acidic electrolytes than in alkaline electrolytes. Alkaline electrolytes can use water as a proton source and non-precious metal anodes for oxygen evolution reaction (OER) and thus suppress the hydrogen evolution reaction. Using water molecules as a source of hydrogen can avoid the problems enable by hydrogen production, storage and transportation (Fig. 4.10) [28]. For commercial electrochemical ammonia synthesis, the requirements include an ammonia production rate $\geq 10^{-6}$ mol/(cm^2 s), a Faraday efficiency $\geq$90%, and an energy utilization efficiency $\geq$60%. Currently, regardless of the type of electrocatalyst used in the electrochemical ammonia synthesis system, its Faraday efficiency is less than 1%. Therefore, the development of high-performance electrocatalysts that can suppress the HER process and simultaneously promote the nitrogen reduction reaction (NRR) is key to electrochemical ammonia synthesis [29].

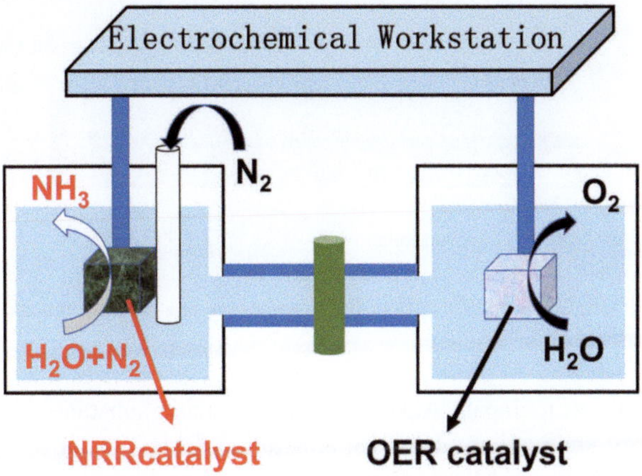

Fig. 4.10 A schematic diagram of electrochemical ammonia synthesis technology

4.2.2.2 Liquid Ammonia Cracking Process

In recent years, ammonia decomposition for hydrogen production, storage and transportation has received widespread attention. This reaction is endothermic, a reverse reaction of ammonia synthesis reaction, and the equilibrium conversion is thermodynamically controlled. At 450 °C, although the equilibrium conversion of ammonia decomposition can reach 99%, the high kinetic energy barrier limits the hydrogen production rate. As a carbon-free fuel, hydrogen fuel cells coupled with ammonia decomposition for hydrogen production can not only reduce carbon emissions, but also avoid carbon monoxide poisoning of the precious metal catalysts in fuel cells. Since Green proposed in 1982 that ammonia decomposition for hydrogen production could serve as a hydrogen source for proton exchange membrane fuel cells [30], Japan and South Korea have made substantial investments in fuel cell research and commercialization since 2009 [31], with a commercialized Japan's FC micro-CHP fuel cell system. Japan is the first country to commercialize fuel cells and bring them to market. As of 2020, its FC micro-CHP fuel cell system has sold more than 300,000 units in Japan and over 10,000 units in Europe, and it is expected to sell a total of 100,000 units by 2022. In 2021, Japan and Russia established a long-term partnership for hydrogen production and ammonia synthesis. By 2030, Japan aims to achieve an ammonia fuel usage rate of 2 million ton/year by establishing co-firing technology in coal-fired power plants and developing alternative fuel markets and supply chains, to achieve the 2050 decarbonization goal. In addition, Professor Nagaoka Katsutoshi of Nagoya University has conducted extensive research on ammonia as a high-density energy and hydrogen storage carrier and has achieved a series of results in the synthesis of ammonia and ammonia decomposition for hydrogen production using ruthenium metal and bimetallic catalyst

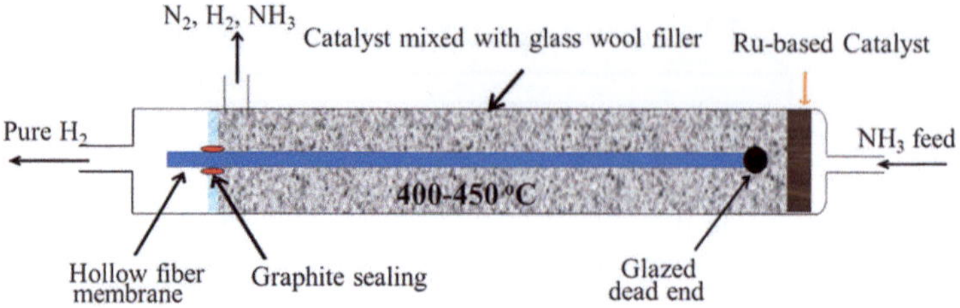

Fig. 4.11 A schematic diagram of a hollow fiber membrane reactor for ammonia decomposition to produce hydrogen [33]

systems. In addition to Japan, Australia has also made substantial investments in ammonia energy applications. In 2018, Australia's Chief Scientist Michael Dolan and his partner Fortescue Metals Group announced an investment of 13.8 million US dollars over 5 years to commercialize ammonia decomposition technology for hydrogen production [32]. Its high-purity hydrogen production process (Fig. 4.11) is divided into two steps [33]:

1. Use ruthenium metal catalyst to decompose ammonia into nitrogen and hydrogen;
2. Use precious metal membrane (Pd/Ag) to efficiently separate hydrogen for compression and use in fuel cells.

In addition, in the UK, Professor Edman Tsang of Oxford University and Professor Laura Torrente of Cambridge University are also seeking to develop new ammonia synthesis and decomposition catalysts under efficient and mild conditions for its commercial applications in fuel cells [34]. Despite the achievements of ammonia fuel cell technologies, the development of low-temperature, low-energy consumption, low-cost, and efficient ammonia decomposition technologies remains challenging, especially requiring the hydrogen separation equipment after ammonia decomposition to limit the ammonia concentration of <100 ppb (parts per billion) in hydrogen gas [22]. The ammonia decomposition technology includes thermal decomposition, electrochemical decomposition, and photochemical decomposition. Whatever the technology is, the development of efficient catalysts remains the key. In general, B5-type active sites, alkali/alkaline earth metal promoters, and strong metal-support interactions are the design guidelines for high-performance ammonia decomposition catalysts [35]. Currently, common ammonia decomposition catalysts are porous carbon or metal oxide supported noble metal catalysts (ruthenium, platinum, palladium, etc.) and non-noble metal catalysts (cobalt, nickel, iron, etc.) (Fig. 4.12) [36]. Traditional ammonia decomposition catalysts are based on the ruthenium (Ru) metal catalyst system supported by carbon nanotubes (CNT), which shows excellent ammonia decomposition performance and cycle stability. Nevertheless, it only exerts a high activity under high ruthenium loading and strong catalyst surface alkalinity, thus the ruthenium consumption is large and the requirement of equipment corrosion resistance is high. Moreover, the cost of CNT supports is high and its biotoxicity is significant, making it difficult to achieve large-scale

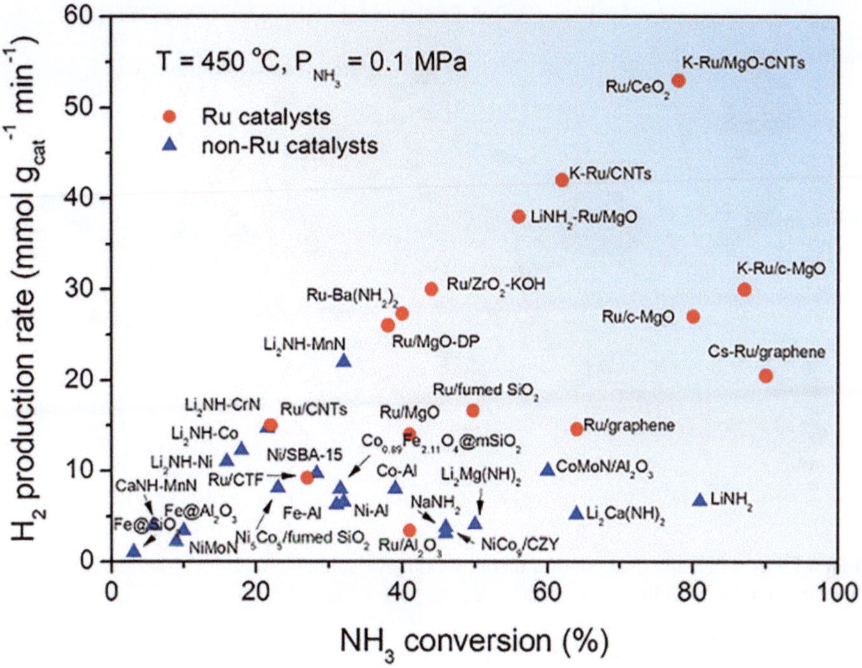

Fig. 4.12 Comparison of ammonia decomposition catalyst performance (450 °C) [22]. Note: The hydrogen production efficiency and ammonia conversion of the catalyst are related to the gas hourly space velocity (GHSV)

industrial applications [37]. Co, Ni, and Fe-based catalysts are abundant in natural reserves, low in prices, easy in preparation, and stable under harsh conditions, which make them the best substitutes for ruthenium-based catalysts. However, their ammonia decomposition performance is lower than that of ruthenium, and their ammonia decomposition reaction temperature is also higher [38]. Therefore, the key to ammonia decomposition technology is to develop green, efficient, low-cost, and long-life ammonia decomposition catalysts.

In addition to traditional fixed-bed reactors, membrane reactors (Fig. 4.11), based on the Le Chatelier's principle, continuously remove the hydrogen product, which breaks the equilibrium of the ammonia decomposition reaction and thereby achieves higher equilibrium ammonia conversion, and the corresponding downstream process does not need to use any hydrogen purification equipment [39]. For membrane reactors, in addition to developing high-performance catalysts, it is also necessary to develop low-cost and high hydrogen-selective membranes and low-cost membrane supports, replacing traditional palladium-silver metal membranes and tantalum metal nanotube supports.

In 2017, Nagaoka and coworkers proposed a thermo-oxidative ammonia decomposition route that uses partial oxidation of ammonia to achieve self-heating (thermo-oxidative ammonia decomposition), as shown in Fig. 4.13 [40]. Compared

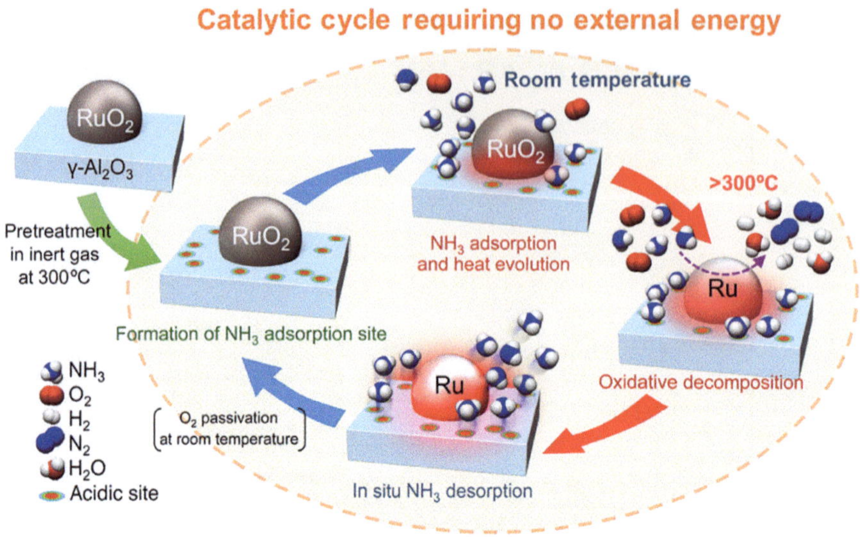

Fig. 4.13 A schematic diagram of a thermos-oxidative ammonia decomposition technology for hydrogen production [40]

to the endothermic reaction property of thermal ammonia decomposition (45.9 kJ/mol), in the presence of oxygen, using the acidic $RuO_2/\gamma\text{-}Al_2O_3$ catalyst allows the adsorption and partial oxidation of ammonia, making the entire thermo-oxidative ammonia decomposition reaction exothermic (−75 kJ/mol). The ammonia decomposition reaction can proceed at "room temperature" without the need for an external heat source (note: this refers to no heating, self-exothermic reactions may also raise the temperature of the reaction system). However, one molecule of hydrogen in the ammonia molecule has been oxidized into a water molecule, resulting in a loss of 33% hydrogen. For the hydrogen storage process, the theoretical hydrogen storage density drops to 11.8 wt%. In addition, the introduction of oxygen makes the reaction system more complex, increases the by-products, complicates the subsequent hydrogen separation process, and thus increases the cost. All of these greatly reduce the potential and practicality of ammonia as a high hydrogen storage vector.

Electro-oxidative ammonia decomposition technology (electro-oxidative ammonia decomposition) for hydrogen production has also been developed to prepare high-purity hydrogen in alkaline aqueous solution [41]. This technology couples the electrochemical oxidation reaction of ammonia and the hydrogen evolution reaction, and theoretically requires a driving potential of only 0.06 V, which is thermodynamically more favorable and far lower than water electrolysis (1.22 V). Nevertheless, it is limited by a low ammonia decomposition efficiency, a high anode overpotential, and a high portion of non-target by-products (NO_X, etc.). Since the ammonia saturation concentration in an aqueous solution at room temperature is only 34.2 wt%, the resultant hydrogen storage density of this technology is limited

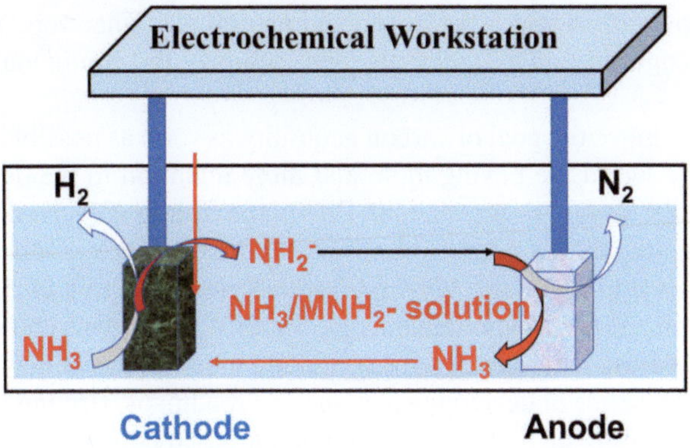

Fig. 4.14 A schematic diagram of direct electrolysis of liquid ammonia for hydrogen production

to 6.1 wt%, thereby reducing the application prospects of ammonia as a high hydrogen storage carrier. In view of this, Ichikawa and coworkers [42] developed a direct liquid ammonia electrolysis technology based on platinum as electrodes and alkali metal amide salts as electrolytes, where ammonia decomposed to form NH_2^- and NH_4^+. On the cathode, ammonia molecules were reduced to evolve hydrogen; on the anode, amide ions were oxidized to form nitrogen molecules (Fig. 4.14). However, this technology uses a large amount of alkali metal amide salts, requires harsh anhydrous and oxygen-free conditions, sets stringent requirements for the electrolysis cell equipment that needs to withstand 100 atm and requires a voltage of 1–2 V, making it complex to operate and not conducive to commercial use.

In addition, ammonia decomposition technologies also include plasma and photocatalytic routes, which are not discussed here due to less research outputs and more complex mechanisms.

4.2.3 Current Status of Ammonia Energy Application

The progress of current utilization and commercialization of hydrogen energy is slow, and hydrogen storage and transportation are both difficult and expensive. Nevertheless, widespread ammonia delivery systems and corresponding operating standards already exist, and the cost of facilities related to ammonia is lower than that of hydrogen. If green hydrogen is used to produce green ammonia, there will be no CO_2 emissions, and liquefying green ammonia for large-scale transportation may be one of the important ways to store and transport hydrogen in the future. In addition, the combustion products of ammonia are water and nitrogen, which will not cause carbon emissions, and the hydrogen industry may progress towards the ammonia energy industry [43]. However, if ammonia is used directly as a fuel, it is necessary to overcome the limitation that ammonia is not easy to burn. The combustion reaction rate of ammonia is much lower than that of hydrogen, and its calorific value is also lower than that of hydrogen and natural gas. Therefore, there remains a huge challenge to ignite ammonia and achieve its continuous and stable combustion.

In order to achieve the goal of carbon neutrality as soon as possible, major countries around the world are paying more and more attention to the development of ammonia energy [44]. At the end of 2020, the fourth EU Hydrogen Online Conference mentioned the need to continuously increase the production of green ammonia. Japan announced the "Green Growth Strategy" action plan, which highlighted the application of ammonia energy. South Korean government announced that 2022 will be the first year of hydrogen and ammonia-based power generation and formulated a series of development plans and roadmaps, striving to become the world's largest country on hydrogen and ammonia-based power generation. In April 2021, the Japanese government planned to make hydrogen and ammonia power generation account for about 10% of Japan's total energy output by 2050, to breakthrough the ammonia-hydrogen mixed combustion technology for coal-fired power plants before 2023, to put 20 wt% ammonia containing fuels into practical applications by 2025, and to achieve the 100% pure ammonia combustion power generation technology by 2040. In November 2021, South Korea planned to complete research and testing of ammonia as a carbon-free power generation fuel by 2027, to achieve the commercialization of ammonia fueled power generation by 2030, and to increase the fuel proportion of ammonia to 3.6% to reduce its dependence on coal and liquefied natural gas in power generation. In September 2020, the Australian Ammonia Energy Association proposed to strengthen the cooperative relationship between the government and the industry, to open safety training courses for ammonia-powered ships. The industry and the government jointly funded the establishment of an ammonia production technology research and development center and established energy security cooperation related to green ammonia with countries, such as Japan and Singapore. In January 2022, the National Development and Reform Commission and the National Energy Administration issued a notice on the implementation of the ≪"Fourteenth Five-Year Plan" for the development of new energy storage≫, mentioning that it is necessary to increase the research and development of key technology equipment to promote diversified technology development, and carry out research on hardcore technologies, equipment, and integrated optimization design in the energy storage systems, including research and application of ammonia energy storage.

In June 2021, Jupiter Ionics, an Australian company, developed a new electrochemical nitrogen reduction method to produce ammonia, which could significantly reduce the greenhouse gas emissions compared with the current Haber-Bosch process (about 2 ton CO_2/ton NH_3). This method uses an electrolyte battery like the lithium battery to produce ammonia gas. Jupiter Ionics' method uses renewable energy, such as solar and wind power, to electrolyze nitrogen that is separated from the air, using lithium metal to reduce it to form lithium nitride, which then is reduced by the hydrogen produced from water electrolysis to produce ammonia through electro-oxidative reduction. The whole process can be seen as "green production"

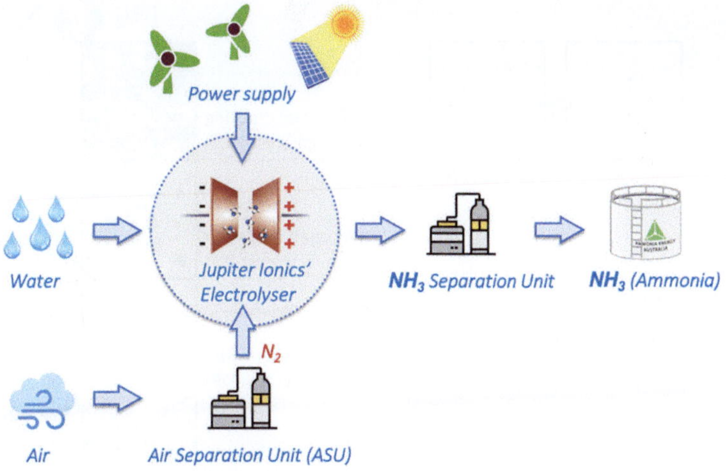

Fig. 4.15 A schematic diagram of the Jupiter Ionics' new green ammonia preparation process

(Fig. 4.15) [45]. Currently, Jupiter Ionics has received an investment of 2.5 million US dollars to expand the commercial production of this green ammonia technology.

In January 2021, Professor Sossina Haile from Northwestern University and researchers from the California energy startup firm, SAFCell, developed an efficient and environmentally friendly method to convert ammonia into hydrogen, overcoming several existing obstacles to producing clean hydrogen from ammonia [46]. The Haile team built a unique electrochemical cell with a proton exchange membrane integrated with an ammonia decomposition catalyst (Ru-Cs/CNT). In brief, ammonia is first decomposed into nitrogen and hydrogen, the hydrogen is then immediately converted into protons, and the protons are subsequently driven through the proton conductive membrane in the electrochemical cell, as shown in Fig. 4.16. By continuously removing the hydrogen, the reaction is driven forward, and the hydrogen produced from the thermal decomposition of ammonia is used for fuel cells.

In December 2021, a research team led by Jiang Lilong at Fuzhou University achieved the industrialization of a new type of low-temperature ammonia decomposition catalysts and explored the hydrogen storage method using ammonia as a hydrogen carrier, which laid a solid foundation for the development of the "ammonia-hydrogen" green energy industry. The first domestic "ammonia-hydrogen" green energy industry innovation platform was also launched in Fujian [47]. The ammonia-hydrogen conversion technology specifically involves an ammonia decomposition catalyst and its preparation method and application in electrodes. The catalyst includes active metal components and catalyst supports, where the active metal components are Ru and/or Ni, and the support materials include barium-based perovskite, zirconium-based rare earth metal oxides, cerium-based rare earth metal oxides, gallium-based perovskite, and alumina [48]. This catalyst can make the thermal expansion coefficient of the catalyst close to that of the electrode material, thereby solving the problem of delamination of the catalyst from the

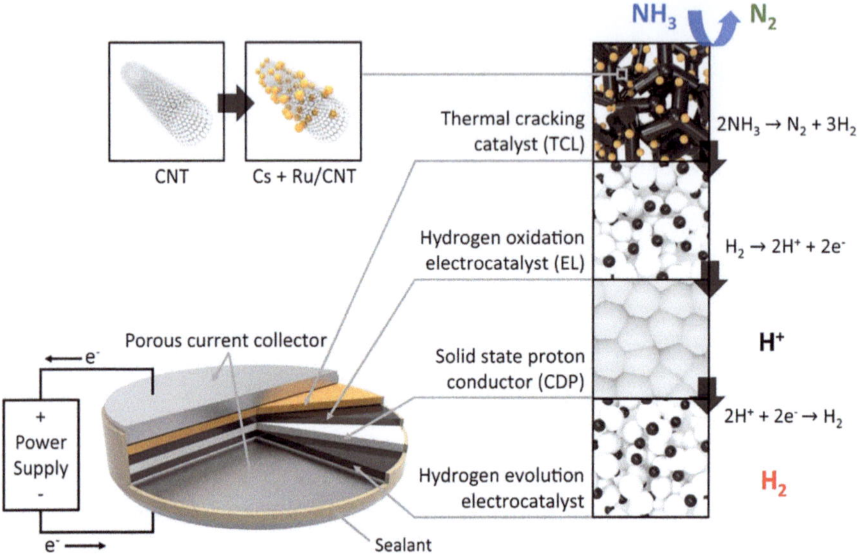

Fig. 4.16 A schematic diagram of the SAFCell acidic electrolysis pool green ammonia hydrogen production process [46]

electrode due to heat. In addition, Fuzhou University, Beijing Sanju Environmental Protection Company, and Zijin Mining Group have jointly invested about 267 million yuan to establish a joint venture company. The three parties will further focus on the "bottleneck" problems in the development of China's ammonia energy industry.

In March 2022, Israel's GenCell energy corporation announced that compared with the traditional ammonia production process commonly used in the world today, they could directly produce green ammonia from water electrolysis at extremely low temperatures and pressures. In addition, GenCell developed a power solution based on zero-emission alkaline batteries and green ammonia power generation technology (GenCell FOX™), allowing uninterrupted power supply to help the world transition from diesel power to clean energy power (Fig. 4.17) [49]. Based on the developed ammonia cracking hydrogen production technology and equipment, GenCell could instantly provide high-purity hydrogen for alkaline-based hydrogen fuel cells for power generation. A storage tank filled with 12 tons of liquid ammonia could provide enough fuels for the fuel cell to operate 24 h per day, 7 days per week, 365 days per year. When installing more than 1000 GenCell FOX™ power generation devices, it could save about 2.5 million US dollars compared to diesel power generation.

Besides the ammonia-hydrogen conversion power generation technology, direct ammonia combustion technology is also an important means of ammonia energy utilization. In 2020, Malaysia International Shipping Corporation, Samsung Heavy Industries of South Korea, Lloyd's Register Group of the UK, and MAN Energy

Fig. 4.17 The GenCell Energy Company's FOXTm ammonia power generation system in Israel

Solutions of Germany collectively reached a cooperation protocol to jointly develop an ammonia fuel oil tanker project in the next 3–4 years. Wärtsilä, a Finnish marine engine manufacturer, Eidesvik, a Norwegian offshore shipbuilder, and Equinor, a Norwegian state-owned energy company, also started collaborating to develop a large zero-emission ship powered by ammonia fuel cells, which was expected to be launched as early as 2024 and commercialized by 2030. MAN ES, a subsidiary of Volkswagen Germany, planned to complete the MANB and WME-LGIP prototype tests by 2024. In October 2021, the ammonia-hydrogen co-firing demonstration project of JERA, a Japanese power giant, was first ignited at a thermal power plant in Hinan City, Aichi County, Japan. According to the plan, the proportion of ammonia fuel co-firing in this project would increase to 20% by 2024, 50% by 2029, and achieve 100% by 2050. The first-generation mixed ammonia low-nitrogen coal powder boiler independently developed by China's National Energy Group has conducted a full-scale mixed ammonia combustion test on the 40 MW-scale Longyuan combustion test platform, with an ammonia conversion of 99.99% and a mixed ammonia combustion ratio of up to 35%, effectively controlling the production of nitrogen oxides [50]. This technological achievement can be applied to coal-fired boilers in the fields of power generation and industry. By carrying out low-cost mixed ammonia combustion modifications to the existing coal-fired boilers, fossil fuels can be replaced to achieve effective CO_2 emission reduction from coal-fired power generation. Green ammonia is crucial for decarbonization in the fields of electricity, shipping, transportation, and agriculture. Many companies around the world have started to performance research or initiate pilot production projects, and the commercialization of the green ammonia industry has begun to prosper.

Global green ammonia market is expected to grow from 16 million USD in 2021 to 540 million USD in 2030, with an annual compound growth rate of about 90% during the forecast period. It is expected that by 2028, the global green ammonia market will generate revenues of 1320 million USD, higher than 850 million USD in 2019, an increase of about 63% during the forecast period. As the number of technologies for producing green ammonia increases, such as alkaline water, solid oxide electrolysis, and proton exchange membranes, and the number of solutions for achieving decarbonization goals using intelligent technology increases, this green ammonia market will also grow. Main companies in the global green ammonia market include BP, OCI N.V., ITM Power, Haldor Topsoe, Thyssenkrupp Industrial Solutions, Yara, Air Liquide, Linde, Air Products, and Chemicals, Hy2gen, etc. Main companies in the domestic green ammonia market include Golden Concord Limited (Group) Holdings, China National Machinery Corporation, Beijing Sanju Environmental Protection Company, Zijin Mining Group, and Ningxia Power Investment Group.

For China, the power sector and the energy chemical sector are the main sources of carbon emissions. If ammonia energy can help replace traditional fossil fuel energy as a new fuel, it will greatly help China achieve its carbon neutrality goal by 2060. China has a very mature ammonia transportation and distribution system, and the same volume of liquid ammonia contains at least 60% more hydrogen than the liquid hydrogen, which has obvious economic advantages. Therefore, hydrogen storage and supply in ammonia and ammonia substitution are possible development trends of hydrogen energy. The ammonia-hydrogen fusion development is a forward-looking, disruptive, and strategic technological development direction in the field of international clean energy. It is an effective way to solve the major storage and transportation bottlenecks of hydrogen energy development, and it is also an important technical route to achieve high-temperature zero-carbon fuel. Although foreign countries have gradually carried out ammonia-hydrogen fusion application projects, domestic research and applications are relatively few, and there remain fundamental bottlenecks and technical challenges in ammonia fuel applications. First, the combustion kinetics and calorific value of ammonia are low, and much lower than hydrogen, which is not conducive to efficient industrial applications. Second, ammonia is not easy to ignite and achieve stable combustion. In addition, to achieve large-scale ammonia-hydrogen conversion, storage and transportation applications, further technical breakthroughs are needed in research areas, such as large-capacity storage and transportation equipment and catalysts.

4.3 Methanol

4.3.1 Principle of Hydrogen Storage Using Methanol

Methanol, also known as xylitol, has the molecular formula CH_3OH and a molecular weight of 32. It is a colorless, volatile, and flammable liquid with a smell similar to ethanol. Under normal pressure, methanol has a boiling point of 64.7 °C and an ignition point of 473 °C. It can dissolve infinitely in water and many organic liquids such as ethanol and ether. It easily absorbs water vapor, carbon dioxide, and some other impurities. Its vapor can form explosive compounds when mixed with air within a certain range, with an explosive range of 6–36.5%. Methanol is the simplest alcohol compound containing C–H, C–O, and O–H bonds. As a liquid hydrogen storage material, methanol has a high energy density (21.6 MJ/kg), gravimetric hydrogen storage density (12.5 wt%), and volumetric hydrogen storage density (99 g/L) [51]. Methanol is a liquid at normal temperature and pressure, easy to store and transport, with abundant and diversified sources. It can be produced from the traditional chemical industry or be prepared from renewable energy. The main product of the direct dehydrogenation reaction of methanol is CO, and the widely used technology in the industry is the low-temperature proton exchange membrane fuel cell (LT-PEMFC) technology. Its fuel must be high-purity hydrogen gas, requiring extremely low impurity, such as CO (5×10^{-6} to 10×10^{-6}), and direct use can cause platinum (Pt) based catalysts to be poisoned and deactivated [52]. Relatively speaking, high-temperature proton exchange membrane fuel cell (HT-PEMFC) technology has more advantages, because its catalyst has strong anti-poisoning ability, and its CO tolerance can reach 1–2% [51].

Considering the needs of hydrogen fuel cell applications, it is more reasonable and effective to release the hydrogen stored in methanol molecules through the catalytic reforming reaction of methanol and water. The temperature range for methanol reforming reaction is 200–300 °C. Compared with other fuel reforming hydrogen production routes, the catalytic reforming reaction of methanol and water is faster and milder. This process can not only effectively reduce the CO content in the produced fuel gas, but also use the reducing ability of hydrogen to further obtain additional hydrogen from water molecules, thereby making the hydrogen storage density of methanol per unit mass breakthrough the theoretical limit to reach 18.75 wt%. Theoretically, 1 kg of methanol and water can produce 187.5 g of hydrogen [53]. This technology, through effective integration with high-temperature proton exchange membrane fuel cell (HT-PEMFC) technology, could successfully realize the "onboard production and utilization" of hydrogen energy, without the need for separation and purification processes. The crude hydrogen directly enters the high-temperature fuel cell stacks, where the chemical energy is directly converted into electrical energy (Fig. 4.18). Among them, overcoming key technologies, such as the integration of high-temperature proton exchange membrane battery stacks and high-temperature methanol reforming reaction systems, has become the key to the market application of methanol reforming-based hydrogen storage technologies.

Another important advantage of the methanol hydrogen storage technology is that it does not require the construction of additional hydrogen refueling stations (high-pressure gaseous hydrogen or liquid/solid hydrogen storage), but can rely on current existing gas refueling stations for simple adjusting and upgrading, turning them into joint refueling stations that can refuel gasoline, diesel, and methanol/water solutions. The cost of building a new methanol refueling station is about 1–3 million yuan/station (excluding land costs), and the cost of converting a gas refueling station into a methanol refueling station is even lower, about 500,000–800,000 yuan/station. Compared with other hydrogen refueling stations with high safety technology thresholds and high investment costs, methanol refueling stations have stronger feasibility and promotability. Methanol, as a hydrogen

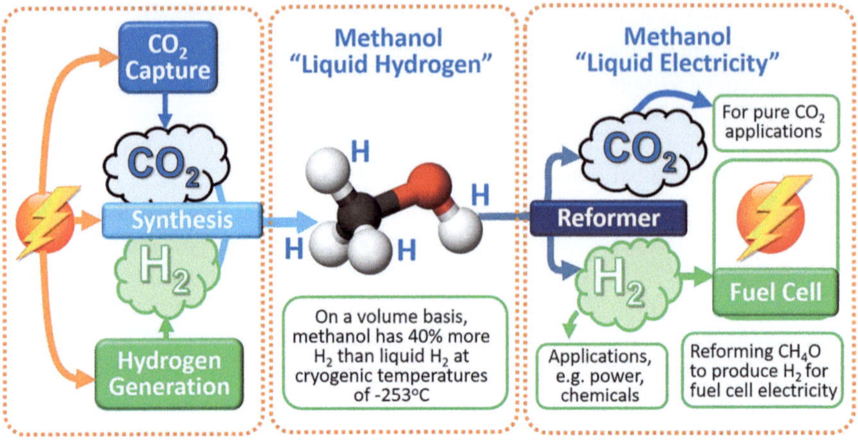

Fig. 4.18 A schematic diagram of the circular methanol economy [52]

storage carrier, has strong competitiveness in terms of economic efficiency in long-distance (>200 km) transportation compared to direct use of hydrogen. In the currently operated "high-pressure gaseous hydrogen transportation-high-pressure hydrogen direct refueling" technological route, the cost of its hydrogen is about 60–80 yuan/kg [51]. Among them, the cost of hydrogen transportation accounts for its high cost. In comparison, with coal-based methanol with an annual output of one trillion tons as raw material, the cost of hydrogen prepared by a methanol steam reforming hydrogen conversion device with a scale of 1000 m^3/h is generally below 2 yuan/m^3, and the cost of methanol reforming is about 20 yuan/kg [51]. Considering various factors, such as H_2 purification, depreciation of various equipment, personnel costs, and profits in the subsequent processes, the terminal H_2 price of a hydrogen refueling station is expected to be about 40–60 yuan/kg.

4.3.2 Methanol Synthesis and Cracking Process

4.3.2.1 Methanol Synthesis Process

The methanol synthesis industry began in 1923, mainly synthesized by compressed synthesis gas under the action of a catalyst. To produce 1 ton of methanol, for the methane steam reforming reaction, it requires 10 MW h equivalent of natural gas and releases about 0.5 ton of CO_2; for the partial oxidation process, it requires about 10.5 MW h of oil and releases about 1.4 ton CO_2 [54]. Of course, methane can also be synthesized through the CO_2 hydrogenation reaction, where the hydrogen comes from water electrolysis, and CO_2 comes from the power plant flue gas or carbon capture and storage (CCS) process from cement plants. However, the CCS technology itself is a high-cost and high-energy consumption process. For example, to

- Spillover theory, ZnO supports serve as a hydrogen storage pool for the CO hydrogenation reaction on the Cu catalyst surface;
- Cu-Zn Alloy theory, ZnO compound partially transfers to the Cu surface, forming the Cu^{+}-O-Zn active sites, meanwhile, part of ZnO dissolves into Cu metal nanoparticles to form Cu-Zn alloy.

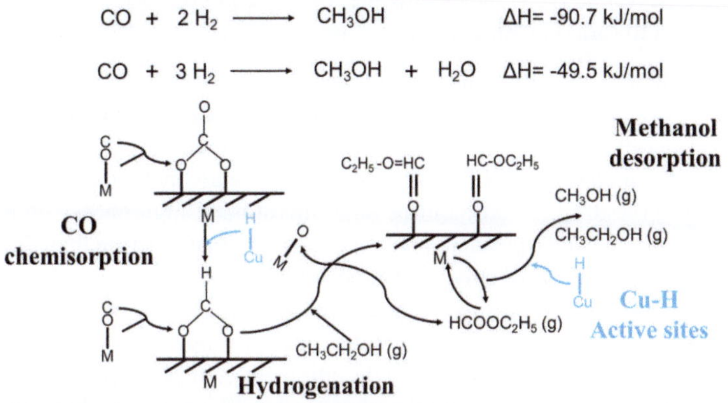

Fig. 4.19 Methanol synthesis reaction equation and mechanism

capture 1 ton CO_2, with an efficiency of 85%, it requires about 44 kW h of compression energy consumption and about 890 kW h of solution regeneration energy consumption. If direct air capture (DAC) of CO_2 is used, to capture 1 ton CO_2, it requires about 300 kW h of electricity consumption and about 1800 kW h of heat consumption [55]. Although the CO_2 hydrogenation synthesis of methanol is feasible, the cost of CCS technology is high and not suitable for large-scale production of methanol.

From the early 1920s to the middle 1960s, all methanol production used zinc-chromium catalysts. In 1966, the British Imperial Chemical Industries (ICI) company successfully developed a copper-based catalyst (Cu/ZnO), using CO_2 containing syngas as raw materials, to synthesize methanol under conditions of 250–300 °C and 8–10 MPa (Fig. 4.19) [56]. Subsequently, researchers studied the composition, activity, various characteristics, and methanol synthesis mechanism of the copper-based catalyst (ZnO as a Cu catalyst promoter). However, the high activity of the copper-based catalyst requires high temperature and pressure, and the CO single-pass conversion is low (15–25%). Therefore, the development of low-temperature and high-efficiency methanol synthesis catalysts is crucial for improving the CO single-pass conversion and reducing the production costs. Tusbaki and coworkers improved the synthesis process using different alcohols as catalyst promoters and solvents, and first achieved methanol synthesis at 150–170 °C [57]. Later, they used a solid-state method to synthesize Cu/ZnO catalyst and studied the effects of different $H_2C_2O_4$/(Cu + Zn) molar ratios and types of chelating agents on the catalyst structure and methanol synthesis performance [58]. Among them, the main role of ZnO is [59]:

- Morphology theory, ZnO supports help to disperse Cu nanoparticles and improve their surface morphology;
- Spillover theory, ZnO supports serve as a hydrogen storage pool for the CO hydrogenation reaction on the Cu catalyst surface;
- Cu-Zn Alloy theory, ZnO compound partially transfers to the Cu surface, forming the Cu^{+}-O-Zn active sites, meanwhile, part of ZnO dissolves into Cu metal nanoparticles to form Cu-Zn alloy.

Currently, main factors affecting the methanol synthesis technology are the synthesis reactor, catalyst, and process flow selection. According to the different synthesis pressures, it can be divided into high-pressure method (19.6–29.4 MPa), medium-pressure method (9.8–19.6 MPa), and low-pressure method (4.9–9.8 MPa) [60]. The high-pressure method requires large investment, high energy consumption, and high cost. The low-pressure method is easier to manufacture than the high-pressure method, with less investment and about a quarter less energy consumption. The medium-pressure method is developed with the scaling up of methanol synthesis equipment, because when the equipment scale increases, if the low-pressure method is used, the process pipelines and equipment will be very large and not compact, so the medium-pressure method has emerged between the high-pressure method and the low-pressure method. Methanol produced by the medium and low-pressure processes accounts for more than 80% of the world's total methanol production. The most typical corporations for methanol production are from the UK's ICI DAVY Process (formerly ICI), Denmark's Haldor Topsoe, and Germany's Lurgi Chemical Company.

Johnson Matthey (DAVY) is one of the main suppliers for large-scale coal-to-methanol synthesis technology in the world. A daily 5000-ton and annual 1.67 million-ton methanol synthesis technology provided by DAVY was successfully put into operation in Trinidad Island in 2005. Its technology is mainly low-pressure synthesis technology. The synthesis process is shown in Fig. 4.20. The process is in

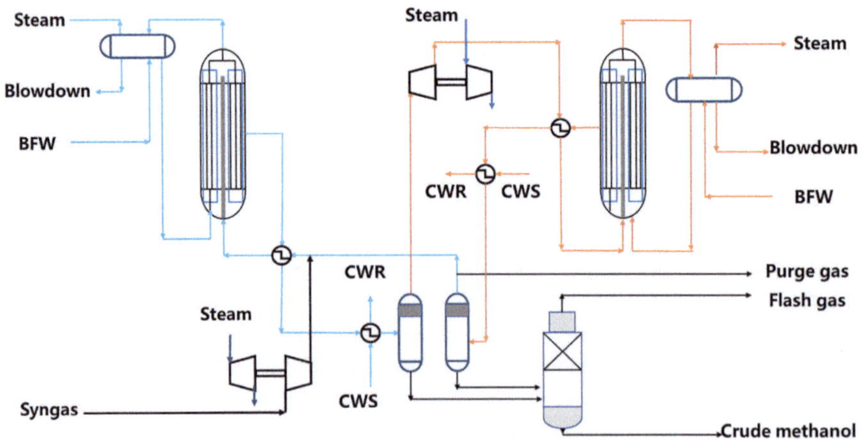

Fig. 4.20 A Davy medium-pressure methanol synthesis process flow diagram

a "series-parallel" mode, and the reaction raw materials are conversion gas ($CO_2 + H_2$) from natural gas conversion or syngas ($CO + H_2$) from coal gasification. Among them, the catalyst uses high-pressure saturated steam and medium-pressure steam at the bottom of the tower to complete physical dehydration and chemical dehydration. When the temperature of the catalyst is higher than 200 °C, fresh gas is added, and crude methanol is obtained through the pre-reactor and synthesis reactor. The synthesis reactor uses a tubular radial flow reactor, and the syngas enters the central tube from both the top and bottom of the reactor, and the reaction is completed under the action of the catalyst evenly distributed on the central tube. DAVY has independently developed a unique reactor catalyst support, which mainly consists of a ring-shaped catalyst container, panels, and bottom plates. It can fill small-particle catalysts, ensuring high heat transfer and low pressure drop while improving the methanol yield [60]. Its compressor adopts a "one-to-three" mode: one gas turbine drives three compressors (low-pressure syngas compressor, high-pressure syngas compressor, and circulating gas compressor), providing a power source for the methanol synthesis reactor. The DAVY methanol synthesis tower can control the uniformity of the bed temperature, and the use of a radial flow reactor can ensure a small pressure drop under large gas flow conditions. By extending the length of the reactor, the production capacity of the reactor can be expanded.

As shown in Fig. 4.20 [60], the methanol feed gas ($CO + H_2$) enters the methanol synthesis unit at a pressure of about 5.1 MPa and a temperature of 30 °C, mixes with the hydrogen-rich gas from the membrane separation unit, and is pressurized to 7.74 MPa by the syngas compressor and heated to 90 °C. The pressurized syngas is preheated in the preheater using the low-pressure steam, and then a small amount of water from the purge gas scrubber (or from the secondary high-pressure boiler feed water in the boundary area) is injected into the syngas to hydrolyze the carbonyl sulfide (COS) impurities into H_2S, which is then removed by the syngas desulfurization to avoid catalyst poisoning. The methanol synthesis loop uses two radial flow steam production synthesis towers (R-SRC) in a both series and parallel configuration, with an initial reaction temperature of the catalyst at ~217 °C, a reaction pressure at ~7.5 MPa, and a final reaction temperature of the catalyst rising to ~243 °C. The syngas flows through the synthesis catalyst located in the shell side, and the reaction heat is removed by the by-product steam inside the tube. The system is equipped with a steam superheater, and the saturated steam from methanol synthesis enters the steam superheater and is superheated to about 320 °C, and then fed into a steam network of about 1.7 MPa. The steam superheater has a separate waste heat recovery system. The high-temperature flue gas discharged from the top of the convection section is introduced into the air preheater for heat exchange with the combustion air, and the cooled flue gas is discharged into the self-standing chimney at the top of the convection section for high-altitude emission.

Resultant desulfurized syngas is divided into two portions. One portion of the syngas is mixed with the circulating gas from the No. 2 crude methanol separator, heated to the reaction temperature by the No. 1 inlet and outlet heat exchanger, and then fed into the No. 1 methanol synthesis tower, where the syngas is catalyzed to synthesize crude methanol. The hot gas leaving the No. 1 synthesis tower is cooled to about 40 °C by passing through the No. 1 inlet and outlet heat exchanger, the No. 1 crude methanol air cooler, and the No. 1 crude methanol water cooler, and then enters the No. 1 crude methanol separator for gas-liquid separation. The liquid phase product is sent to the crude methanol flash tank, and the gas phase product is mixed with the other portion of the syngas, which are subsequently compressed to about 8.0 MPa by the circulating gas compressor and then fed into the No. 2 methanol synthesis tower after passing through the No. 2 synthesis tower inlet and outlet heat exchanger. The gas after the reaction in the No. 2 methanol synthesis tower is cooled to 40 °C by passing through the No. 2 synthesis tower inlet and outlet heat exchanger, the No. 2 crude methanol air cooler, and the No. 2 crude methanol water cooler, and then sent to No. 2 crude methanol separator for gas-liquid separation. The liquid phase methanol is sent to the crude methanol flash tank, and the gas phase product is circulated into the No. 1 methanol synthesis tower. A small portion of the vent gas is discharged from the top of the gas pipeline of the No. 2 crude methanol separator to control the content of inert gas in the circuit. The crude methanol separated from the No. 1 and No. 2 crude methanol separators enters the methanol distillation process of the methanol-to-olefins (MTO) technology, and soluble gases are removed by vacuum flash evaporation and washing in the crude methanol flash tank. The flash vapor from the crude methanol flash tank is sent to the internal steam superheater as a fuel, and the liquid obtained after separation is sent to the stabilizer tower for recovery.

Like DAVY company, Germany's LURGI company is also one of the world's main suppliers of methanol synthesis technology. It successfully developed the LURGI low-pressure methanol synthesis technology in the 1970s. Its unique process design can achieve faster reaction rates and higher conversion, improve the single-pass conversion, reduce the circulation gas volume, and save the power consumption of the circulation gas compressor. At the same time, because it uses a water-cooled reactor, the operating temperature is higher, which makes the pressure of the by-product steam higher than the DAVY process, conducive to the efficient recovery and utilization of the by-product steam. However, the use of a water-cooled reactor will cause a large pressure drop. The methanol synthesis reactor from Danish TOPSOE is similar to the water-cooled reactor of LURGI company. The catalyst is filled in the tube bundle, and the shell side is boiler feed water. The heat released by the reaction is transferred to the boiler water between the tubes through the outer wall of the tube bundle, generating about 4.0 MPa steam. Unlike the DAVY and LURGI processes, the TOPSOE process uses two or several reactors in parallel to achieve an increase in production scale.

4.3.2.2 Methanol Cracking Process

From molecular level analysis, the essence of cracking methanol to produce hydrogen is the process of releasing all the hydrogen atoms in the molecule, mainly involving the dissociation of chemical bonds, such as C–H and O–H, and the carbon atoms from low valence undergo multiple reactions and get oxidized to CO_2. In the methanol-to-hydrogen reaction, the addition of an oxidant will have a significant impact on the thermodynamics of the hydrogen production reaction, the efficiency of hydrogen production, and the optimization of the reactor design and reaction conditions. Traditional methanol cracking technologies include (Table 4.2): direct decomposition (DE), partial oxidation (POX), methanol steam reforming (MSR), and autothermal reforming (ATR).

Traditional methanol decomposition (DE) technology is mature and has some applications in small and medium scale hydrogen production needs. The key technological development focuses on the catalyst optimization, improvement, and reaction coupling to achieve lower reaction temperatures, higher hydrogen selectivities, and higher yields. Direct methanol cracking for hydrogen production is an endothermic reaction. High temperatures are beneficial for complete methanol conversion, but excessively high temperatures can lead to high energy consumption and poor thermal stability of the catalyst [52]. The catalyst for direct methanol cracking should have the following properties: a high selectivity for H_2, good thermal stability, and a high low-temperature activity. Similar to the catalytic methanol synthesis process, copper-based catalysts are also considered as good methanol decomposition catalysts. Early research focused more on Cu-based catalysts, with Cu/ZnO and Cu/Cr binary catalysts being the most studied. Since Cu-ZnO catalysts are excellent methanol synthesis catalysts, this catalyst has become one of the earliest catalysts studied in the methanol decomposition catalyst systems [51]. However, its activity in the methanol cracking process is poor, its stability is not high, and the methanol conversion is low. Although the Cu/Cr catalyst system has better activity and stability, its selectivity is not high and there are pollution problems. The key to direct methanol cracking for hydrogen production lies in the development of high-performance and high-selectivity catalysts to achieve complete methanol conversion within the range of 250–300 °C.

Methanol steam reforming (MSR) for hydrogen production is an endothermic reaction, with the reaction temperature generally between 250–300 °C. Transition metal catalysts, such as Cu and Ni, especially PT catalysts, are commonly used [61]. Methanol steam reforming can be seen as an integration of the methanol cracking reaction and CO steam conversion reaction, obtaining an additional 50% of hydrogen from water molecules, thus the hydrogen content in the product is high, approaching 75%. Its reaction process is simple, and the products are easy to separate. As a front-end hydrogen source for hydrogen refueling stations, MSR hydrogen production technology (Fig. 4.21) has a high H_2 content and mature technology, making it the best choice for current hydrogen production reactions [62]. In the research of methanol steam reforming catalysts, Xiaodong Wen from Institute of Coal Chemistry, Chinese Academy of Sciences, and Ding Ma from Peking University developed a new single-atom Pt/α-MoC dual-function catalyst based on the characteristics of methanol-water hydrogen production reaction, achieving efficient hydrogen production at low temperatures (150–190 °C) [63]. In 2021, Ding Ma and coworkers developed a low-cost and widely sourced non-precious metal nickel-based catalyst 2 wt% Ni/α-MoC, with a reaction temperature of 240 °C, and

Table 4.2 Traditional methanol cracking hydrogen production technologies

Types	Reaction formula	Advantages	Disadvantages
Decomposition	$CH_3OH \rightarrow CO + 2H_2\ \Delta H = 90.7$ kJ/mol	Fast at high temperatures	Require additional heating and produce CO product
Steam reforming	$CH_3OH + H_2O \rightarrow CO + 3H_2\ \Delta H = 49.5$ kJ/mol	High hydrogen production yield, low temperature	Require full heat management systems to supply heat
Partial oxidation	$CH_3OH + \frac{1}{2}O_2 \rightarrow CO + 2H_2\ \Delta H = -192.2\,\text{kJ / mol}$	Mild and fast	Low hydrogen yield
Autothermal reforming	$CH_3OH + (1-p)H_2O \rightarrow CO_2 + (3-p)H_2$ $\Delta H = (49.5 - 241.8\text{p})\,\text{kJ / mol}$	Easy lo start, and good for heat management	Low hydrogen yield, exquisite heat management system

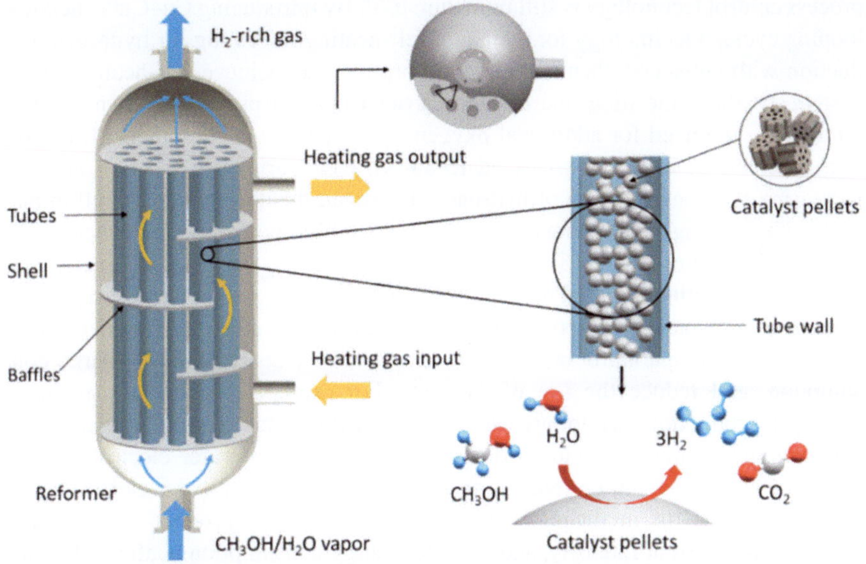

Fig. 4.21 A schematic diagram of methanol steam reforming processes for hydrogen production [62]

its activity was 6 times higher than that of the platinum-based catalyst 2 wt%Pt/Al_2O_3 [64].

Partial oxidation methanol cracking (POX) refers to the methanol reforming technology where oxygen partially or completely replaces water as the oxidant. Oxygenation can significantly change the reaction thermodynamics of methanol hydrogen production. When the concentration of molecular oxygen in the reaction atmosphere exceeds 12.5% of the water concentration, the methanol hydrogen production reaction becomes exothermic [65]. Hydrogen production by partial oxidation of methanol is an exothermic reaction, the reaction rate is high, the by-product is CO_2, the CO content is extremely low, and no additional heating equipment is needed. Based on the stoichiometry of the chemical reaction, each molecule of methanol can produce 2–3 molecules of hydrogen in the autothermal reforming process. When pure oxygen is used as the oxidant, the concentration of hydrogen in the product can reach 66%, but an air separation unit is required; when air is used as the oxidant, the concentration of hydrogen in the product is 41%, and the high nitrogen content increases the difficulty of subsequent separation. Currently, the catalyst system for this technology is not ample enough, and the intense exothermic reaction is difficult to control.

Autothermal reforming (ATR) of methanol for hydrogen production is an integration of partial oxidation of methanol and steam reforming of methanol. The overall reaction is slightly exothermic, with the temperature in the range of 300–500 °C, and the catalysts are based on Cu, Zn and other oxides. This method has a higher reaction rate and hydrogen production yield, but the development of its catalyst and process control technology is still immature [66]. By introducing Cu–CuO chemical looping cycle, a technology for methanol self-heating reforming for hydrogen production with enhanced chemical chain absorption, can achieve self-heating of the system. At the same time, the oxygen carrier CuO can provide its own oxygen, eliminating the need for additional oxygen or air separation equipment. The addition of a carbon carrier is expected to absorb CO_2 during the reaction process, increasing the concentration of hydrogen. However, this technology is still immature, with challenges including the development of high-performance cyclic oxygen carriers and reactor design [67].

Compared with traditional methanol cracking technologies for hydrogen production, new methanol decomposition technologies mainly aim to achieve reactions under normal temperature and pressure, improve conversion, reduce energy consumption, and reduce the use of catalysts. For example, electrolytic methanol hydrogen production technology can produce hydrogen under normal temperature and pressure. Compared with hydrogen production from water electrolysis, the power consumption can be reduced from 5.5 kW h/m^3 H_2 of water electrolysis to 1.2 kW h/m^3 H_2 of methanol electrolysis. The hydrogen production is linearly related to the current intensity, and the total energy consumption is affected by the working temperature and the properties of the anode material. The development of suitable anode materials is expected to significantly reduce the cost of hydrogen production. Ultrasonic methanol cracking technology uses ultrasound as an inducing factor to trigger the methanol cracking reaction without adding other external conditions, producing hydrogen at room temperature, and avoiding the high temperatures required by traditional methanol tracking processes. However, the chemical reactions under ultrasonic radiation are extremely complex, and the specific reaction mechanism is still unknown. Plasma methanol cracking technology uses highly active particles, such as electrons and excited state substances, to provide energy for the reaction process, increase the reaction speed, and avoid the use of heterogeneous catalysts. Experiments demonstrates that methanol shows significantly higher reactivity than water molecules in the cathode plasma layer, and the hydrogen content in the products can reach 95%. However, the energy consumption of the plasma conversion process is too high. Among them, sliding arc plasma and glow discharge plasma can control the energy consumption at 3 kW h/m^3 H_2, which has certain market development potential. Photochemical methanol-to-hydrogen technology selects suitable photochemical catalysts and uses specific light sources to catalyze the production of hydrogen from the methanol-water system. The reaction occurs at room temperature and is still in the preliminary research stage.

In summary, among various methanol-to-hydrogen production methods, the methanol-water reforming hydrogen production reaction is a relatively mature field in catalyst synthesis and process development due to its high hydrogen production rate and easy control. From an engineering perspective, the energy required for the initial start-up of methanol reforming can be solved by coupling with small energy storage batteries.

4.3.3 Current Status of Methanol Energy Applications

In the context of high costs in hydrogen production, storage, transportation, and refueling, and lagging infrastructure construction, the methanol economy may promote cost reduction and efficiency improvement in the hydrogen industry chain and unblock the industry "bottleneck". Tao Zhang, the deputy dean and academician of Chinese Academy of Sciences, once stated that using renewable energy to generate electricity to produce green hydrogen, and then combining it with CO_2 to produce green methanol that is easy to store and transport, is an important pathway to zero carbon emissions. Green methanol, as the hub of energy transformation, can solve the cleanliness problem of energy throughout the carbon footprint process, and achieve multiple benefits, such as expanding the hydrogen energy application industry chain, reducing carbon emissions, and realizing carbon utilization. In 2019, the Ministry of Industry and Information Technology issued the "Guidance of Developing Methanol Vehicles Applications in Some Parts of China", pointing out that enterprises should be encouraged and supported to research and develop methanol fuel cells, and accelerate the transformation and industrial applications of scientific research results of methanol fuel cell vehicles.

Since 2012, China Hydrogen Energy Technology Company has started the technical development and market research of methanol reforming based hydrogen fuel cells, and in 2018, it passed the "China's first methanol reforming hydrogen fuel cell powered logistics vehicle" and "China's first silent mobile power station MFC30" projects approved by the 303 batch of the Ministry of Industry and Information Technology (Fig. 4.22) [68].

Fig. 4.22 The first silent mobile power station developed by China Hydrogen Energy Technology Company

The methanol-to-hydrogen fuel cell system developed by Shanghai Palcan New Energy Technology Co., Ltd. (Fig. 4.23) uses methanol and water as the fuel, uses medium-high temperature proton exchange membranes to reduce the system's requirements for hydrogen purity, and the rich hydrogen gas produced by methanol reforming generates electricity within the module [69]. The entire methanol reforming process is mild, hydrogen is produced and used instantly without storage and transportation, and it exerts the advantages of low noise, a high degree of automation, a small size, a high efficiency, and strong continuous working ability. Importantly, the power generation of this system is stable for a long time, as long as enough methanol and water fuel can provide stable power outputs, meeting the power needs in most mobile scenarios, so it can be widely used in fuel cell powered systems, mobile charging pile systems, military-civilian integration power stations and other fields, and can also be applied to 5G base stations, emergency rescue, island off-grid power supply, and other special scenarios.

In September 2019, the world's largest methanol reforming fuel cell (Blue World Technologies) production base broke ground in Aalborg Port, Denmark, with an annual output expected to be 750 MW, equivalent to 50,000 sets of fuel cell outputs. The methanol reforming fuel cell system designed and developed by Blue World Technologies in the United States (Fig. 4.24) has zero harmful emissions. From the perspective of the entire industry chain from the mine to the wheel, carbon dioxide has zero increase in emissions, focusing on applications in the automotive and mobile fields. This technology can provide an effective solution to global air pollution and climate change. With decades of professional experience in the fuel cell field, Blue World Technologies uses a mixed configuration system of methanol reforming fuel cells and lithium batteries. A complete system includes a methanol reformer for fuel conversion, a DC-to-DC power converter, and a fuel cell stack for power generation. The fuel cell control unit controls the fuel cell system and interacts with the vehicle system.

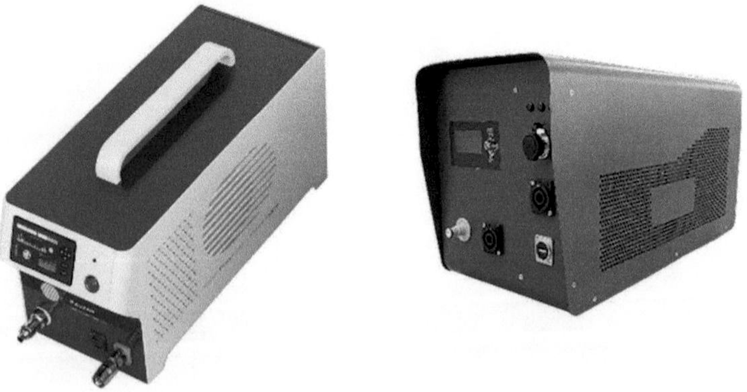

Fig. 4.23 A photo of Shanghai Palcan New Energy Technology Co., Ltd. 's 300 W portable methanol reforming hydrogen fuel cell system

Fig. 4.24 The high temperature methanol reforming fuel cell released by Blue World Technologies in the United States

Fig. 4.25 The world's first light-duty truck based on methanol reforming hydrogen fuel cells, jointly produced by Dongfeng Special Vehicle Co., Ltd. and Suzhou Qingjie Power Supply Technology Co., Ltd.

In October 2018, the world's first light-duty truck based on methanol reforming hydrogen fuel cells was officially put into commercial operation in Kunshan. This vehicle was jointly produced by Dongfeng Special Vehicle Co., Ltd. and Suzhou Qingjie Power Supply Technology Co., Ltd. (Fig. 4.25) [70, 71]. The methanol reforming hydrogen fuel cell system was provided and organized for mass production by Shanghai Palcan New Energy Technology Co., Ltd., the parent company of Suzhou Qingjie Power Supply Technology Co., Ltd. This model is a type of fuel cell electric vehicle, but its fuel is methanol, which is a new solution proposed for the current problems of lagging construction of hydrogen refueling stations and high costs in China. The hydrogen supply of this hydrogen fuel cell system is different from the traditional high-pressure tank hydrogen storage method. It obtains hydrogen through the methanol-water mixture reforming chemical reaction, which enters the fuel cell stacks to generate electricity, achieving "on-demand" hydrogen production with a high efficiency and low cost.

In January 2020, the world's first 1000-ton scale solar energy green hydrogen methanol demonstration project led by the academician Can Li from Dalian Institute of Chemical Physics was successfully tested in Lanzhou New Area, achieving the first step in the industrial production of liquid methanol fuels from solar energy (Fig. 4.26) [72]. This project consists of three basic units: solar photovoltaic power generation, water electrolyzer to produce green hydrogen, and CO_2 hydrogenation to synthesize methanol. It is led by Hualu Engineering & Technology Co., Ltd. in project design, with a total power of 10 MW photovoltaic power station built to provide electricity for two 1000 Nm^3/h water electrolyzers for hydrogen production. The academician Can Li 's team developed a new type of water electrolysis catalysts with China's independent intellectual property rights, which can manufacture large-scale (1000 Nm^3/h) alkaline water electrolysis equipment, reducing the unit hydrogen production energy consumption to 4.0–4.2 kW h/m^3 H_2, significantly reducing the cost of water electrolysis based hydrogen production, and is currently the highest efficiency of large-scale alkaline water electrolysis based hydrogen

Fig. 4.26 The world's first 1000-ton scale solar fuel synthesis demonstration project developed by the Dalian Institute of Chemical Physics

production in the world. At the same time, the solid solution bimetallic oxide catalyst (ZnO-ZrO_2) independently developed by Can Li's team achieved a high selectivity and high stability of CO_2 hydrogenation to synthesize methanol, with a single pass methanol selectivity greater than 90%, and the catalyst performance decays less than 2% after 3000-h operation, laying the foundation for industrial applications.

The storage, transportation, and distribution of methanol can use current existing infrastructures and logistics systems, and green methanol can be made from various raw materials, such as renewable electricity, biomass, biogas, and urban solid waste. Methanol is a good liquid energy carrier, and China has abundant methanol resources. As the world's largest producer and consumer of methanol, China's methanol production capacity accounts for more than 50% of the world. As of 2017, China's methanol production capacity reached 83.51 million tons/year, and it is still increasing the construction of the methanol industry, with an expected additional capacity of 20 million tons/year in the future. China's transportation sector has a relatively high dependence on crude oil and natural gas, and the use of methanol fuel cell systems is a "fuel cell solution with Chinese characteristics". Based on pure electric trucks, a methanol reforming hydrogen fuel cell power generation system is added, using liquid methanol as raw materials, producing hydrogen, generating electricity, driving vehicles, or charging the batteries, keeping the battery in the best power state, extending the battery life, and reducing the battery load. The main bottleneck of the current hydrogen storage technology based on methanol reforming is: (1) the efficiency of green methanol production is low, the selectivity is low, the preparation cost is high, and the energy consumption is large; (2) the efficiency of methanol reforming reactors is low, and the high-purity hydrogen separation equipment is expensive, with high operating costs and limited lifespan; (3) issues with the integrated development of hydrogen fuel cells, including hydrogen supply purity, CO catalyst poisoning, and operation lifespan.

4.4 Comparison of LOHC, Ammonia, and Methanol Hydrogen Storage Technologies

After a detailed introduction of LOHC, ammonia, and methanol as three hydrogen storage systems and their technological processes, this section will focus on analyzing the cost of the hydrogen refueling process (synthesis of H18-DBT, NH_3, CH_3OH) of these three systems. These hydrogen storage media are currently the most widely studied and applied, but there is little research on their large-scale dehydrogenation. Nevertheless, researchers estimate that for the NH_3 hydrogen storage system, producing 200 tons/day of hydrogen gas would cost 510 million US dollars; for the LOHC hydrogen storage system, producing 200 tons/day of hydrogen gas would cost 24 million US dollars [5]. Typically, in addition to hydrogen production by cracking, additional separation equipment is needed to separate the hydrogen gas from the by-products. Moreover, the energy consumption of N_2 hydrogen refueling (ammonia synthesis) and CO_2 hydrogen refueling (methanol synthesis) processes is greater than that of the LOHC hydrogen refueling process. If storing 500 tons of hydrogen gas per day, with an estimated 10% loss, this is equivalent to daily cracking about 3200 tons of ammonia, about 4500 tons of methanol, and about 8700 tons of H18-DBT. The electricity and heat consumption of the hydrogen refueling process is shown in Fig. 4.27 [5]: for the NH_3 hydrogen storage system, the endothermic ammonia decomposition hydrogen production requires about 100 MW of electricity, while the exothermic ammonia synthesis process generates a thermal power of about 153 MW; for the methanol hydrogen storage system, methanol based on the CO_2 DAC technology requires the heat and electricity consumption of about 500 MW and about 151 MW, respectively, while generating about 92 MW heat and about 7 MW electric power; for the DBT/H18-DBT hydrogen storage system, the electricity consumption is about 0.35 MW, which can be ignored, and it generates a thermal power of about 220 MW. In addition, for the ammonia and LOHC systems, the generated thermal power can be stored for later use in the dehydrogenative hydrogen release reaction, especially in the case of stationary factory sites.

In addition to the heat and electricity consumption of the hydrogen refueling process, the capital costs of hydrogen storage materials and related systems also need to be considered. Typically, chemical plants need to store hydrogen in two parts of the manufacturing process: first, a day-night storage unit for storing about 8 h of reactants to maintain operations without shutting down the entire facility; second, in the event of a supply chain disruption, all products and reactants can be stored for 30 days. Since the synthesis or hydrogen refueling processes of ammonia, methanol, and LOHC are continuous and will not be interrupted, sufficient on-site product storage space is necessary. Therefore, only the daily storage of the final products (i.e., 3200, 4500, and 8700 tons of ammonia, methanol, and H18-DBT, respectively) and storage containers for 30 days are considered here. The approximate capital costs of storage containers, along with other major components, are shown in Fig. 4.27. For the ammonia hydrogen storage system, in addition to the storage containers, there are also facilities including air separation units (ASUs) for nitrogen separation, mechanical compression for concentrating the synthesis gas mixture, and the synthesis loop for ammonia conversion. Among them, the synthesis loop for ammonia conversion has the highest cost because it includes a series of compressors, heat exchangers, pumps, reactors, and flash tanks. For the methanol hydrogen storage system, the DAC device for CO_2 capture, the synthesis loop for methanol conversion, and storage tanks are the main components. The synthesis loop for methanol is also composed of different components, such as reactors, boilers, a series of heat exchangers, flash separators, distillation towers, and compressors. For methanol hydrogen storage, the largest cost comes from the DAC process for capturing CO_2. If the cost of CO_2 capture is reduced, the cost of synthesizing methanol will also decrease. It is worth noting that some studies predict that the system cost of methanol is lower than that of ammonia and LOHC, because it is assumed that DAC is more expensive than CCS [73]. For the LOHC hydrogen storage system, the material cost is the main cost for operating for 30 days. Overall,

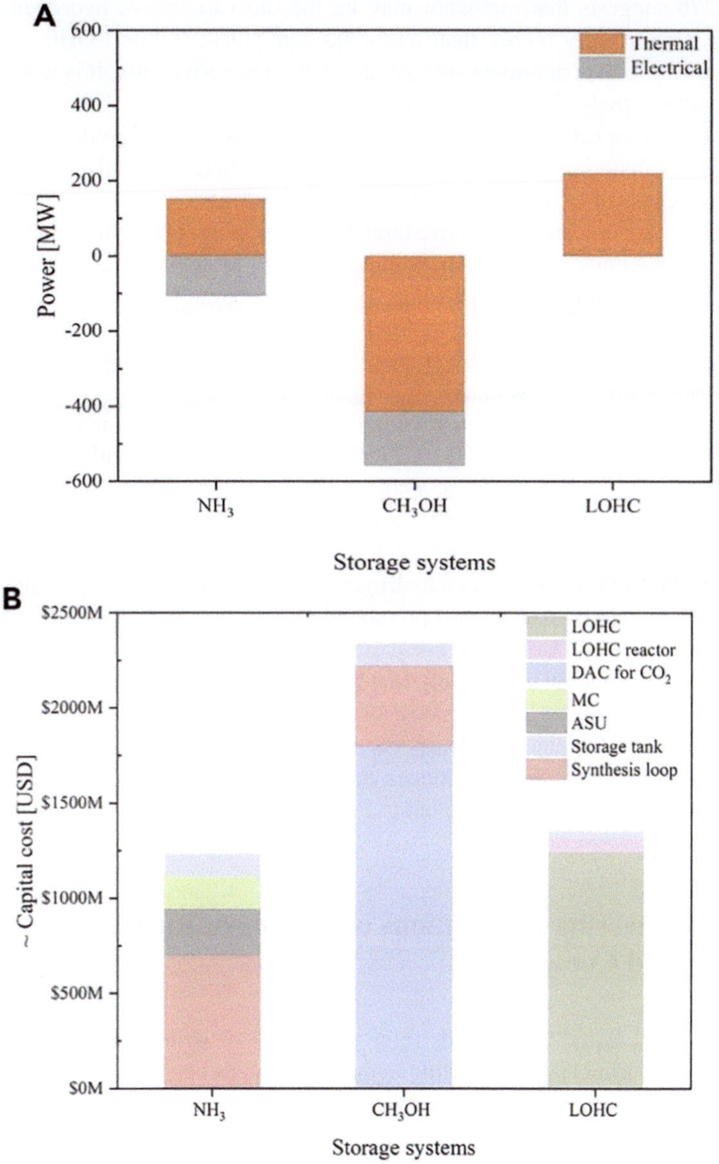

Fig. 4.27 Comparison of energy consumption and capital cost of hydrogen storage process for LOHC, ammonia, and methanol systems [5]. (**a**) predicted energy consumption for storing 500t of hydrogen gas per day, capital cost, US dollars, hydrogen storage system, reactor, direct air capture, hydrogen storage tank, synthesis chain; (**b**) predicted capital cost of hydrogen storage materials and main system components for storing 500t of hydrogen gas per day, calculated for 30 days for the LOHC hydrogen storage tank

Fig. 4.27b suggests that methanol may be the most expensive hydrogen storage method, significantly higher than ammonia and LOHC, while LOHC is slightly more expensive than ammonia storage due to high material costs. It is worth noting that the above analysis assumes a new LOHC material cost of 5 USD/L. However, for large-scale storage and transportation systems, due to mass production and use, the price of materials may significantly decrease, thereby greatly enhancing the competitiveness of LOHC hydrogen storage technology.

Catalysts are crucial to the above three hydrogen storage systems, and the cost of catalysts is not considered as a major component of system capital costs because it is a consumable and therefore belongs to operating costs. For the LOHC hydrogen storage system, the U.S. Office of Energy Efficiency and Renewable Energy initially assumed that 1 kg of catalyst, priced at 150 $, could hydrogenate 500 tons of LOHC. Some studies show that the hydrogen refueling capacity of 0.3 wt%Pt/Al_2O_3 catalyst is 3 gH_2/gPT-min or 0.54 kg H_2/kgcatAlysT-hour, and the price of 0.5–5 wt%Pt/Al_2O_3 catalyst is 110–2100 $ [74]. Calculating with a 0.3 wt% Pt/Al_2O_3 catalyst priced at 110 $/kg, a conservative estimate of the catalyst's lifespan is 14,000 h. To hydrogenate 15,000 tons in 30 days, approximately 1984 kg of catalysts or 220,000 $ are needed. For large-scale hydrogen refueling reactions, assuming 5 wt%Pt/Al_2O_3 with the same hydrogen refueling capacity of 3 gH_2/gPT-min or 0.54 kg H_2/kgcatAlysT-hour, the corresponding cost is about 4.17 million US dollars or 2100 $/kg [5]. Relatively speaking, the cost of catalysts for the synthesis of ammonia and methanol is relatively low because they do not require precious metal catalysts. Despite this, the high cost of catalysts has not reduced the economic attractiveness and application prospects of the LOHC hydrogen storage system. If the cost of LOHC catalysts continues to decrease, the overall cost of the LOHC hydrogen storage system is expected to further decrease.

4.5 Engineering Applications of Hydrogen-Rich Liquid Compounds

Hydrogen-rich liquid compounds refer to the use of hydrogen-rich liquids as a medium for storing and transporting hydrogen, such as benzene, N-ethylcarbazole, liquid ammonia, methanol, etc. Hydrogen-rich liquid compound storage and transportation technology has the characteristics of a high hydrogen storage density, high safety, and a low transportation cost. However, its hydrogen refueling and dehydrogenation need to be carried out under the action of precious metal catalysts for high-temperature catalytic synthesis and decomposition, which results in high energy consumption, catalysts are prone to decay, and the decomposition process is often accompanied by side reactions. The released hydrogen is impure and needs to be separated and purified, which leads to the high cost of hydrogen-rich liquid compound storage and transportation technology. Using hydrogen-rich liquid compounds as hydrogen carriers, its transportation network is mature, standardized, and highly flexible, and is also considered a choice for large-scale hydrogen storage and transportation. The application process is shown in Fig. 4.28.

1. Application of liquid ammonia storage and transportation technology

The infrastructures and related technologies for the synthesis and storage and transportation of liquid ammonia are mature. Ammonia has a low activity, and its combustion and explosion hazards are lower than other gases and liquids, but it is toxic and harmful to human health. Ammonia has the following characteristics: anhydrous ammonia is not easily combustible, but ammonia vapor is easily combustible in the air and will explode when ignited; ammonia decomposes when heated and releases toxic gases, requiring the wearing of face masks and liquid-sealed protective clothing; ammonia corrodes alloys, such as copper and zinc, and reacts with oxidants, halogens, acids, etc. Preventing ammonia leakage during transportation is a key issue to be solved. If a liquid ammonia leak occurs, it can be handled according to HG/T 4686—2014 "Treatment and Disposal Method for Liquid Ammonia Leakage" [75]. Therefore, the toxicity and corrosiveness of ammonia are one of the main problems faced by liquid ammonia as a hydrogen transport carrier.

Liquid ammonia is mainly filled in tank trucks, transported by vehicles, trains, etc., or filled in large liquid ammonia transport ships for transportation, as shown in Fig. 4.29. According to the Ammonia Energy Association report, the global production level of ammonia is currently close to about 200 million tons per year, almost 98% of the raw materials for ammonia production come from fossil fuels, of which 72% use natural gas as raw materials. In the future, ammonia as a carrier for hydrogen transportation needs to use green hydrogen to produce ammonia to reduce global carbon emissions. In 2020, NYK, JMU, and NK of Japan jointly developed the application of ammonia as ship fuels and liquid ammonia transportation ships, planning to synthesize green hydrogen in Southeast Asia into ammonia, and transport it back to Japan on a large scale by sea for hydrogen production or direct combustion of ammonia. However, the hydrogen produced

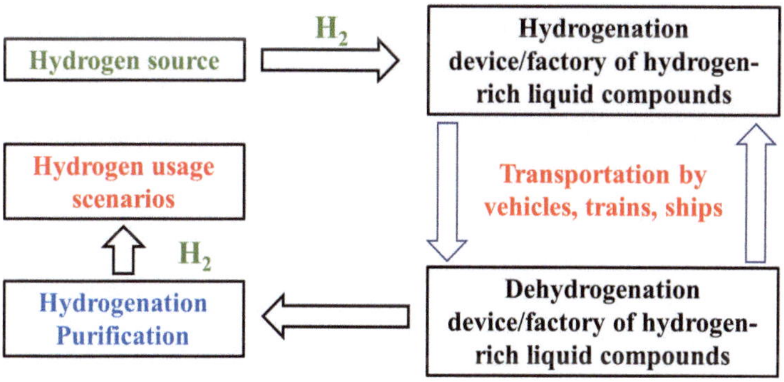

Fig. 4.28 A schematic diagram of the hydrogen storage, transportation and utilization application of hydrogen-rich liquid compounds

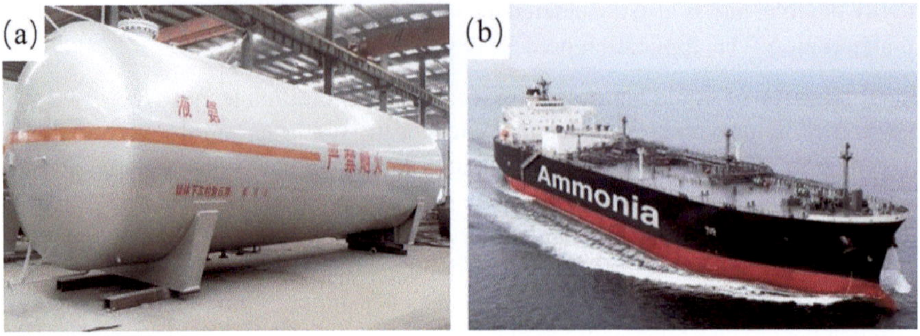

Fig. 4.29 Liquid ammonia transportation methods. (**a**) a liquid ammonia tank; (**b**) a liquid ammonia transportation ship

by ammonia decomposition will contain a small amount of ammonia impurity. If hydrogen gas containing 1×10^{-6} L/L ammonia is used for fuel cells for a week, it will significantly reduce the lifespan of fuel cells. This is mainly because the ammonia molecules could enter the proton exchange membrane, get irreversibly adsorbed by the H^+ in the membrane, and accumulate in the form of ammonium ions, resulting in compromised fuel cell performance. According to the requirements of GB/T37244—2018 " Hydrogen Fuels for Proton Exchange Membrane Fuel Cell Powered Vehicles" [76], the ammonia concentration in hydrogen should be $\leq 0.1 \times 10^{-6}$. This requires that the hydrogen produced from ammonia, after being separated from nitrogen, needs to be further deammoniated.

2. Application of LOHC hydrogen storage and transportation technology

Organic liquids (mainly including cyclohexane, pyrazole, indole, etc.) serve as hydrogen storage and transportation media. They are transported by cars, ships, and trains under normal temperature and pressure, similar to the transportation of oil/gasoline, and can use existing oil/gasoline infrastructures. The key issues mainly include the following points:

(a) Toxicity of organic liquids. Cyclohexane, pyrazole, indole, etc. have slight toxicity. Developing non-toxic organic liquid hydrogen storage and transportation media is one of the main tasks for the safe use of hydrogen energy.
(b) Economic issues of long-distance transportation. Although the method of organic liquid storage and transportation is similar to the existing oil/gasoline transportation, there is no concept of empty-return vehicles/ships. The dehydrogenated organic liquid hydrogen storage medium still needs to be returned to the factory with the vehicle for hydrogen refueling. Such round trips are heavy-load transportation, and the economy of long-distance transportation needs to be considered.
(c) Hydrogenation and dehydrogenation conditions are still stringent. The common organic liquid absorption/desorption hydrogen characteristics are seen in Sect. 4.1. Taking N-ethylpyrazole as an example, it needs to be hydrogenated at 200 °C and 6 MPa, and dehydrogenated at 230 °C and 0.1 MPa.

(d) Dehydrogenation reactions often accompany side reactions. Organic liquids often undergo side reactions during dehydrogenation, resulting in impure hydrogen release, with gases such as CO, CH_4 that are toxic to fuel cells.
(e) The price of the catalyst is still high. To achieve excellent hydrogenation/dehydrogenation performance, precious metal catalysts are usually required, and the lifespan of the catalyst is relatively short, which increases the system cost.

In summary, the development of high-capacity, non-toxic, fully hydrogen-absorbing/desorbing organic liquids at low temperatures and cheap catalysts is the main research direction for promoting the large-scale applications of organic liquid hydrogen storage methods. Chiyoda Corporation of Japan planned to transport hydrogen from overseas to Japan using methylcyclohexane as a hydrogen storage medium. Its typical hydrogenation/dehydrogenation plant is shown in Fig. 4.30 [77]. The company built a methylcyclohexane synthesis plant in Brunei in 2019, which could transport hydrogen produced in Brunei to Japan by ship. The scale of hydrogen transportation could reach 210 tons/year, which was the world's first global hydrogen supply chain project based on organic liquid hydrogen storage. Germany's Hydrogenious Technologies used dibenzyltoluene as a hydrogen storage medium, transporting green hydrogen produced by wind power in northern Germany's Heligoland via organic liquids to Germany, with an estimated hydrogen transport capacity of 100 tons/year by 2030 [78]. The organic liquid dehydrogenation device is shown in Fig. 4.31. Meanwhile, Wuhan Hydrogen Energy Technology Co., Ltd., based on the N-ethylcarbazole organic liquid hydrogen storage technology of China University of Geosciences (Wuhan), jointly developed the world's first ambient temperature and pressure liquid organic hydrogen fuel cell powered engineering prototype bus "Taige", medium-sized fuel cell bus "Xingrui", upgraded hydrogen oil fuel cell bus "Qingyang", and logistics fuel cell vehicle "Xinqingka". The company has now realized the world's first set of annual 1000-ton N-ethylcarbazole production device and hydrogenation/dehydrogenation catalyst production line. In 2021, they realized the application demonstration of the whole hydrogen energy industry chain in Fangshan, Beijing, including liquid organic

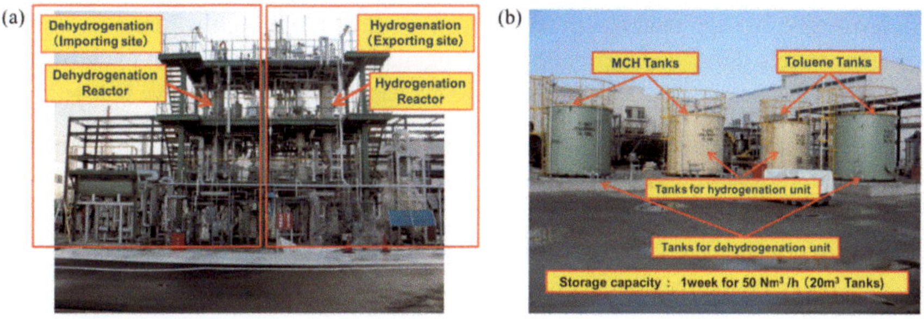

Fig. 4.30 The Chiyoda Corporation's toluene/methylcyclohexane hydrogen refueling/dehydrogenation demonstration plant (**a**) and storage tanks (**b**)

Fig. 4.31 The Hydrogenious Technologies' 1.5t/day dehydrogenation device [78]

hydrogen storage technology to produce hydrogen oil [79]. Overall, organic liquid hydrogen storage and transportation technology is currently in the transition stage from laboratory to industrial production.

Exercises

1. Briefly describe the concept of LOHC and its advantages and disadvantages in hydrogen storage.
2. Compare the advantages and disadvantages of LOHC, ammonia, and methanol as hydrogen storage technology.
3. Briefly describe the Le Chatelier's principle and its application in the field of ammonia based chemical hydrogen storage.
4. 2-Methylbiphenyl ($C_{14}H_{14}$, abbreviated as H0-BT) is a widely studied liquid hydrogen storage material in recent years, which has a high volumetric hydrogen storage density and excellent hydrogenation/dehydrogenation cycle stability. Please answer the following questions based on what you have learned:

   (a) Refer to the literature and calculate the corresponding product of H0-BT hydrogenation and the theoretical mass and volumetric hydrogen storage density.
   (b) According to the basic principle of LOHC hydrogen storage, briefly describe the hydrogen storage process of H0-BT.

5. Please describe the SPERA hydrogen storage technology process developed by Chiyoda Corporation and its advantages and disadvantages.

6. Briefly describe the advantages and disadvantages of ammonia synthetic methods (Haber-Bosch ammonia synthesis, chemical looping ammonia synthesis, electrochemical ammonia synthesis).
7. The ammonia decomposition reaction is an important part of the ammonia hydrogen storage technology. Please briefly describe the current bottlenecks of the ammonia decomposition technology and the methods to solve these bottlenecks.
8. Ammonia has a high gravimetric hydrogen storage density of up to 17.6 wt%, but the ammonia decomposition reaction requires high temperature, high pressure, and precious metal catalysts. Suppose there is a compound (H_2A) that can recombine with NH_3 at room temperature. Assuming the atomic mass of A is 10 g/mol, and A can form a stable NA_2 compound with N, please answer the following questions:

    (a) Write out the recombination reaction equation of H_2A with NH_3.
    (b) Calculate the theoretical gravimetric hydrogen storage density of NH_3 after its recombination with H_2A.

9. Methanol is an important chemical raw material and bioenergy source. Please briefly describe the advantages and disadvantages of four methanol cracking hydrogen production technologies and explain the technical bottlenecks and solutions for methanol hydrogen storage based on the current technological status.
10. The gravimetric hydrogen storage density of methanol is 12.5 wt%, but the catalytic reforming reaction with water can further obtain additional hydrogen from water, thereby breaking the theoretical upper limit of the hydrogen storage density of methanol per unit mass to reach 18.75 wt%. Please answer the following questions:

    (a) Please prove by calculations that the gravimetric hydrogen storage density with water reforming is 18.75 wt%.
    (b) Calculate the mass of hydrogen that can be obtained from the reaction of 10 L methanol with water based on the density of methanol (0.79 g/mL).

References

1. Teichmann D, Arlt W, Wasserscheid P et al (2011) A future energy supply based on liquid organic hydrogen carriers (LOHC). Energy Environ Sci 4(8):2767–2773
2. Andersson J, Grönkvist S (2019) Large-scale storage of hydrogen. Int J Hydrog Energy 44(23):11901–11919
3. Modisha PM, Ouma CNM, Garidzirai R et al (2019) The prospect of hydrogen storage using liquid organic hydrogen carriers. Energy Fuel 33(4):2778–2796
4. Collins L (2020) World's first international hydrogen supply chain realized between Brunei and Japan
5. Abdin Z, Tang C, Liu Y et al (2021) Large-scale stationary hydrogen storage via liquid organic hydrogen carriers. iScience 24(9):102966

6. CHIYODA (2016) What is “SPERA HYDROGEN” system [Z]. https://www.chiyodacorp.com/en/service/spera-hydrogen/innovations/
7. Zhang Y, Zhao J, Lu X et al (2016) Research progress on organic liquid hydrogen storage materials. Chem Ind Prog 35:2869–2874
8. Zhao L, Zhang J, Zhu W et al (2019) Research progress on liquid organic hydrogen storage technology. Chem Reagents 41:47–53
9. Biniwale RB, Kariya N, Ichikawa M (2005) Dehydrogenation of cyclohexane over Ni based catalysts supported on activated carbon using spray-pulsed reactor and enhancement in activity by addition of a small amount of Pt. Catal Lett 105(1):83–87
10. Xia Z, Lu H, Liu H et al (2017) Cyclohexane dehydrogenation over Ni-Cu/SiO_2 catalyst: effect of copper addition. Catal Commun 90:39–42
11. Okada Y, Shimura M (2013) Development of large-scale H_2 storage and transportation technology with liquid organic hydrogen carrier (LOHC). In: Proceedings of the technical paper at joint GCC-JAPAN environment symposia. [s. n.: S. v.]
12. Gora A, Tanaka DAP, Mizukami F et al (2006) Lower temperature dehydrogenation of methylcyclohexane by membrane-assisted equilibrium shift. Chem Lett 35(12):1372–1373
13. Suttisawat Y, Sakai H, Abe M et al (2012) Microwave effect in the dehydrogenation of tetralin and decalin with a fixed-bed reactor. Int J Hydrog Energy 37(4):3242–3250
14. Yang M, Dong Y, Fei S et al (2014) A comparative study of catalytic dehydrogenation of perhydro-N-ethylcarbazole over noble metal catalysts. Int J Hydrog Energy 39(33):18976–18983
15. Sotoodeh F, Smith KJ (2013) An overview of the kinetics and catalysis of hydrogen storage on organic liquids. Can J Chem Eng 91(9):1477–1490
16. Jorschick H, Bösmann A, Preuster P et al (2018) Charging a liquid organic hydrogen carrier system with H_2/CO_2 gas mixtures. ChemCatChem 10(19):4329–4337
17. Green Car Congress (2016) Hydrogenious Technologies partners with United Hydrogen Group (UHG) to bring novel LOHC H_2 storage system to US market[Z]. https://www.greencarcongress.com/2016/05/20160504-hydrogenious.html
18. Yadav V, Sivakumar G, Gupta V et al (2021) Recent advances in liquid organic hydrogen carriers: an alcohol-based hydrogen economy. ACS Catal 11(24):14712–14726
19. Service RF (2018) Ammonia—a renewable fuel made from sun, air, and water—could power the globe without carbon. Science 361(6398):120–123
20. Suryanto BHR, Matuszek K, Choi J et al (2021) Nitrogen reduction to ammonia at high efficiency and rates based on a phosphonium proton shuttle. Science 372(6547):1187
21. Service RF (2018) Liquid sunshine. Science 361(6398):120–123
22. Chang F, Gao W, Guo J et al (2021) Emerging materials and methods toward ammonia-based energy storage and conversion. Adv Mater 33(50):2005721
23. Li Y, Wang H, Priest C et al (2021) Advanced electrocatalysis for energy and environmental sustainability via water and nitrogen reactions. Adv Mater 33(6):2000381
24. Foster SL, Bakovic SIP, Duda RD et al (2018) Catalysts for nitrogen reduction to ammonia. Nat Catal 1(7):490–500
25. Rayment T, Schlögl R, Thomas JM et al (1985) Structure of the ammonia synthesis catalyst. Nature 315(6017):311–313
26. Morgan ER (2013) Techno-economic feasibility study of ammonia plants powered by offshore wind. [s. n.]: University of Massachusetts Amherst
27. Gao W, Guo J, Wang P et al (2018) Production of ammonia via a chemical looping process based on metal imides as nitrogen carriers. Nat Energy 3(12):1067–1075
28. Zhang L, Ji X, Ren X et al (2018) Electrochemical ammonia synthesis via nitrogen reduction reaction on a MoS_2 catalyst: theoretical and experimental studies. Adv Mater 30(28):1800191
29. Hu L, Xing Z, Feng X (2020) Understanding the electrocatalytic interface for ambient ammonia synthesis. ACS Energy Lett 5(2):430–436
30. Green L (1982) An ammonia energy vector for the hydrogen economy. Int J Hydrog Energy 7(4):355–359
31. Bia Energy. Fuel cell micro-CHP in 2020: the news from the fuel cell review. 17 Apr 2022. https://biaenergyconsulting.com/2021/04/29/fuel-cell-micro-chp-in-2020-the-news-from-the-fuel-cell-review/

32. CROLIUS S (2018) CSIRO partner revealed for NH_3-to-H_2 technology[Z]. https://ammoniaenergy.org/articles/csiro-demonstrates-ammonia-to-hydrogen-fueling-system/
33. Jiang J, Dong Q, McCullough K et al (2021) Novel hollow fiber membrane reactor for high purity H_2 generation from thermal catalytic NH_3 decomposition. J Membr Sci 629:119281
34. MacFarlane DR, Cherepanov PV, Choi J et al (2020) A roadmap to the ammonia economy. Joule 4(6):1186–1205
35. Mukherjee S, Devaguptapu SV, Sviripa A et al (2018) Low-temperature ammonia decomposition catalysts for hydrogen generation. Appl Catal B Environ 226:162–181
36. Le TA, Do QC, Kim Y et al (2021) A review on the recent developments of ruthenium and nickel catalysts for COx-free H_2 generation by ammonia decomposition. Korean J Chem Eng 38(6):1087–1103
37. Wang D, Jiang W, Zhao Z et al (2018) Research of industrial hydrogen production at home and abroad. Ind Catal 26(5):26–30
38. Qiu S, Ren T, Li J (2018) Progress in the modification of ammonia decomposition hydrogen production catalysts. Chem Ind Prog 37(3):1001–1007
39. Zhang Z, Liguori S, Fuerst TF et al (2019) Efficient ammonia decomposition in a catalytic membrane reactor to enable hydrogen storage and utilization. ACS Sustain Chem Eng 7(6):5975–5985
40. Nagaoka K, Eboshi T, Takeishi Y et al (2017) Carbon-free H_2 production from ammonia triggered at room temperature with an acidic RuO_2/γ-Al_2O_3 catalyst. Sci Adv 3(4):e1602747
41. Vitse F, Cooper M, Botte GG (2005) On the use of ammonia electrolysis for hydrogen production. J Power Sources 142(1):18–26
42. Hanada N, Hino S, Ichikawa T et al (2010) Hydrogen generation by electrolysis of liquid ammonia. Chem Commun 46(41):7775–7777
43. Zhang Y, Liu H, Li J et al (2022) Life cycle assessment of ammonia synthesis in China. Int J Life Cycle Assess 27(1):50–61
44. Wu S, Salmon N, Li MM-J et al (2022) Energy decarbonization via green H_2 or NH_3? ACS Energy Lett 7(3):1021–1033
45. ELLIS J (2021) Here's why agrifoodtech VC tenacious ventures backed a green ammonia startup[Z]. https://agfundernews.com/jupiter-ionics-tenacious-ventures-backed-green-ammonia-startup
46. Lim D-K, Plymill AB, Paik H et al (2020) Solid acid electrochemical cell for the production of hydrogen from ammonia. Joule 4(11):2338–2347
47. Office of the People's Government of Fujian Province (2021) The first domestic "ammonia-hydrogen" green energy industry innovation platform is launched in Fujian
48. Chen C, Wu K, Ren H et al (2021) Ru-based catalysts for ammonia decomposition: a mini-review. Energy Fuel 35(15):11693–11706
49. ATCHISON J. GenCell to roll out its ammonia-fed, off-grid power solution[Z]. (2022) https://ammoniaenergy.org/articles/gencell-to-roll-out-its-ammonia-fed-off-grid-power-solution/
50. Chen D, Li J, Huang H et al (2020) Progress in the study of ammonia combustion and reaction mechanism. Chem Bull 83:508–515
51. Fan S, Wu Y, Li X et al (2021) Catalytic research on methanol-H_2 energy system: progress and challenges. Chem Bull 84:21–30
52. Shih CF, Zhang T, Li J et al (2018) Powering the future with liquid sunshine. Joule 2(10):1925–1949
53. Olah GA (2005) Beyond oil and gas: the methanol economy. Angew Chem Int Ed 44(18):2636–2639
54. Pérez-Fortes M, Schöneberger JC, Boulamanti A et al (2016) Methanol synthesis using captured CO_2 as raw material: techno-economic and environmental assessment. Appl Energy 161:718–732
55. Fasihi M, Efimova O, Breyer C (2019) Techno-economic assessment of CO_2 direct air capture plants. J Clean Prod 224:957–980
56. Graaf GH, Sijtsema PJJM, Stamhuis EJ et al (1986) Chemical equilibria in methanol synthesis. Chem Eng Sci 41(11):2883–2890

57. Shi L, Yang G, Tao K et al (2013) An introduction of CO_2 conversion by dry reforming with methane and new route of low-temperature methanol synthesis. Acc Chem Res 46(8):1838–1847
58. Chen F, Zhang P, Zeng Y et al (2020) Vapor-phase low-temperature methanol synthesis from CO_2-containing syngas via self-catalysis of methanol and Cu/ZnO catalysts prepared by solid-state method. Appl Catal B Environ 279:119382
59. Chen F, Zhang P, Xiao L et al (2021) Structure-performance correlations over Cu/ZnO interface for low-temperature methanol synthesis from syngas containing CO_2. ACS Appl Mater Interfaces 13(7):8191–8205
60. Zhao Z (2021) Development and analysis of coal-to-methanol synthesis technology. Chem Ind Manag 33:111–113
61. Yu KMK, Tong W, West A et al (2012) Non-syngas direct steam reforming of methanol to hydrogen and carbon dioxide at low temperature. Nat Commun 3(1):1230
62. Zhu J, Araya SS, Cui X et al (2020) Modeling and design of a multi-tubular packed-bed reactor for methanol steam reforming over a $Cu/ZnO/Al_2O_3$ catalyst. Energies 13(3)
63. Lin L, Zhou W, Gao R et al (2017) Low-temperature hydrogen production from water and methanol using Pt/α-MoC catalysts. Nature 544(7648):80–83
64. Lin L, Yu Q, Peng M et al (2021) Atomically dispersed Ni/α-MoC catalyst for hydrogen production from methanol/water. J Am Chem Soc 143(1):309–317
65. Sun Z, Zhang X, Li H et al (2020) Chemical looping oxidative steam reforming of methanol: a new pathway for auto-thermal conversion. Appl Catal B Environ 269:118758
66. Pan L, Wang S (2004) Research on methanol autothermal reforming for hydrogen production in plate reactors. J Fuel Chem 32(03):362
67. Yan WRZ, Wang HN, Lu SF et al (2021) Advancement toward reforming methanol high temperature polymer electrolyte membrane fuel cells. Chem Ind Eng Prog 40(6):2980–2992
68. Zhongqing New Energy (2022) Methanol to hydrogen, methanol reforming to hydrogen fuel cell technology gradually enters the public eye
69. Shanghai Bohydro New Energy Technology Co., Ltd, 300W portable product (2022) http://www.palcan.com.cn/product_2.html
70. MoHydrogen Technology Co., Ltd. (2021) A global first! MPT Long-life methanol fuel cell power station delivered to the tower base station, successfully generating electricity
71. Zhongqing New Energy Technology Co., Ltd. (2018) China's first batch of methanol reforming hydrogen fuel cell logistics vehicles officially put into commercial operation
72. Zhongqing New Energy Technology Co., Ltd (2018) China's first batch of methanol reforming hydrogen fuel cell logistics vehicles officially put into commercial operation [Z]. https://m.chinatrucks.org/news/8307.html
73. Raab M, Maier S, Dietrich R-U (2021) Comparative techno-economic assessment of a large-scale hydrogen transport via liquid transport media. Int J Hydrog Energy 46(21):11956–11968
74. Rüde T, Bösmann A, Preuster P et al (2018) Resilience of liquid organic hydrogen carrier based energy-storage systems. Energ Technol 6(3):529–539
75. Ministry of Industry and Information Technology of the People's Republic of China (2014) Treatment and disposal methods for liquid ammonia leakage: HG/T 4686—2014. China Standard Quality Inspection Publishing House, Beijing
76. General Administration of Quality Supervision, Inspection and Quarantine of the People's Republic of China, China National Standardization Management Committee (2018) Proton exchange membrane fuel cell vehicle fuel hydrogen gas: GB/T 37244—2018. China Standards Publishing House, Beijing
77. Okada Y, Yasui M (2015) Large scale H_2 storage and transportation technology. Hyomen Kagaku 36(11):577–582
78. Hydrogenious LOHC Technologies (2022) Projects [Z]. https://hydrogenious.net/worldwide-novelty-hydrogenious-supplies-hydrogen-filling-station-in-erlangen-germany-via-liquid-organic-hydrogen-carriers/
79. Wuhan Hydrogen Energy Co., Ltd (2022) Company history [Z]. https://www.hynertech.com/en/nd.jsp?fromColId=2&id=170#_np=2_690

Chapter 5
Material-Based Solid Hydrogen Storage and Transportation

Many solid-state materials are potential reversible hydrogen storage carriers. From the perspective of hydrogen storage mechanisms, solid-state hydrogen storage materials can be divided into physical adsorption types and chemical hydride types. Physical adsorptive hydrogen storage is a method of hydrogen storage that relies on the interaction of van der Waals forces between the material and hydrogen molecules for adsorption and desorption. During the adsorption and desorption process, hydrogen gas exists in the form of H_2 molecules. Since physical adsorption is usually an exothermic process, and the binding force between hydrogen and the material is weak, physical adsorptive hydrogen storage materials generally have high hydrogen adsorption capacities under low temperature conditions (generally at the boiling point of liquid nitrogen, 77 K). Physical adsorptive hydrogen storage materials include carbon materials, metal-organic framework (MOF) materials, zeolite materials, etc. Chemical hydride materials store hydrogen by forming a stable compound with hydrogen, which can be mainly divided into metal hydrides, complex hydrides, and ammonia boranes and their derivatives.

In this chapter, firstly, we discuss the hydrogen storage properties of metal hydrides, that is, metal elements or alloys can absorb gaseous hydrogen to form binary or multinary metal hydrides to store hydrogen gas; secondly, introduce complex hydrides, where the central atom and the hydrogen atom covalently form anionic coordination groups, which simultaneously coordinate with the metal ions to form complex hydrides; thirdly, introduce the most basic B-N-H compound,

J. Zou, *Hydrogen Storage and Transportation*,
https://doi.org/10.1007/978-981-96-2875-9_5

ammonia borane, and its derivatives; finally, discuss physical adsorptive hydrogen storage materials. In practical applications, based on different application scenarios and operating conditions, different hydrogen storage materials are selected to achieve efficient hydrogen storage.

5.1 Alloys and Metal Hydrides for Hydrogen Storage

Metal hydrides are binary or multinary, formed by metal elements or alloys absorbing gaseous hydrogen. Hydrogen molecules dissociate into hydrogen atoms on the metal surface, and the hydrogen atom diffuses into the bulk phase through the interstitial positions in the material lattice to form metal hydrides. Hydrogen storage alloys are intermetallic compounds composed of metal element A (e.g., La, Ce, Zr, Ti, V, etc.) that easily form stable hydrides, and transition metal B (e.g., Fe, Co, Ni, Cu, Mn, etc.) with weak affinity for hydrogen [1–3]. Among them, component A easily reacts with hydrogen, has a large hydrogen absorption capacity, and the formation of strong-bond hydrides with hydrogen is an exothermic reaction ($\Delta H < 0$); component B has a small affinity for hydrogen, and hydrogen can easily move in it, usually does not form hydrides, and the hydrogenation process is an endothermic reaction ($\Delta H > 0$). The former determines the amount of hydrogen absorption, while the latter controls the reversibility of hydrogen absorption/desorption processes and can adjust the thermodynamics and kinetics of hydrogen absorption/desorption processes of the hydrogen storage alloy. Currently, most common and developed hydrogen storage alloys include rare-earth AB_5-type hydrogen storage alloys, Mg-based hydrogen storage alloys, AB_2-type Laves phase hydrogen storage alloys, AB-type Ti alloys, V-based solid-solution hydrogen storage alloys, and rare-earth magnesium-nickel hydrogen storage alloys. The types and properties of common alloys and their hydrides are shown in Table 5.1. This section mainly introduces the hydrogen storage principles, hydrogen absorption and desorption characteristics, and preparation technology of various metal elements and alloys.

Table 5.1 Main hydrogen storage alloys and their hydride properties

Type	Typical alloys or metals	Hydrides	Hydrogen absorption (wt%)	Enthalpy of hydride formation/(kJ/mol)
AB_5	$LaNi_5$	$LaNi_5H_6$	1.4	−30.1
AB_2	$TiMn_2$	$TiMn_2H_{2.5}$	1.8	−28.5
AB	TiFe	$TiFeH_2$	1.8	−23.0
Mg-based alloy	Mg_2Ni	Mg_2NiH_4	3.6	−64.5
V-based alloy	Ti-V	Ti-V-H_4	2.6	–

5.1.1 Principle of Hydrogen Storage in Metals (Alloys)

Metals (alloys) possess a special atomic structure arrangement, and the lattice gaps can serve as vacant sites for storing hydrogen atoms. Under certain temperature and hydrogen pressure conditions, hydrogen storage materials can "absorb" a large amount of hydrogen gas, that is, react with hydrogen to form metal hydrides, while releasing heat. When heated, the metal hydride will decompose and release the stored hydrogen. The "absorption" and "release" process of hydrogen gas is reversible and can be repeated cyclically, as shown in Fig. 5.1.

Releasing high-purity hydrogen and absorbing heat, Endothermic, Metal (alloy), Cycle, Metal hydride, Exothermic, Absorbing low-purity hydrogen and releasing heat.

The hydrogen absorption reaction of metals can be divided into the following four steps [4, 5]:

1. Under the effect of van der Waals forces, hydrogen gas is firstly adsorbed on the metal surface, and the surface metal atom dissociates H_2 into H atoms.
2. H atoms diffuse from the surface to the interior of the metal, entering the lattice gaps of metals.
3. With the continuous increase of H atom concentration in the bulk phase, an α solid solution starts to form in the metal lattice.
4. As the hydrogen atom concentration continues to increase, after its saturation in the α "solid solution", a β phase change occurs, forming β-phase metal hydrides, and the hydriding reaction is completed.

The above are the main steps for metals (alloys) to absorb hydrogen to form hydrides. Since most hydrogen absorption reactions of metal hydrides are reversible reactions, the dehydrogenation reaction steps are the reverse of the above steps.

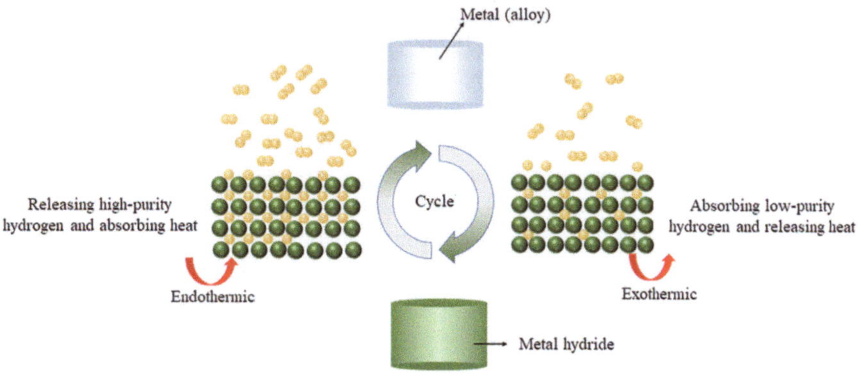

Fig. 5.1 Working principle of hydrogen storage by metals (alloys)

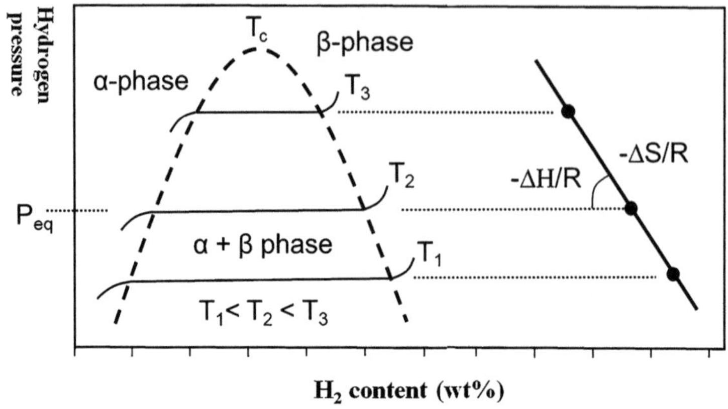

Fig. 5.2 Pressure-composition-temperature (PCT) curves in typical metal hydride hydrogen storage materials (left part), showing the solid solution (α phase), hydride (β phase), and coexisting phase of the two (α + β phase), and the right part is its corresponding van 't Hoff curve

5.1.1.1 Thermodynamic Principles

The hydrogen absorption and release of metal hydride hydrogen storage materials is a dynamic equilibrium process of three phases, hydrogen, metal, and corresponding hydrides. As can be seen from the left part of Fig. 5.2, the hydrogen storage material absorbs very little hydrogen when it forms a solid solution (α phase) at the beginning of hydrogen absorption. Under isothermal conditions, when the hydrogen pressure and hydrogen solubility rise to a specific value, the solid solution (α phase) begins to transform into the hydride (β phase). The β-phase hydride nucleates and grows, and this process does not change the hydrogen pressure as the phase change proceeds, until the α phase completely transforms into the β phase, the hydrogen pressure continues to rise, and the hydriding reaction is completed. However, as the temperature rises, the phase transition platform pressure of the same material is also higher; when the temperature exceeds the critical temperature (T_c), the platform disappears. As such, the hydrogen pressure (P), composition (C), and temperature (T) are the decisive factors affecting the phase equilibrium. Therefore, the curve obtained by calibrating the pressure-composition-temperature (PCT) equilibrium points during the hydrogen absorption and release process of metal hydride hydrogen storage materials, that is, PCT curve, a series of fundamental thermodynamic data, such as the platform pressure, maximum hydrogen absorption and release capacity, and hydrogen absorption and release enthalpy, can be obtained.

Hydrogen pressure, α phase, β phase, α + β phase, H_2 content (wt%).

The relationship between the platform pressure (P) in the PCT curve and the temperature (T) can be described by the van 't Hoff equation:

$$\ln\left(\frac{P}{P_0}\right) = \frac{\Delta H}{RT} - \frac{\Delta S}{R} \tag{5.1}$$

In the formula, P_0 is atmospheric pressure (1.01 × 10^5 Pa); ΔH and ΔS are the enthalpy change and entropy change of the hydrogen absorption and desorption reaction, respectively; T is the absolute temperature of the hydrogen absorption and desorption reaction.

The enthalpy change and entropy change of the hydrogen absorption reaction of most metal hydrides are negative, so the hydrogen absorption process is exothermic, and the hydrogen desorption process is endothermic. The enthalpy change of the hydrogen absorption and desorption reaction (ΔH) is an important indicator of the strength of the M-H bond and also an important criterion for designing the thermal management system of metal hydride hydrogen storage. The larger the absolute value of ΔH, the stronger the binding force of the M-H bond, the more difficult the failure of hydrogen storage systems, and the more difficult the hydrogen release. ΔS represents the trend of the reaction to form metal hydrides. In the same type of hydrogen storage alloys, the larger the ΔS value, the lower the equilibrium partial pressure, and the more stable the formed metal hydrides. Through the isobaric data on the left side of Fig. 5.2, by fitting Ln*P* vs 1000/T, the corresponding slope and intercept values from the linear relationship are used to calculate the ΔH and ΔS data of this hydrogen storage material and systematically characterize its thermodynamic performance.

5.1.1.2 Kinetic Principle

The hydrogen storage performance of materials depends on both the thermodynamics and kinetics of absorption and desorption processes. The kinetic performance focuses on evaluating the reaction rate of hydrogen absorption and desorption, which is mainly determined by the structure of the material itself and the mechanism of hydrogen absorption and desorption under specific conditions. The kinetic performance of the metal hydride hydrogen storage system can be expressed by the Arrhenius relationship:

$$v(t) \propto \exp\left(-E_a / \mathrm{k_B T}\right) \quad (5.2)$$

In the formula, v represents the reaction rate; E_a represents the reaction activation energy; k_B represents the Boltzmann constant (1.381 × 10^{-23} J/K).

Therefore, when the system has a higher reaction temperature and a smaller reaction activation energy, the reaction rate is higher. As such, in order to obtain a higher reaction rate at a lower temperature, it is necessary to reduce the reaction activation energy of the system.

Figure 5.3 is a schematic diagram of the activation energy barrier in the reaction process of the metal hydride hydrogen storage system. Before the metal + H_2 system transitions from the initial state to the final state of the metal hydride with a lower energy state, the reaction process needs to overcome the energy barriers of the decomposition of H_2 on the metal surface, the diffusion of H atoms in the metal, and

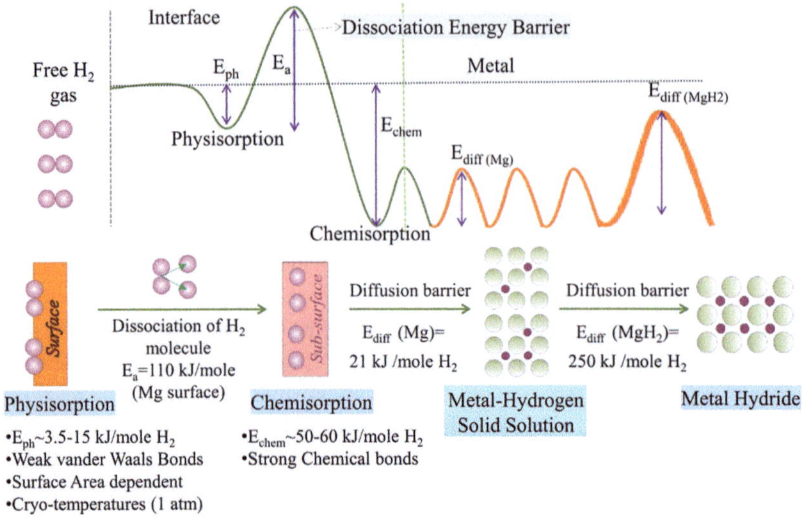

Fig. 5.3 Schematic diagram of the energy barrier change from the initial state to the final state through the excited state in the hydrogen absorption and desorption process of the metal hydride hydrogen storage system

the diffusion of H atoms in the metal hydride, etc., collectively referred as the reaction activation energy (E_a). The reaction activation energy (E_a) as an important parameter of kinetic performance, can directly reflect the difficulty of the phase transition process, thus the higher the activation energy, the more difficult the reaction proceeds.

Interface, Decomposition energy, Within the phase.

Physical adsorption, Chemical adsorption, Metal, Metal hydride.

Physical adsorption, H_2 decomposition, Chemical adsorption, H diffusion in metal, Solid solution state, H diffusion in hydride, Metal hydride.

5.1.2 Synthesis of Hydrogen Storage Metals (Alloys)

The preparation methods of hydrogen storage metals (alloys) can be divided into the following categories: induction melting method, mechanical alloying method, arc plasma evaporation method, and hydriding combustion method, according to different routes and yields.

Induction melting method [6] uses the high-frequency current generated by the high-frequency induction power supply flowing across the induction water-cooled copper coil, where an induction current is generated in the metal furnace charge due to electromagnetic induction. This induction current flows in the metal furnace to generate heat, causing the metal furnace to be heated and melted. At the same time, the melt is stirred by the electromagnetic induction, and the liquid metal rolls along the direction of the paramagnetic force line, causing the melt to be fully mixed and homogeneously melted, resulting in homogeneous metal alloys. Such induction melting methods possess the advantages of simple operation, high production efficiency, stable and facile-control temperature field. It is a melting method suitable for titanium-based, vanadium-based, and magnesium-based hydrogen storage alloys in industrial production, with a melting scale ranging from several kilograms to several tons.

Mechanical alloying method [7–9] is a process of forming an alloy from the metal powder by repeatedly squeezing and deforming different powders through high-energy ball milling, and then fracturing, impacting, cold welding, inter-atomic diffusion, and breaking. The high-energy ball milling method is simple to operate and can alloy immiscible metals, while ball milling can simultaneously introduce the dislocation, stacking fault, vacancy, and other defective sites, thereby accelerating the diffusion and transport of hydrogen. In addition, the metal (alloy) materials synthesized by such high-energy ball milling methods will possess more active phases/interfaces, thereby significantly improving the hydrogen absorption and desorption kinetic performance. The schematic diagram of the planetary ball mill and the high-energy ball milling method is shown in Fig. 5.4. However, the metal alloys prepared by the high-energy ball milling method usually suffer from the problem of uneven particle size, which is not conducive to the regulation of hydrogen absorption and release performance, and is prone to agglomeration in repeated cycles, leading to a decrease in hydrogen storage performance. It is therefore necessary to add auxiliary materials or adjust process parameters to improve the cycle stability of the metal alloys.

Arc plasma evaporation method [10–12] using the arc plasma generated between the metal and the electrode to melt and evaporate the metal under certain atmospheric conditions (Ar, H_2, CH_4, N_2, NH_3, etc.), and the vapor condenses in the atmosphere to generate hydrogen storage material powder. This method is suitable for materials with low melting points and boiling points, and has the advantages of

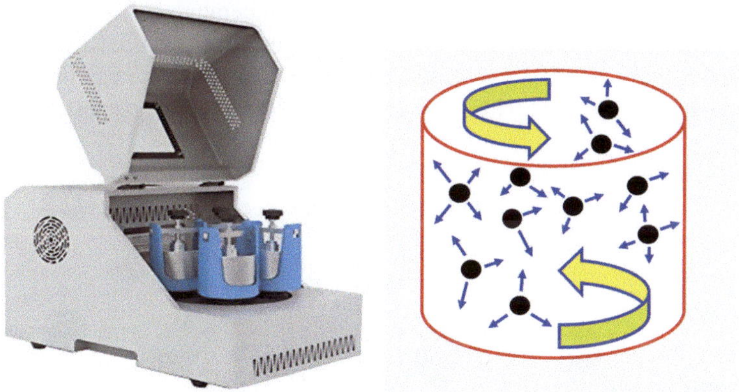

Fig. 5.4 Schematic diagram of planetary ball mill and high-energy ball milling method

high efficiency, high purity, tunable particle size, and achievable gas phase alloying. Meanwhile, adding reactive gases is a way to obtain some special structures or compounds. For example, adding CH_4 to the arc plasma evaporation process of magnesium can lead to the formation of Mg@C nano-composite hydrogen storage material with a core-shell structure, and adding NH_3 to evaporate magnesium provides hollow structure $Mg(NH_2)_2$.

Hydriding combustion is a new method for preparing magnesium-based hydrogen storage alloys [13]. The mixed powder of magnesium and nickel is placed under a high-pressure gaseous hydrogen environment, and the magnesium-based hydride is directly obtained through a one-step hydriding combustion, fully utilizing the heat released by the reaction between Mg, transition metal Ni, and H_2 to promote the completion of the reaction, and the resultant product does not need any activation treatment. At present, this method has been successfully used to prepare Mg_2NiH_4, Mg_2FeH_6, Mg_2CoH_5 and other magnesium-based hydrides, and the synthesized products have the advantages of high purity, high cycle stability, and high reaction activity.

5.1.3 Rare Earth AB_5-Type Hydrogen Storage Alloys

The research on AB_5-type hydrogen storage alloy was started very early, and the A side is composed of a single rare earth element of La, or multinary rare earth elements composed of Ce, Pr or Nd. Meanwhile, the B side is made up of non-hydrogen absorbing metals, such as Ni, Co, Mn, Al, represented by $LaNi_5$ [14, 15]. The crystal structure of the AB_5-type alloys is usually a $CaCu_5$-type crystal, with a space group of P6/mmm. The crystal structure of its typical representative $LaNi_5$ is shown in Fig. 5.5. The hydrogen absorption product of $LaNi_5$ is the hexagonal structured $LaNi_5H_6$, with hydrogen atoms located in the octahedral interstices, and the mass hydrogen storage density is about 1.38 wt%. The hydrogen absorption enthalpy change of $LaNi_5$ is −30.1 kJ/mol H_2, and the hydrogen release platform pressure at 25 °C is about 0.2 MPa, suitable for use at room temperature.

The advantages of $LaNi_5$ alloy are mild hydrogen absorption/release conditions, fast kinetics, easy activation, insensitive to impurities, moderate equilibrium pressure, and small hysteresis. However, due to the 24% volume expansion of $LaNi_5$

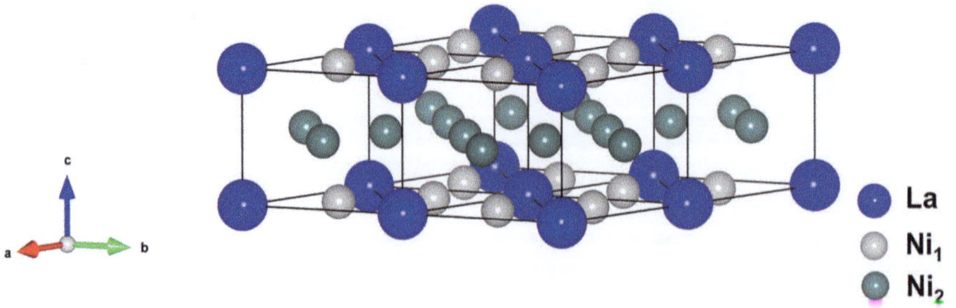

Fig. 5.5 Crystal structure of $LaNi_5$

alloy after hydrogen absorption, the material is prone to pulverizing, leading to the poor cycle performance. In addition, the high cost also limits the practical application of $LaNi_5$ alloy. Currently, the hydrogen storage performance of AB_5-type rare earth alloy can be improved by various methods, such as composition optimization, structure control, and adjustment of stoichiometric ratios [15].

5.1.3.1 Alloy Composition Optimization

Alloy composition optimization is the most widely used method to improve the hydrogen storage performance of AB_5-type alloy, including optimization of the A side and B side.

A-Side Optimization Elements, such as Ce, Pr, Nb, Sm, Gd, with different physical and chemical properties can partially or jointly replace La, forming mixed rare earth $MmNi_5$ hydrogen storage alloys. The presence of a small amount of Ce improves the toughness of the alloy, increasing the alloy's resistance to pulverization and cycling performance. Meantime, because the atomic radius of Ce is smaller than Ni, the increase in Ce content will reduce the cell volume, thus decreases the maximum hydrogen absorption, reducing the alloy volume expansion. Therefore, the pulverization rate of the alloy can be reduced. However, when the cell volume is too small, the distortion of the lattice will accelerate the pulverization process of the alloy. Excessive Ce content will cause Ce segregation, leading to a decrease in the alloy stability. In addition, beside Ce element, Pr and Nd rare earth elements also have similar functions. Research has shown that in La-rich $MmNi_5$ alloy, $MmNi_5$ alloy [w(La + Nd) = 70%] not only retains the excellent characteristics of $LaNi_5$ alloy, but also has better hydrogen storage capacity and kinetic characteristics than those of $LaNi_5$, and is thus of more practical values [14].

B-Side Optimization Partial replacement of Ni with third component elements, such as Al, Cu, Mn, Si, Ca, Zr, V, Co, Ag, can improve the hydrogen storage performance of $LaNi_5$. Among them, Co is the most effective element to reduce volume expansion during hydrogen absorption, improving the resistance to pulverization, and thereby improve the cycle life. Co can reduce the microhardness of the hydrogen storage alloy, and reduce the volume expansion of the alloy after hydrogenation, thus increase the flexibility to enhance the resistance to pulverization of the alloy. During the re-/dehydrogenation cycles, Co forms a protective film on the surface of the hydrogen storage alloy, which can inhibit the dissolution of Al, Mn and other elements on the alloy surface, thereby reducing the corrosion rate of the alloy and improving the cycle life of the alloy. Since the atomic radius of Mn is larger than Ni, after partial replacement of Ni by Mn, the cell volume of the alloy increases, reducing volume expansion. However, addition of excessive amounts of Mn will reduce the cycle life of the alloy. Mn also plays a role in reducing the equilibrium hydrogen pressure of the hydrogen storage alloy and reduces the degree of hysteresis of hydrogen absorption and release. Partial replacement of Ni by Al can reduce the equilibrium hydrogen pressure of the hydrogen storage alloy, but as the replacement amount gradually increases, the hydrogen storage capacity of the alloy will decrease. The most mature and commercialized alloys are $Mm(NiCoMnAl)_5$ (rich in cerium) and $Ml(NiCoMnAl)_5$ (rich in lanthanum) alloys, which have been widely used in domestic and foreign MH-Ni batteries.

5.1.3.2 Microstructure Control

The lattice stress of AB_5-type hydrogen storage alloy is high, and segregation and defects are prone to occur in the alloy. Rapid quenching and annealing processes can evolve the alloy towards nanoscale or amorphous structures, which can better eliminate structural stress and component segregation, thus homogenize the alloy, and improve the cyclic stability of the alloy. By controlling the cooling rate according to the alloy composition, the alloy structure can be improved, which can further enhance the overall hydrogen storage performance of the alloy [16]. In addition, by nanosizing the hydrogen storage alloy, due to its unique surface effects and small size effects, hydrides are more likely to form, which can greatly improve the hydrogen storage performance.

5.1.3.3 Non-stoichiometric Ratio Research

Non-stoichiometric hydrogen storage alloys refer to alloys in which the ratio of A and B components is not in accordance with the stoichiometric ratio, that is, alloys formed by insufficient or excessive A/B ratios. In the La-Ni binary system, when the Ni content is too poor or too rich, segregation will occur beyond this region, producing a second phase. The composition, quantity, morphology, size, and distribution of the second phase often affect the alloy's structure, phase composition, equilibrium hydrogen pressure, and cycle life. When the second phase is evenly distributed in the main phase of the alloy, it will show good performance, such as the cycle life of $LaNi_{5.2}$ alloy is higher than $LaNi_5$, and the anti-powdering property of $LaNi_{4.27}Sn_{0.24}$ is excellent. After 1000 hydrogen absorption/desorption cycles, its hydrogen storage capacity remains unchanged.

5.1.4 Magnesium-Based Hydrogen Storage Alloys

Magnesium-based hydrogen storage materials have a broad application prospect due to their high hydrogen storage capacity (gravimetric hydrogen density, MgH_2 ~ 7.6 wt%, Mg_2NiH_4 ~ 3.6 wt%, Mg_2FeH_6 ~ 5.5 wt%), abundant reserves in the earth's crust, and low cost. Magnesium-based hydrogen storage materials are based on the MgH_2/Mg hydrogen storage system. The chemical reaction formula for the hydrogen absorption and release of the MgH_2/Mg system is as follows:

$$Mg + H_2 \rightleftharpoons MgH_2 \tag{5.3}$$

$$MgH_2 + 2H_2O(l) \rightarrow Mg(OH)_2 + 2H_2 \uparrow \tag{5.4}$$

$$MgH_2 + H_2O(g) \rightarrow MgO + 2H_2 \uparrow \tag{5.5}$$

The most common stable structure of MgH_2 is β-MgH_2. The space group of β-MgH_2 is P42/mnm, each magnesium atom coordinates with six hydrogen atoms to form a deformed octahedron (Fig. 5.6), and the charge distribution is $Mg^{1.91}$ + $H^{0.26-}$ [17]. This unique structure makes the MgH_2 a high structural stability, with an enthalpy change of 75 kJ/mol H_2 for the dehydrogenation reaction, and a decomposition temperature of 280 °C under 100 kPa H_2 pressure [18]. In addition to thermal decomposition to release hydrogen, magnesium hydride can also undergo a hydrolysis reaction to release hydrogen, with a theoretical hydrogen production amount of up to 15.2 wt% (mass fraction).

5.1.4.1 Thermal Decomposition Hydrogen Release Performance and Modification Methods of Magnesium-Based Hydrogen Storage Materials

At present, the magnesium-based hydrogen storage system mainly improves the hydrogen storage performance of the material through forming alloys, adding catalysts, optimizing the structure, and combining with light metal complex hydrides.

1. **Alloying.** The thermodynamic stability of MgH_2 is high, and the decomposition reaction is not easy to occur. One of the methods to improve the thermodynamic performance of MgH_2/Mg is to form alloys with transition elements, rare earth elements, and some main group elements. By forming a thermodynamically more stable alloy phase, the reaction path can be changed, thereby reducing the reaction enthalpy change. Researchers have prepared intermetallic compounds using Mg and alloying elements, such as Ni, Ti, In, Al, Ag, Si, Ga, La, Cd [19–26], forming alloys of specific compositions, as shown in Table 5.2. It can

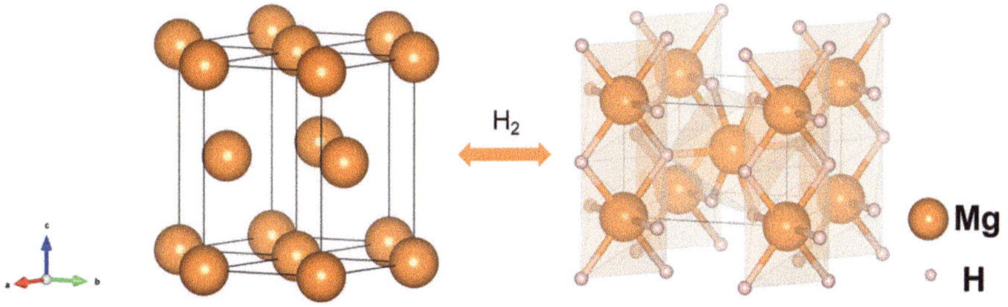

Fig. 5.6 Crystal structure of Mg and β-MgH_2

Table 5.2 Hydrogen storage performance of magnesium-based alloys

Alloy	Hydride formation enthalpy ΔH/(kJ/mol H_2)	Mass hydrogen storage density (wt%)
Mg_2Ni	−64.5	3.6
Mg_3Cd	−65.5	2.8
Mg_2Si	−36.4	5.0
Mg_5Ga_2	−68.7	5.7
Mg_3Ag	−68.2	2.1
$Mg_2Fe(H_6)$	−87	5.5
MgH_2-Ti	−75.2	6.7

be seen that, except for La, Fe, and Ti, the enthalpy change of the hydrogen absorption and release reactions of most alloys is significantly reduced, especially after the addition of Si, the formation enthalpy change of the Mg_2Si hydride is further reduced to 36.4 kJ/mol H_2. Among all magnesium-based alloys, Mg_2Ni is one of the most representative Mg-based hydrogen storage alloys, with a mass hydrogen storage density of 3.6 wt%, and the enthalpy change of the hydrogen absorption and release reaction is ±64.5 kJ/mol H_2 [26]. Mg_2FeH_6 has the highest volumetric hydrogen storage density (150 kg/m^3), and the mass hydrogen storage density is 5.5 wt% [27]. The enthalpy change of Mg_2FeH_6's dehydrogenation is 87 kJ/mol H_2. Since the Mg_2Fe phase is not stable, the formation of Mg_2FeH_6 requires a long process of metal atom migration, and high-purity Mg_2FeH_6 can be prepared using high-energy ball milling and hydriding combustion methods [28, 29].

2. **Catalyst addition.** The addition of catalysts can improve the kinetic performance of magnesium-based hydrogen storage materials. The dissociation energy of hydrogen on the surface of magnesium is high (1.15 eV), and the addition of catalysts can reduce the dissociation energy of hydrogen molecules on the surface of magnesium, improving the kinetic performance of the hydrogen absorption reaction [30]. At the same time, some metals/compounds will in situ form MgM_XH_y or MH_X and other catalytic components, such as Mg_2NiH_4, TiH_2, providing a fast channel for H dissociation/desorption through the "hydrogen pump" effect, thus reducing the activation energy (E_a), and greatly improving the kinetic performance of magnesium-based hydrogen storage materials. In addition, the effect of the catalyst is not only related to its own properties, but also its morphology, particle size, dispersion, and other factors also significantly affect its catalytic activity. The mechanism of improving the kinetic performance of MgH_2 by adding catalysts is shown in Fig. 5.7.

Catalysis, Hydrogen release, Hydrogen absorption, Reduction of activation energy, Catalyst.

Additives that improve the hydrogen storage performance of the MgH_2/Mg system can be classified as metal and intermetallic compounds, metal oxides, other metal compounds, carbon materials, metal-based and carbon-based composite catalysts, etc. Transition metals usually have a positive effect on enhancing the

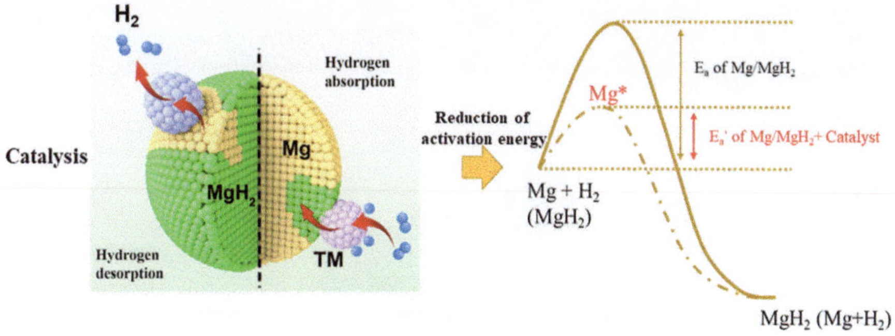

Fig. 5.7 Mechanism of catalyst improving the kinetic performance of magnesium hydrogen absorption and release

hydrogen absorption and release reaction of the MgH_2/Mg system. The decomposition energy of hydrogen molecules on Pd, Cu, Ni, and Fe transition metals is reduced to 0.39, 0.56, 0.06, and 0.03 eV, respectively, which is lower than the 1.15 eV of Mg [30]. Metal based catalysts can cause H_2 molecules to quickly dissociate and undergo hydrogen absorption reactions, which is beneficial to the absorption/release kinetics of the Mg/MgH_2 system. Cui et al. [31] prepared Mg-Tm (Tm = Ti, Nb, V, Co, Mo, Ni) composite materials with a core-shell structure through a direct ball milling reduction method. The in situ generated catalytic components were in close contact with Mg, and the hydrogen release temperature had been significantly reduced. The hydrogen storage kinetic performance of Mg-Tm was ranked as Mg-Ti > Mg-Nb > Mg-Ni > Mg-V > Mg-Co > Mg-Mo, and structural analysis shows that Ti/TiH_2, Nb, Mg_2Ni/Mg_2NiH_4, V, Co, Mo are catalytic phases.

Among other types of additives, MXene is a type of two-dimensional layered materials that effectively improve the hydrogen absorption and release performance of Mg-based materials. Zhu W et al., [32] reported the impact of self-assembled Ni nanoparticle loaded Ti_3C_2 MXene (Ni@Ti-MX) additives on improving the hydrogen storage performance of MgH_2. During the hydrogen absorption and desorption processes, the MgH_2/Ni@Ti-MX composite formed multi-component catalytic phases (Mg_2Ni, TiO_2, metallic Ti, amorphous carbon) and constituted a synergistic catalytic effect, significantly enhancing the adsorption/desorption of H atoms on the Mg surface, as shown in Fig. 5.8. In addition, by introducing carbon materials and metal-based catalysts into the MgH_2/Mg system, the growth/aggregation of Mg/catalyst components can be prevented due to the isolation/wrapping effect of carbon materials. Addition of a small amount of metal-based catalysts combined with carbon materials could significantly enhance the cyclic stability of magnesium-based composite materials [33, 34].

Amorphous carbon, Interface of catalytic phase, Hydrogen pump, Hydrogen channel, Cross-section.

3. **Nanostructure regulation.** Nanosizing of Mg-based materials is an effective method capable of simultaneously improving both thermodynamic and kinetic

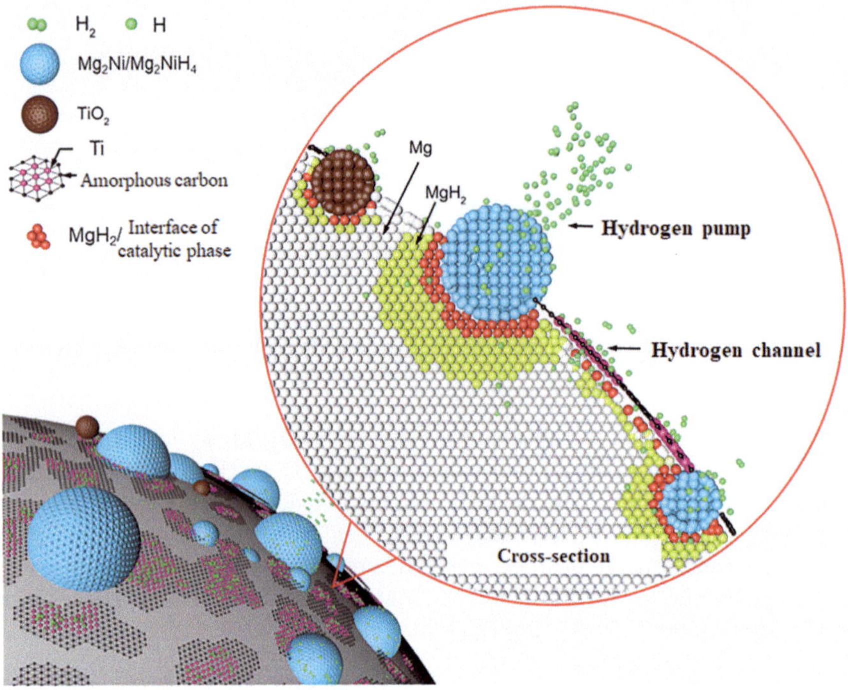

Fig. 5.8 Schematic diagram of the catalytic hydrogen absorption and desorption mechanism of the MgH_2/Ni@Ti-MX system

performance of the MgH_2/Mg system for hydrogen absorption and desorption reactions. When the particles of MgH_2/Mg hydrogen storage materials are reduced to a sufficient extent, most atoms will be exposed on the surface. Consequently, the material will possess a higher specific surface area, a higher interface area, and more active sites, causing thermodynamic instability of MgH_2/Mg and making the hydrogen absorption/desorption reaction more likely to occur. Nanosized MgH_2/Mg usually requires template materials to confine MgH_2 nanoparticles, thereby effectively maintaining the structural stability of Mg-based particles. Confinement materials can be divided into 1D carbon nanotubes, 2D graphene, 3D porous carbon materials, and metal-organic framework (MOF) materials, etc. The synthesis methods of such nanostructures include high-energy ball milling method, chemical reduction method, hydriding method, vapor deposition method, and melting method, etc. The mechanism of nanosizing improving the thermodynamic and kinetic performance of MgH_2 is shown in Fig. 5.9. Nanosizing can simultaneously change the enthalpy change (ΔH) and activation energy (E_a).

Porous carbon, Graphene, Carbon nanotubes.

High-energy ball milling has been widely used for the preparation of nanoscale magnesium-based materials, but it cannot precisely control the particle size dis-

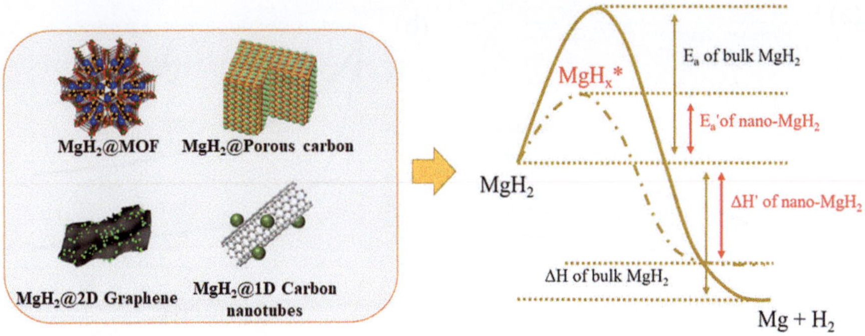

Fig. 5.9 Mechanism of nanosizing in improving the thermodynamic and kinetic performance of MgH_2

tribution within the required narrow range. Chemical reduction method obtains metallic Mg nanocrystals by reducing Mg^{2+} precursors [35, 36]. Hydriding method is way to directly hydriding magnesium salts into MgH_2 under a high H_2 pressure (3–8 MPa), for example, the reaction of dibutylmagnesium with hydrogen gas can be represented as:

$$Mg(C_4H_9)_2(s) + 2H_2(g) \rightarrow MgH_2(s) + 2C_4H_{10}(g) \tag{5.6}$$

The hydrogen pressure of the hydriding reaction is 3–8 MPa, and the reaction temperature is 170–200 °C. Xia et al., [37] synthesized graphene-supported MgH_2 and Ni nanoparticles via the hydriding method under mild reaction conditions (200 °C and 3.5 MPa H_2). The prepared MgH_2 nanoparticles are small (only 2–5 nm) and highly dispersed, and the mass hydrogen storage density after 100 cycles of hydrogen absorption and desorption remains 5.35 wt%. Zhu et al., [38] constructed a three-dimensional folded structure of $Ti_3C_2T_x$(MXene) using the electrostatic interaction between cetyltrimethylammonium bromide (CTAB) and $Ti_3C_2T_x$, and uniformly loaded MgH_2 nanoparticles. It can release about 3.0 wt% H_2 within 2.5 h at 150 °C, with excellent cycle stability, as shown in Fig. 5.10. Arc plasma evaporation method is based on the high temperature generated by the arc to instantly evaporate the metal. Under the hydrogen gas environment, metal atoms undergo a series of processes, such as evaporation, nucleation, growth, and aggregation to generate nano-sized metal particles. It is worth noting that the arc plasma method can prepare Mg nanopowder and nanowires with diameters of 30–170 nm [39].

Hydrogen adsorption, Hydrogen absorption, Commercially available, Time

(a) Isothermal hydrogen absorption curve
Hydrogen adsorption, Time
(b) Isothermal hydrogen desorption curve
Hydrogen desorption, Time, Cycle number

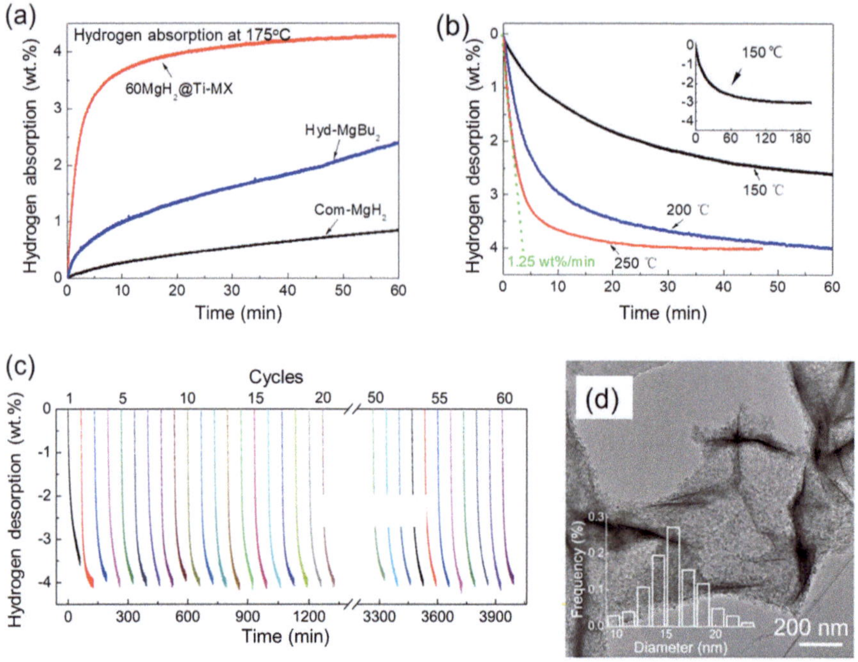

Fig. 5.10 Performances of the MgH_2@Ti-MX hydrogen storage composite material at different temperatures for hydrogen absorption (**a**), desorption (**b**), cycling (**c**) and morphology after cycle tests (**d**)

(c) Hydrogen desorption cycle performance
(d) Morphological characterization

The performance of nanoscale magnesium-based hydrogen storage materials has been significantly improved, and they have wide application prospects in hydrogen storage and transportation, distributed power generation, fuel cell powered vehicles, etc. However, the introduction of template materials could introduce "dead mass" that does not adsorb or desorb hydrogen, which affects the hydrogen storage capacity of magnesium-based materials, and their scale-up production process is yet to be broken through.

4. **MgH_2-based composite hydride system.** The dehydrogenation/hydrogenation reaction pathway of MgH_2 can be changed with the addition of light metal complex hydride composite materials, resulting in a decrease in thermodynamic stability and thus improving the thermodynamic/kinetic performance of magnesium-based hydrogen storage composite materials. At the same time, the hydrogen storage capacity of light metal complex hydrides is high, which can significantly increase the hydrogen storage capacity of the composite system. Among all these composites, MgH_2-$LiBH_4$ composite material is one of the most studied systems in recent years [40] owing to its high theoretical storage capac-

ity. The dehydrogenation reactions of MgH_2-$LiBH_4$ composite materials are carried out at 300 and 350 °C, respectively. During the hydrogenation process, a more stable intermediate phase of MgB_2 is formed, which significantly reduces the hydrogen absorption and desorption enthalpy change to 46 kJ/mol H_2. The formation of these intermediate compounds changes the reaction pathway, thereby significantly improving the thermodynamic performance of the composite material. However, its kinetic performance remains sluggish and needs to be improved by adding catalysts.

5.1.4.2 Hydrolysis Performance of Magnesium-Based Hydrogen Storage Materials

During the hydrolysis hydrogen release process of MgH_2, the product $Mg(OH)_2$ gradually wraps on the surface of MgH_2, thus inhibits the progress of the hydrolysis reaction. The hydrolysis performance of MgH_2 can be improved by changing the hydrolysis environment, adding catalysts, and reducing the particle size to increase the active surface area.

Since the $Mg(OH)_2$ passivation layer is an alkaline precipitate, adding strong acids, such as HCl, H_2SO_4, HNO_3, etc., can help adjust the PH value of the solution. As the PH value decreases, the hydrolysis rate significantly increases [41]. However, adding too much acids will lead to the excessive consumption of acids, reducing the overall economic benefits. Besides, adding salts, such as $FeCl_3$, $TiCl_3$, $MgCl_2$, $ZrCl_4$, etc., through the hydrolysis reaction of metal ions, could help generate $M(OH)_X$ precipitate and H^+, H^+ could help dissolve the surface $Mg(OH)_2$ and thus help enhance the hydrolysis performance of MgH_2 [42]. In addition, the mentioned approaches, such as arc plasma and high-energy ball milling, adding transition metal elements, oxides, halides, etc., can be directly used to modify the surface of MgH_2 [43, 44]. Metal salts can hydrolyze in water to form insoluble hydroxide precipitates and generate H^+ ions, reducing the pH of the hydrolysis environment; adding metals can form primary batteries with Mg substrate in the solution, thus boosting the hydrolysis reaction. Such hydrolysis processes could increase the fresh surface of the MgH_2, thereby continuously promoting the hydrolysis reaction of MgH_2.

A typical fuel cell system powered by the hydrogen released via hydrolysis of magnesium hydrides is shown in Fig. 5.11. The fuel cell system uses the hydrolysis of magnesium hydrides to produce hydrogen, which is then introduced into the fuel cell to generate electricity. It possesses several advantages: high energy density, high safety, and the non-toxic and recyclable attribute of $Mg(OH)_2$ byproduct. The main challenges of developing hydrolysis-based magnesium hydride hydrogen fuel cell system are: the control of the hydrolysis reaction kinetics and device integration, among which the problems such as large water consumption, unstable hydrogen production rate, and the unexpected side reactions. At present, both overseas and domestic researchers adopt material optimization and precision-control of water addition rate to realize stable and controllable hydrogen generation, and use

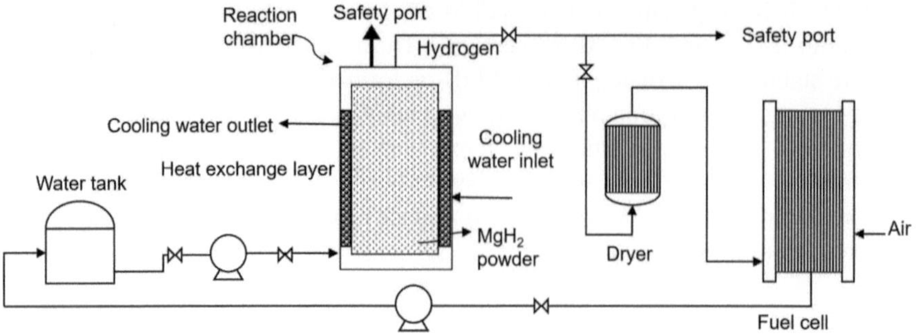

Fig. 5.11 A typical fuel cell system powered by the hydrogen released via hydrolysis of magnesium hydrides

the hydrogen purification system to further improve the purity of hydrogen supply to diversify engineering applicability, making it suitable for kilowatt-level and medium-sized backup power, drones, underwater vehicles, etc. [45].

Water tank, Reaction chamber, Safety port, Hydrogen, Cooling water outlet, Heat exchange layer, Cooling water inlet, Powder, Dryer, Safety port, Fuel cell, Air.

5.1.5 *Other Metal Alloys*

5.1.5.1 Rare Earth-Magnesium-Nickel Hydrogen Storage Alloys

In recent years, a new type of rare earth-magnesium-nickel hydrogen storage alloy has been developed [46]. Because its crystal structure is formed by stacking [A_2B_4] subunits and [AB_5] subunits along the c-axis, its general formula can be expressed as:

$$AB_y = A_2B_4 + x(AB_5)(x = 1,2,3,4) \tag{5.7}$$

Typical alloy superlattice structure types include AB_3, A_2B_7, A_5B_{19}, AB_4, etc., for example:

$$\text{The } AB_3 \text{ structure can be expressed as } A_2B_4 + AB_5 = A_3B_9 = 3(AB_3)(x = 1) \tag{5.8}$$

$$\text{The } A_2B_7 \text{ structure can be expressed as } A_2B_4 + 2(AB_5) = 2(A_2B_7)(x = 2) \tag{5.9}$$

$$\text{The } A_5B_{19} \text{ structure can be expressed as } A_2B_4 + 3(AB_5) = A_5B_{19}\,(x = 3) \tag{5.10}$$

$$\text{The } AB_4 \text{ structure can be expressed as } A_2B_4 + 4(AB_5) = 6(AB_4)(x = 4) \tag{5.11}$$

This type of alloy combines the high-capacity characteristics of AB_2-type alloys and the easy-to-activate advantages of AB_5-type alloys.

The superlattice structure of each type of the rare earth-magnesium-nickel alloy can be divided into two categories: when the [A_2B_4] subunit is a $MgZn_2$ type structure, the formed superlattice structure is a P63/mmc space group structure, that is, 2H-type; and when it is a $MgCu_2$ type structure, the formed superlattice has an R-3m space group structure, that is, 3R-type. Therefore, each type of superlattice alloy exists in 2H-type and 3R-type: AB_3-type alloys can be divided into $CeNi_3$ type (2H-type) and $PuNi_3$ type (3R-type); A_2B_7-type alloys can be divided into Ce_2Ni_7 type (2H-type) and Gd_2Co_7 type (3R-type); A_5B_{19}-type alloys can be divided into Pr_5Co_{19} type (2H-type) and Ce_5Co_{19} type (3R-type). In the preparation process of rare earth-magnesium-nickel superlattice hydrogen storage alloys, due to the differences in thermodynamic stability of different superlattice structures and their selectivity toward valence electrons, different chemical compositions and heat treatment conditions of these alloys often lead to different crystal structures, thus forming multiple phases. Under certain conditions, different phases will coexist and transform into each other, therefore, the prepared alloys usually contain multiple-phase structures. In addition, the superlattice structure is formed through the inclusion reaction, and non-superlattice structures, such as the $CaCu_5$-type AB_5 phase will also appear in the alloy crystal structure.

The study of rare earth-magnesium-nickel superlattice hydrogen storage alloys began with $PuNi_3$-type alloys. In the early stage of the study, Kadir and his collaborators prepared $PuNi_3$-type $REMg_2Ni_9$ (RE is a rare earth element) [47, 48]. The $LaMg_2Ni_9$ alloy only absorbs 0.33 wt% of hydrogen at 30 °C, with a plateau pressure of 2 atm (2.02×10^5 Pa). However, compared with $LaNi_3$ alloy, the $REMg_2Ni_9$ alloy does not turn amorphous after hydrogen absorption, indicating that the structural stability of the alloy has been enhanced, and the reversibility of hydrogen absorption/release process has also been significantly improved. Kohno and others [49] prepared alloys such as La_2MgNi_9, $La_5Mg_2Ni_{23}$, and La_3MgNi_{14}, among which the $La_5Mg_2Ni_{23}$ alloy showed the highest hydrogen absorption capacity up to 1.1 wt%. In order to further increase the hydrogen storage capacity of this type of alloy, researchers adopt metals with smaller atomic masses to partially replace heavier rare earth elements and prepare some superlattice hydrogen storage alloys with a higher hydrogen storage capacity [49]. Among them, a reversible hydrogen storage capacity of 1.48 wt% is obtained for $CaMg_2Ni_9$ alloy at 273 K, while using Y to partially replace Ca and Ca to partially replace Mg results in a reversible hydrogen storage capacity of 2 wt% for $(Y_{0.5}Ca_{0.5})(MgCa)Ni_9$ alloy at 263 K. In addition, the hydrogen absorption capacity of $(La_{0.65}Ca_{0.35})(Mg_{1.32}Ca_{0.68})Ni_9$ alloy is significantly increased, reaching 1.87 wt%, and the alloy has good cycle stability. After 2000 cycles of hydrogen absorption and desorption processes, the hydrogen storage capacity remains at ~1.61 wt%.

5.1.5.2 AB_2-Type Laves-Phase Hydrogen Storage Alloys

There are two main types of AB_2-type Laves-phase hydrogen storage alloys, Ti-based and Zr-based, with a crystal structure being hexagonal C14 (P63/mmc) or cubic C15 (Fd/3m) [50, 51]. American ECD Ovonic company is the first corporation to develop and apply AB_2-type hydrogen storage alloys. The company developed a multicomponent Ti-Zr-V-Ni-Cr-Co-Mn-Al-Sn hydrogen storage alloy composed of the Laves phase, with a cycle life of up to 1000 times, showing great application prospects. AB_2-type alloys have a very wide hydrogen pressure plateau. For example, the hydrogen absorption enthalpy change of ZrV_2 is $\Delta H = -78$ kJ/mol H_2, and its hydrogen absorption plateau pressure at room temperature is as low as 10^{-4} Pa, while $ZrFe_2$ has a hydrogen absorption plateau pressure exceeding 30 MPa ($\Delta H = -21$ kJ/mol H_2) at room temperature [52], and $TiFe_2$ has a hydrogen absorption plateau pressure of about 4000 MPa ($\Delta H = -3.6$ kJ/mol H_2) at room temperature [52]. The modulation of the plateau pressure of AB_2-type hydrogen storage alloys can be achieved by adjusting the Ti/Zr ratio and doping Fe, Mn, Cr, V and other alloy elements, among which a low content of V, Cr and Mn is beneficial to reduce the plateau pressure, whereas Fe is beneficial to increase the plateau pressure. In addition, AB_2-type hydrogen storage alloys are easily poisoned and require surface treatment or element doping to enhance their stability [52]. In terms of structure control, rapid quenching or annealing processes can be used to achieve uniformly structured alloying materials. Rapid quenching by rapidly cooling the alloy melt in a very short time could suppress the element segregation, making the alloy structure uniform and the grains well-defined, thereby improving the alloy properties. During the preparation process, annealing treatment is to put the alloy ingot into a high-temperature furnace, heat the alloy to a certain temperature, and maintain them under the protection of inert gas. The purpose is to homogenize the alloy, thereby reducing the internal stress of the alloy. Such a method is particularly effective for improving the hydrogen absorption capacity and cycle life of the resultant alloys.

5.1.5.3 AB-Type Ti-Based Hydrogen Storage Alloys

TiFe alloy is a typical representative of AB-type hydrogen storage alloys. After absorbing hydrogen, the formed hydrides exert a TiFeH phase (orthorhombic system, a = 0.298 nm, b = 0.455 nm, c = 0.442 nm) and $TiFeH_2$ phase (orthorhombic system, a = 0.704 nm, b = 0.623 nm, c = 0.283 nm). The absorbed hydrogen is located at the center of the body-centered cubic octahedron, surrounded by four Ti atoms and two Fe atoms [53, 54]. After activating the TiFe alloy, it can reversibly absorb and release hydrogen at room temperature, with a theoretical hydrogen storage capacity of 1.86 wt%, an equilibrium pressure at room temperature of 0.3 MPa, and a reversible hydrogen storage capacity of 1.5 wt%. It demonstrates a reversible hydrogen storage characteristic within a practical application range, and the price of raw materials is low and resources are abundant, which shows certain advantages in industrial applications.

TiFe-based hydrogen storage alloy has the following advantages [55]:

1. diverse applications: TiFe-based hydrogen storage alloy can be used as the cathode material for nickel-hydrogen power batteries for hybrid electric vehicles, pure electric vehicles; it can also be made into low-pressure metal hydride hydrogen storage tanks as a hydrogen supply source for hydrogen fuel cells, hydrogen internal combustion engine vehicles;
2. the reversible hydrogen storage capacity of TiFe alloy is higher than that of $LaNi_5$;
3. the TiFe alloy can absorb and release hydrogen at room temperature;
4. the cycle life is up to 2000 times;
5. the cost of raw materials for TiFe-based alloys is lower than that of $LaNi_5$.

The main disadvantages of TiFe alloy are: (i) it is easy to form a dense TiO_2 layer, which makes the corresponding activation difficult, and the alloy activation requires a high temperature of 400 °C and a high pressure of 5 MPa H_2; (ii) it has weak resistance to impurities, easy to be poisoned, and easily affected by H_2O and O_2 and other gas impurities, leading to the fact that the hydrogen storage performance decreases after repeated absorption and release of hydrogen cycles, and the cycle life is thus short [56–58]. At present, element substitution is a common processing technology to improve the hydrogen storage performance of TiFe alloys. Using transition elements, such as Mn, Cr, Zr, and Ni to replace part of Fe in TiFe alloys, one can effectively improve the hydrogen storage performance of the alloy and facilitate its activation.

5.1.5.4 V-Based Solid-Solution Hydrogen Storage Alloy

Solid-solution alloys mainly refer to one or several hydrogen-absorbing metal elements dissolved in another metal to form a solid solution alloy, which is different from intermetallic compounds. Solid solution alloys do not necessarily follow a strict stoichiometric ratio or a ratio close to the stoichiometric one. From the perspective of hydrogen storage, V-based solid-solution alloys exhibit good hydrogen storage performance. V-based solid-solution alloys mainly refer to the solid solution (BCC structure) formed by the alloying between V-Ti and Ni, Cr, Mn [59, 60]. In vanadium-based body-centered cubic structured solid solution alloys, hydrogen atoms can stably exist in the tetrahedral lattice interstices (a coordination number of 4) and octahedral lattice interstices (a coordination number of 6). When absorbing hydrogen, most hydrogen atoms enter into the tetrahedral interstices. Since there are 12 tetrahedral interstices in each unit cell, this provides more interstitial positions for the entry of hydrogen atoms, so the theoretical hydrogen storage capacity is high, and the hydrogen content of VH_2 can reach 3.8 wt% [60]. The temperature of the vanadium hydriding reaction is relatively low, and a large amount of hydrogen can be absorbed at room temperature, while as the temperature rises, the solubility of hydrogen in vanadium decreases. The vanadium hydriding process can be divided into four stages:

1. hydrogen molecules are adsorbed on the alloy surface and dissociate into hydrogen atoms;
2. hydrogen atoms dissolve in the vanadium alloy to form a solid-solution phase, and the surface-adsorbed hydrogen atoms diffuse into the alloy body;
3. hydrogen reacts further with the saturated solid solution phase to form a hydride phase layer;
4. hydrogen atoms diffuse further into the alloy body through the hydride layer to react continuously.

Vanadium absorbs hydrogen to first form the β1 phase (V_2H, the low-temperature phase). As hydrogen absorption proceeds, the β1 phase transforms into the β2 phase (VH, the high-temperature phase). Then, after complete hydrogen absorption, it transforms into the γ phase (VH_2). Therefore, there are two plateaus in the PCT curve of the V-H system. The first plateau corresponds to the coexistence of the α phase (hydrogen solid solution phase) and the β1 phase:

$$2V(\alpha)+1/2H_2 \rightleftharpoons V_2H(\beta1) \tag{5.12}$$

The β1 phase is very stable, and the equilibrium hydrogen pressure corresponding to the first plateau is 0.1 Pa at 353 K. Therefore, it is difficult for the hydrogen release reaction from the β1 phase to occur under medium temperature conditions.

The second plateau corresponds to the coexistence of the β2 phase (VH) and the γ phase:

$$VH(\beta2)+1/2H_2 \rightleftharpoons VH_2(\gamma) \tag{5.13}$$

In summary, at room temperature, the hydride formed by the hydriding of the V-based solid solution alloy is too stable, making the hydrogen release reaction difficult to proceed. In fact, the reversible hydrogen storage capacity of the alloy is about 1.9 wt%.

Introduction of Ti into the V-Ti solid solution alloy can adjust its plateau hydrogen pressure to 0.02–1 MPa at 60 °C, significantly improving the applicability of the V-based solid solution alloy. The addition of 3d transition metals (Cr, Mn, Fe, Co, Ni) can enhance the hydrogen absorption and desorption kinetics and cycle stability of the V-Ti solid solution alloy [59–61]. Among them, Mn and Co enhance the surface activity of the alloy, Cr, Al, and Fe enhance the corrosion resistance of the alloy, and Ni enhances the kinetic activity. The effective hydrogen storage capacity of the $Ti_{43.5}V_{49.0}Fe_{7.5}$ alloy is 2.4 wt%, while that of $Ti_{0.32}Cr_{0.43}V_{0.25}$ alloy is 2.3 wt%, and this alloy has good cycle stability. After 1000 hydrogen absorption and desorption cycles, the reversible hydrogen storage capacity remains around 2 wt%. Currently, a V-Ti-Cr-Fe quaternary alloy system [62] has been developed by Hope Clean Energy (Group) Co., Ltd. (China) in cooperation with Sichuan University, and its vanadium content can vary between 20–60 wt%. These types of alloys do not require additional activation treatment and can directly absorb and release hydrogen at room temperature. The hydrogen absorption of the V-Ti-Cr-Fe alloy generally exceeds 3.6 wt% within 6 min at 298 K; when the cut-off pressure is 0.01 MPa, some alloys could release more than 2.5 wt% of hydrogen at 298 K, which can be used for proton exchange membrane (PEM) fuel cell-driven portable power supplies, uninterruptible power supplies, fuel cell powered bicycles and tricycles.

5.2 Complex Hydrides for Hydrogen Storage

Complex hydrides, including metal aluminohydrides, boron hydrides, and amides, have already been introduced in Chap. 1. These hydrides have a high theoretical mass hydrogen storage density and can serve as good hydrogen storage materials. Table 5.3 lists their theoretical hydrogen storage capacities. Pure metal complex hydrides undergo a series of decomposition reactions to release hydrogen, but due to their good thermodynamic stability and poor kinetic performance, the temperature for reversible hydrogen absorption and release is high, and the reaction rate is slow. This section introduces the hydrogen storage performance of complex hydrides and their modification methods.

5.2.1 Principle of Hydrogen Storage in Light Metal Borohydrides

The general formula for metal borohydrides is $M(BH_4)_n$ (n is the valence state of metal), where hydrogen and boron are connected by covalent bonds to form $[BH_4]^-$, and the negative charge is compensated by metal cations, such as Li, Na, K, Be, Mg, Ca, etc. Because $LiBH_4$, $Mg(BH_4)_2$, and $NaBH_4$ have high mass hydrogen storage

Table 5.3 Hydrogen storage performance of typical complex hydrides

Material	Density/ (g/cm^3)	Mass hydrogen storage density (wt%)	Volumetric hydrogen storage density/(kg/m^3)	Hydride formation enthalpy/(kJ/mol H_2)
$LiBH_4$	0.66	18.36	122.5	−194
$Mg(BH_4)_2$	0.989	14.82	146.5	
$NaBH_4$	1.07	10.57	113.1	−191
$LiAlH_4$	0.917	10.50		−119
$NaAlH_4$	1.28	7.41		−113
$Mg(AlH_4)_2$		9.27	72.3	
$LiNH_2$	1.18	8.78	103.6	−179.6
$Mg(NH_2)_2$	1.39	7.15	99.4	

densities, they have wide application prospects, and this section will focus on these three borohydrides.

5.2.1.1 $LiBH_4$ and Its Composite System

$LiBH_4$ is in the form of a white salt-like powder at room temperature, with a density of 0.66–0.68 g/cm^3, a melting point of 275–278 °C, and a high mass hydrogen storage density of up to 18.36 wt% [63]. Commercially available $LiBH_4$ is synthesized through the exchange reaction between $NaBH_4$ and halogenated lithium (LiCl, LiBr) in diethyl ether or isopropylamine solution [64], and its reaction formula is:

$$NaBH_4 + LiX \longrightarrow LiBH_4 + NaX(X = Cl, Br) \tag{5.14}$$

The structure of $LiBH_4$ at room temperature belongs to the orthorhombic system, with a space group of Pnma [65, 66], and lattice constants of a = 7.18 Å, b = 4.43 Å, c = 6.80 Å. Each $[BH_4]^-$ anion group is surrounded by 4 lithium ions (Li^+) in a tetrahedral form, and each Li^+ is also surrounded by 4 $[BH_4]^-$ groups in a tetrahedral form. At around 105 °C, $LiBH_4$ undergoes a crystal transition from orthorhombic to hexagonal (P63mc, a = 4.28 Å, c = 6.95 Å) one, accompanied by a small amount of hydrogen release, the structures of which are shown in Fig. 5.12.

The decomposition of $LiBH_4$ to release hydrogen involves multiple steps. The diagram of the enthalpy change between different phases during the decomposition of $LiBH_4$ is shown in Fig. 5.13. The phase transition of $LiBH_4$ at 118 °C is endothermic with an enthalpy change of 4.18 kJ/mol, transitioning to a hexagonal crystal structure, releasing about 0.3 wt% H_2 at 100–200 °C; the hexagonal system further melts at 277 °C, accompanied by a latent heat of 7.56 kJ/mol and 0.5–1.0 wt% H_2 release. The main hydrogen release starts at about 400 °C, subsequently decomposing into LiH and B [67]. The decomposition temperature of LiH is above 700 °C, with $\Delta H = -90.7$ kJ/mol H_2, being very stable so that the hydrogen release process is generally stopped at the product of LiH. The hydrogen release process of $LiBH_4$

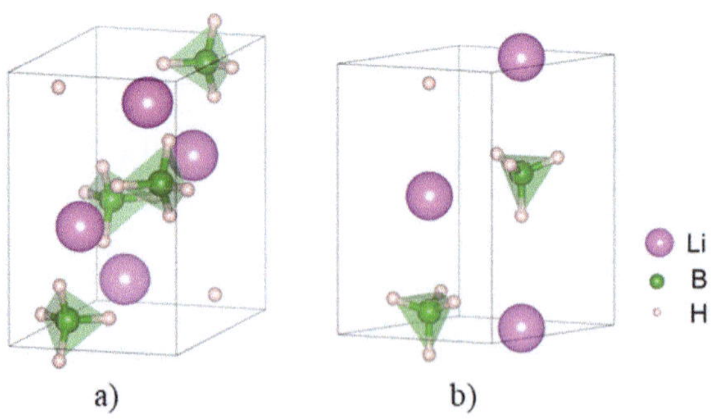

Fig. 5.12 Crystal structures of $LiBH_4$ at low and high temperatures. (**a**) Low temperature. (**b**) High temperature

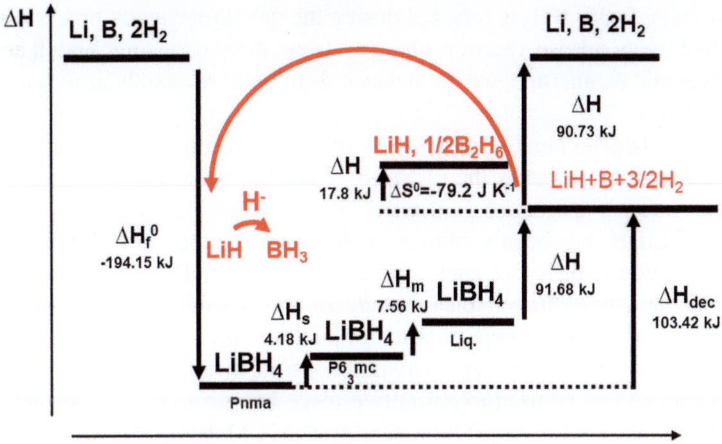

Fig. 5.13 A diagram showing the enthalpy change between different phases generated during the decomposition of $LiBH_4$

is reversible, but due to the high energy barrier required to break the B-B bond in the product B during the hydrogen absorption process, the decomposition products LiH and B can only be regenerated into $LiBH_4$ under high temperature and high pressure conditions (600 °C and 35 MPa hydrogen pressure for 12 h or 727 °C and 15 MPa hydrogen pressure for 10 h) [68, 69]. In view of these, reducing the hydrogen release/absorption reaction temperature and increasing the hydrogen release/absorption reaction rate are the main research directions for the industrialization of $LiBH_4$-based hydrogen storage energy materials.

$$LiBH_4 \longrightarrow LiH + B + 3/2H_2 \tag{5.15}$$

The ways to improve the hydrogen storage performance of $LiBH_4$ are as follows:

1. Thermodynamic control using element substitution and composite hydrides;
2. Catalytic doping or in-situ generated active substances, providing sufficient active reaction points to improve reaction kinetics;
3. Nanostructuring by reducing the particle size, optimizing material structure, and nano-confinement, to simultaneously improve the thermodynamic and kinetic performances of the system.

The $LiBH_4$ can be destabilized by adding another reactants, causing it to react with the additive during the reaction to change the reaction pathway, resulting in thermodynamically more stable products and thus modifying the thermodynamic performance of $LiBH_4$. Pinker-Ton et al., [70] first prepared the $LiBH_4/2LiNH_2$ hydrogen storage material of $Li_3BN_2H_8$ by compounding $LiNH_2$ with $LiBH_4$ at a molar ratio of 2:1. This system has a hydrogen storage capacity of 11.9 wt%, which melts at 190 °C, and starts to release hydrogen at 250 °C, where the hydrogen release can reach 10 wt% and the hydrogen release product is Li_3BN_2. However, at the same time, 2–3% NH_3 is released during the dehydrogenation process, and the $LiBH_4$ hydrogen release reaction changes from thermodynamic endothermic to thermodynamic exothermic, which makes it difficult to reversibly absorb hydrogen after desorption.

The use of MgH_2 can also improve the hydrogen absorption and release performance of $LiBH_4$ [71], that is, the Li-Mg-B-H system, which possesses a maximum hydrogen storage capacity of 11.4 wt%. After adding MgH_2, the decomposition enthalpy of $LiBH_4$ has been significantly decreased to about 46 kJ/mol H_2, lower than the reaction enthalpy of each single-phase system. This is because MgB_2 is more stable than the hydrogen release products, B or Mg of the single-phase system, and this system has good cycle stability. In addition, the stability of metal borohydrides is related to the electronegativity of the central metal atoms, and high electronegative cations can be used to partially replace Li ions to reduce the stability of $LiBH_4$. Currently, cation substitution can form $Li_{1-X}M_XBH_4$ (M = Mg, Ca, Cu, Sc, Mn, Al, Zn) and other systems, which will significantly reduce the hydrogen release enthalpy change and temperature of the system [72].

Catalyst doping is an important technical means to improve the hydrogen absorption and release kinetic performance of $LiBH_4$, and typical additives include oxides, halides, metal elements, and carbon-based materials. Züttel and others [63] first discovered the performance improvement effect of SiO_2 on $LiBH_4$, and $LiBH_4$ with 25 wt% SiO_2 can start to release hydrogen at 200 °C, with a final hydrogen release amount of 9 wt%. In addition to ordinary oxides, oxides with porous structures will confine $LiBH_4$, making $LiBH_4$ to achieve better hydrogen absorption and release performance under the joint effect of catalytic active substances and porous structures. For example, by adding porous TiO_2 microrods to form $LiBH_4$@3TiO_2 (mass ratio of 1:3) [73], the initial hydrogen release temperature can be reduced to 100 °C, and the apparent activation energy is also reduced to 121.9 kJ/mol H_2. As such, the kinetic performance has been improved, and due to the appearance of Li/Ti/O ternary compounds in the product, the hydrogen release pathway of $LiBH_4$ changes, and the thermodynamic performance changes.

Controlling the particle size of $LiBH_4$ at the nanometer level can significantly improve the hydrogen absorption and desorption performance. The methods to reduce the particle size of $LiBH_4$ mainly include solvent evaporation method, melting method, and in-situ reaction method [74–78]. Solvent evaporation method can be used to achieve nano $LiBH_4$ by dissolving $LiBH_4$ in organic solvents, such as tetrahydrofuran, and then evaporating the solvent. Direct melting method is based on melting $LiBH_4$ at 277 °C to confine it in a substrate material, and the mobility of all products in the hydrogen absorption and release cycle process is thus suppressed, maintaining good contact between the products, and there will be no obvious enrichment of elements. Therefore, better cycle stability can be achieved. Carbon materials with porous structures and high specific surface areas, and porous TiO_2 or two-dimensional layered Ti_2C_3 with catalytic effects are usually adopted as the substrate material. Zhang et al., [79] used n-butyllithium and triethylamine borane as reaction precursors, graphene as the substrate, and under certain hydrogen pressure and temperature conditions, they in-situ generated graphene-supported Ni nanoparticle catalyzed ultrafine $LiBH_4$ based hydrogen storage composite material ($LiBH_4$/Ni@G). The content of $LiBH_4$ could reach as high as 70%, and its morphology and hydrogen storage performance are shown in Fig. 5.14. The $LiBH_4$/Ni@G composite material started to release hydrogen when heated to 130 °C. Isothermal absorption and desorption results revealed that $LiBH_4$/Ni@G could release 9.2 wt% hydrogen after maintaining at 300 °C for 150 min and it could completely absorb hydrogen after maintaining at 300 °C and 10 MPa H_2 for 200 min. Hydrogen absorption and release cycle tests at 300 °C showed that after 100 cycles the hydrogen release capacity remained close to 8.5 wt%, and the capacity retention rate was as high as 92.4%, which was the best cycle stability performance in current borohydride research.

Temperature, Capacity, Cycle times.

5.2.1.2 $Mg(BH_4)_2$ and Its Composite System

The mass hydrogen storage density of $Mg(BH_4)_2$ is as high as 14.8 wt%, and the electronegativity of Mg element is higher ($\chi_p = 1.31$) than that of Li ($\chi_p = 0.98$) and Na ($\chi_p = 0.93$), making it possess a lower hydrogen release temperature. As a result, $Mg(BH_4)_2$ can release hydrogen at 300 °C [80]. However, the hydrogen absorption

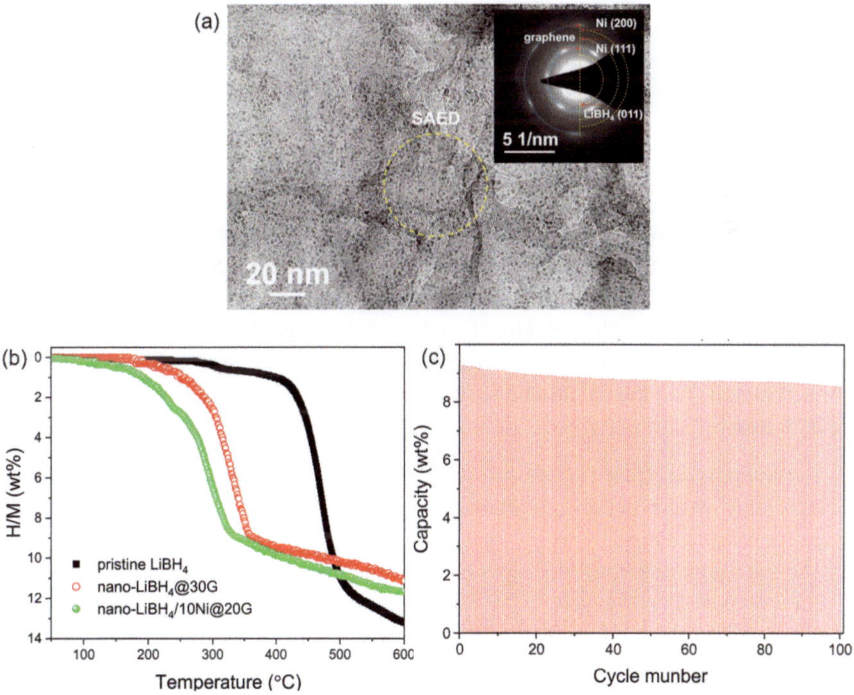

Fig. 5.14 Morphology, hydrogen storage performance, and cycle stability of $LiBH_4$/Ni@G sample [89]

and release temperature of $Mg(BH_4)_2$ is still high, and the hydrogen absorption and release rate is low. Indeed, the reaction kinetic and thermodynamic performances are still poor, which seriously affects its practical application in the field of hydrogen energy.

$Mg(BH_4)_2$ is usually synthesized using wet chemical methods. In 1957, Köster first discovered that MgH_2 could react with triethylamine borane complex at 100 °C to generate a compound composed of $Mg(BH_4)_2$ and triethylamine, then after removing the triethylamine at 100–170 °C under vacuum, $Mg(BH_4)_2$ was obtained, with a yield as high as 95%. Another commonly used method for synthesizing $Mg(BH_4)_2$ is to use $MgCl_2$ and $LiBH_4$ or $NaBH_4$ to perform ion exchange reactions in dimethyl ether solution to prepare $Mg(BH_4)_2$:

$$MgCl_2 + 2MeBH_4 = Mg(BH_4)_2 + 2MeCl(Me = Li,Na) \tag{5.16}$$

Because the LiCl or NaCl produced by this method is insoluble in dimethyl ether while $Mg(BH_4)_2$ can be dissolved, it is thus possible to separate the product and by-product at the same time to obtain a compound composed of $Mg(BH_4)_2$ with coordinated dimethyl ether, while dimethyl ether can be effectively removed under vacuum conditions above 200 °C. Because the production cost of $MgCl_2$ and $NaBH_4$ is lower, this ion exchange method to prepare $Mg(BH_4)_2$ is widely used.

The thermal hydrogen release reaction of $Mg(BH_4)_2$ is:

$$6Mg(BH_4)_2 \rightleftharpoons 5MgH_2 + Mg(B_{12}H_{12}) + 13H_2 \tag{5.17}$$

$$5MgH_2 \rightleftharpoons 5Mg + 5H_2 \tag{5.18}$$

$$5Mg + Mg(B_{12}H_{12}) \rightleftharpoons 6MgB_2 + 6H_2 \tag{5.19}$$

The main crystal configurations of $Mg(BH_4)_2$ at room temperature are α, β, and γ phases. According to the density functional theory (DFT) calculations [81], the reaction enthalpy change can be determined to be −38–54 kJ/mol H_2 for these phases, indicating that the release of hydrogen by $Mg(BH_4)_2$ at a temperature of 20–75 °C is feasible. However, the actual hydrogen release temperature of $Mg(BH_4)_2$ is much higher than the calculated temperature. Main reasons for this phenomenon are as follows.

1. The hydrogen release process of $Mg(BH_4)_2$ requires the breaking of B-H covalent bonds, which has a high kinetic barrier, and the B-H bond cannot be broken at low temperatures.
2. The hydrogen desorption process of $Mg(BH_4)_2$ is complex, with many intermediate phases. For example, γ-$Mg(BH_4)_2$ undergoes a transition process from γ phase to unstable ε phase and to β phase at 150–200 °C, while there exist multiple reactions in the hydrogen desorption process, making the hydrogen desorption difficult.

The DSC spectrum of γ-$Mg(BH_4)_2$ is shown in Fig. 5.15. As shown, the reversible hydrogen storage performance of $Mg(BH_4)_2$ is also poor. The hydrogen release product of MgB_2 re-absorbs hydrogen to generate $Mg(BH_4)_2$, which requires a high temperature of 400 °C and a super high hydrogen pressure of 95 MPa H_2.

γ phase, ε phase, β phase, Temperature.

The improvement of the hydrogen absorption and release performance of $Mg(BH_4)_2$ should focus on the following two aspects. First, to destabilize the B-H bond to reduce the kinetic barrier of the hydrogen release process of $Mg(BH_4)_2$. It is necessary to change the hydrogen release pathway of $Mg(BH_4)_2$, and reduce the production of complex intermediate products, thereby improving its reversible hydrogen storage performance. Combining with other hydrides to change the hydrogen release pathway is an important way to improve the hydrogen storage performance of $Mg(BH_4)_2$. For example, by adding LiH, $Mg(BH_4)_2$ can start to release hydrogen from around 150 °C. The interaction between Li^+ and Mg^{2+} changes the pathway of hydrogen circulation in the system, and thus stabilizes the generation of $[B_3H_8]^-$ groups during the cycle, which serves as the main reversible hydrogen storage phase [82]. Such a $Mg(BH_4)_2$-xLiH (x = 0–0.8) system can release more than 10 wt% of hydrogen at 250 °C, and maintain a cycle capacity of 3.6 wt% at 180 °C for 20 cycles without significant decay. Besides, by adding $NaAlH_4$, the $Mg(BH_4)_2$ system can start releasing hydrogen from 101 °C, with a hydrogen release amount reaching 9.1 wt% [83]. This system can reversibly absorb 6.5 wt% hydrogen at 400 °C, and the re-absorbed products are $NaBH_4$, MgH_2, and Al. At the same time, the destabilization modification method of positive-negative hydrogen coupling reaction can effectively reduce the hydrogen release temperature of $Mg(BH_4)_2$ and accelerate its hydrogen release kinetics, but the products after hydrogen release are irreversible. For example, the $Mg(BH_4)_2$ and $LiNH_2$ composite with a molar ratio of

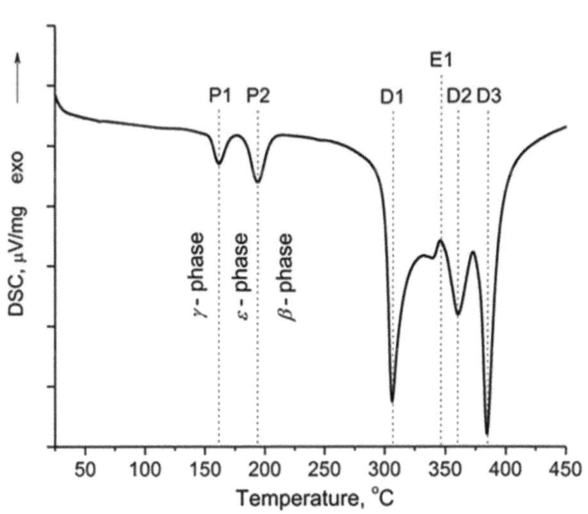

Fig. 5.15 DSC profile of γ-$Mg(BH_4)_2$

1:1 [84], can start releasing high-purity hydrogen from 160 °C, and 7.2 wt% of hydrogen can be released at 300 °C. The hydrogen release products are Li-Mg alloy and amorphous B-N compound. The additives to improve the $Mg(BH_4)_2$ hydrogen storage system mainly include transition metal elements, transition metal oxides (chlorides/fluorides), and carbon-based materials. Transition metal atoms can strongly attract a hydrogen atom in the $[BH_4]^-$ group, thereby destroying its structural stability and reducing the hydrogen dissociation energy. Among various metal doping, Ti atoms can easily replace Mg atoms in the $Mg(BH_4)_2$ structure, and the doping of various metal atoms can all help reduce the bond energy of B-H bonds and facilitate the diffusion and escape of hydrogen atoms [85].

In addition, nanosizing is also an important method to modify the hydrogen storage performance of $Mg(BH_4)_2$. At present, nanosizing modification of $Mg(BH_4)_2$ mainly include melting method, solvent evaporation method, and in-situ synthesis method. Solvent evaporation method is to dissolve $Mg(BH_4)_2$ in an organic solvent, mix the solution containing $Mg(BH_4)_2$ with confined materials, and then being heated or treated under vacuum to evaporate the solvent and precipitate the solute to obtain nanosized $Mg(BH_4)_2$. Compared with the melting method, this solution method has a lower operating temperature and does not introduce other borohydrides. Common solvents for dissolving $Mg(BH_4)_2$ include tetrahydrofuran and dimethyl ether. For example, tetrahydrofuran has been used as a solvent to dissolve $Mg(BH_4)_2$ and confine it to porous Cu_2S hollow nanospheres through this solution method [86]. The obtained $Mg(BH_4)_2$@Cu_2S confined hydrogen storage material can reduce the initial hydrogen release temperature of $Mg(BH_4)_2$ to about 50 °C, and improve the hydrogen release kinetics of the system to completely release hydrogen within 300 °C. Hydrogen absorption tests show that under the conditions of 300 °C and 6 MPa hydrogen pressure, this system can reabsorb 0.5 wt% of hydrogen, showing partial reversibility. However, this nanosizing method also shows distinct disadvantages, such as low effectiveness loading, low actual hydrogen release capacity, difficult removal of solvent residues, and complex operation process.

5.2.1.3 $NaBH_4$ and Its Composite System

$NaBH_4$ is a white crystalline powder that decomposes easily when exposed to moisture. $NaBH_4$ belongs to the orthorhombic system and has a regular tetrahedral crystal structure. The Na^+ and $[BH_4]^-$ groups are bonded by ionic bonds, with four Na^+ surrounding each $[BH_4]^-$ group, and each Na^+ is surrounded by four $[BH_4]^-$ groups. All in all, $NaBH_4$ is a typical ionic-covalent compound and has relatively strong chemical bonding so that it has high stability.

1. Thermal decomposition of $NaBH_4$ to release hydrogen

The mass hydrogen storage density of $NaBH_4$ is 10.57 wt%, and the thermal decomposition temperature to release hydrogen is >500 °C. The hydrogen release reaction formula is:

$$NaBH_4 = Na + B + 2H_2 \tag{5.20}$$

$NaBH_4$ has strong stability, but it also suffers from serious application bottleneck problems, such as severe heat absorption and release, poor kinetic performance, etc. Nanoscale confinement, catalyst doping, and cation substitution are common methods to improve the hydrogen release performance of $NaBH_4$.

For $NaBH_4$, an effective destabilizer is rare-earth metal fluorides (LnF_3) [87, 88]. Rare earth metal fluorides not only reduce the hydrogen release temperature of $NaBH_4$ and accelerate the kinetics of hydrogen release, but also play a significant role in improving hydrogen absorption reversibility. For example, PrF_3 as a destabilizer can reduce the peak hydrogen release temperature of $NaBH_4$ to 419 °C, while NdF_3 can reduce it to 413 °C, which is about 104 °C lower than that of pure $NaBH_4$ (517 °C) under the same condition. Compared with pure $NaBH_4$, the hydrogen absorption reversibility is poor, and the product after complete hydrogen release of $3NaBH_4$-NdF_3 can reach a hydrogen absorption capacity of 3.27 wt% within 1.9 h under relatively mild conditions of 360 °C and an initial hydrogen pressure of 3.2 MPa. $3NaBH_4$-GdF_3 maintains constant hydrogen absorption kinetics after 4 cycles at 400 °C and 4 MPa hydrogen pressure, indicating its high cycle stability.

Regarding the nanoscale confinement of $NaBH_4$, it can be encapsulated in graphene at the nanoscale through a simple wet chemical method [89]. This composite exhibits a unique morphology, in which Na borohydride nanoparticles are completely wrapped by a single layer of graphene nanosheets. Its hydrogen release process is a one-step release, with a broad endothermic peak at 400 °C, significantly lower than the 534 °C of pure $NaBH_4$, and it can achieve a reversible hydrogen storage capacity of 7 wt% under 350 °C and 4 MPa H_2. A series of core-shell nanostructures <30 nm $NaBH_4$@Ni/Co/Cu/Fe/SN, etc. [90] were obtained by anti-solvent precipitation of $NaBH_4$ particles in transition metal precursor solutions. Due to the catalytic and destabilizing effects of Ni nanoparticles dispersed in the $NaBH_4$ phase, the core-shell structured $NaBH_4$@Ni starts to release H_2 at temperature as low as 50 °C, and can maintain a stable structure at high temperatures, demonstrating excellent cycle stability.

2. **Hydrolysis of $NaBH_4$ to release hydrogen**

In addition to thermal decomposition to release hydrogen, $NaBH_4$ is also a commonly used material for hydrolysis-facilitated hydrogen release. $NaBH_4$ can directly react with water without a catalyst at room temperature and neutral conditions to produce hydrogen gas and sodium metaborate. It is generally believed that the hydrolysis reaction rate of $NaBH_4$ is related to its concentration, temperature, pH value, and catalyst, and the rate of hydrolysis hydrogen release reaction can be adjusted by optimizing these parameters.

$$NaBH_4 + 2H_2O = NaBO_2 + 4H_2 \; \Delta H = -217\,\text{kJ/mol} \tag{5.21}$$

The product of the reaction between the sodium borohydride with water, $NaBO_2$, can be re-generated to sodium borohydride by the principle of electrochemical reduction, which can reduce the cost of the raw material $NaBH_4$ and achieve resource recycling.

$$\text{Cathode reaction}: BO_{2^-} + 6H_2O + 8e^- = BH_{4^-} + 8OH^- \tag{5.22}$$

$$\text{Anode reaction}: 2H_2O = O_2 + 4H^+ + 4e^- \tag{5.23}$$

Daimler-Chrysler launched the Millennium Cell in 2000, and has developed a hydrogen fuel cell powered car that uses a hydrogen production device by $NaBH_4$ hydrolysis, with a mileage range of up to 480 km.

The main advantages of producing hydrogen through sodium borohydride hydrolysis are:

(a) High hydrogen storage efficiency: in practical applications, $NaBH_4$ needs to be stored in an alkaline solution; for example, with a 35 wt% $NaBH_4$ solution, the theoretical hydrogen storage density is 7.4 wt%.
(b) Easy control of reaction kinetics: at room temperature, $NaBH_4$ can be stored in a strong alkaline solution for a long time, and only releases hydrogen when it comes to contact with a suitable catalyst; by controlling the amount of $NaBH_4$ solution flowing over the catalyst bed or the amount of catalyst (surface area) in contact with the $NaBH_4$ solution, the amount and speed of hydrogen production can be controlled.
(c) High safety: $NaBH_4$ solution can be stored and transported in conventional plastic containers, with a high safety factor; as a vehicle fuel, $NaBH_4$ solution is non-flammable, making it much safer than gasoline when loaded onto a vehicle.

The main industrial applications of hydrolysis-based sodium borohydride as a hydrogen source are to solve the problems related to the structure of hydrogen generators, hydrogen purification, and compatibility with fuel cells.

(a) $NaBH_4$ hydrolysis generator. $NaBH_4$ hydrolysis generator mainly considers the effective use of catalysts, the collection of by-product sodium metaborate, and material volume expansion. Figure 5.16 shows a scheme of single-chamber $NaBH_4$ fuel-waste integrated generator. The single-chamber structure is initially filled with $NaBH_4$ solution, which is pumped into the single-tube reactor from the bottom of the storage tank by the inlet pump. The hydrogen and $NaBO_2$ solution produced by the reaction, along with any residual material that may accompany it, after passing through the porous filter material, return to the fuel tank. The solution remains in the fuel tank, while the hydrogen is led out from the hydrogen outlet at the top. As the

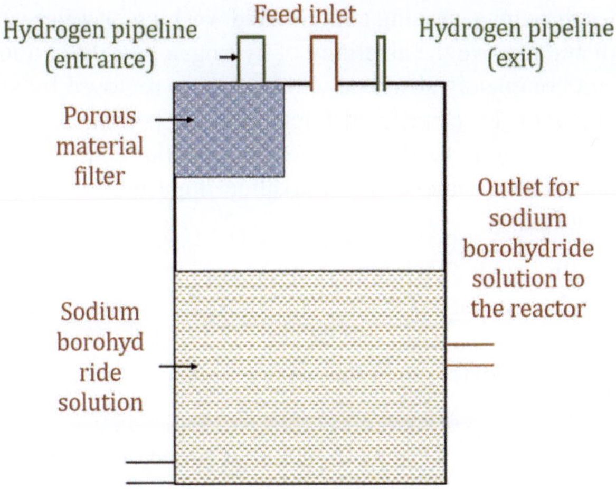

Fig. 5.16 A scheme of single-chamber sodium borohydride aqueous solution fuel-waste integrated generator

reaction consumes water, the volume of the product solution is smaller than the volume of the consumed solution, and the remaining space in the fuel tank can accommodate the hydrolysis products. The advantage of this hydrolysis reactor is that it solves the problem of large volume of the dual-chamber structure of the fuel chamber and waste chamber, which leads to a decrease in volumetric energy density. The disadvantage is that the fuel concentration in the fuel tank becomes lower and lower, leading to a decrease in hydrogen production efficiency, and the remaining $NaBH_4$ solution also causes fuel waste.

Hydrogen inlet, Feed inlet, Hydrogen outlet, Porous filter material, Sodium borohydride solution, Outlet for sodium borohydride solution to the reactor.

(b) Purification of hydrogen produced by $NaBH_4$ hydrolysis. Theoretically, $NaBH_4$ hydrolysis only produces hydrogen and sodium metaborate, but in real reactions, due to the intense exothermic reaction, the reaction system temperature reaches 70–90 °C. Under these conditions, a lot of salt-containing steam is brought out with the rapidly flowing hydrogen. This solution accounts for more than 5 wt% of the total solution. The solute in the steam makes the gas alkaline, which can affect the performance of the acidic proton exchange membrane (PEM) fuel cell, and it also contains Na^+ cations and BO_2^- and BH_4^- anions, which may irreversibly damage the proton exchange membrane, reducing the service life of the fuel cell. Therefore, it is necessary to remove these impurities from the hydrogen gas to improve its purity and extend the service life of the PEM fuel cell. Currently, methods, such as membrane separation, acid washing, gas-liquid separators, and desiccants, have been used to remove these impurities. For example, the gas first goes through a washing tank for acid washing or water washing, which can cool and remove the alkalinity of hydrogen gas; the residual alkali mist that is not completely absorbed will be further removed by an inertial gas-liquid separator by centrifugal forces; finally, porous absorbents (such as activated carbon) are used to remove the alkali liquid mixed in the hydrogen gas. This method removes most alkaline impurities in the hydrogen gas, ensuring its purity.

(c) Compatibility with fuel cells. A hydrogen fuel cell system supplied by $NaBH_4$ hydrolysis includes a fuel ($NaBH_4$) tank, a catalyst reactor, a heating and cooling device, a waste ($NaBO_2$) tank, a hydrogen gas tank, a hydrogen gas purification device, and other devices, such as pumps, valves, and pressure sensors.

For low-power fuel cell systems, to reduce volume, a single chamber reactor that integrates the fuel chamber and the waste chamber can be used. The produced hydrogen gas and sodium metaborate are returned to the fuel chamber together. After initial condensation, the fuel liquid is supplied for fuel cell uses. Its characteristics are small volume, light weight, and high spatial utilization; but there are also problems, such as the thermal decomposition of sodium borohydride due to the increase in temperature in the fuel chamber, the insufficient purity of hydrogen gas leading to a decrease in the service life of the PEM fuel cell. Meantime, the change in fuel concentration with the progress of the reaction causes a decrease in the hydrogen production rate of the reactor and the output power of the fuel cell as well as the increased difficulty in the control of system operation.

For high-power fuel cell systems, the fuel tank is separated from the hydrolysis reaction generator. The residual liquid and hydrogen gas after the reaction flow back to the fuel tank together and undergo gas-liquid separation in the fuel tank. The hydrogen gas required by the fuel cell is transported to the proton exchange membrane fuel cell through a filter. The water produced by the cathode is collected and returned to the fuel tank. The excess hydrogen gas produced will enter the hydrogen storage tank for storage. Devices, such as the hydrogen storage tank and cathode water collector, are more suitable for use in high-power fuel cell systems. The advantage of this system is that the fuel tank is filled mainly with solid $NaBH_4$, and water can be recycled.

Although it is relatively mature in the engineering development for hydrogen production by $NaBH_4$ hydrolysis at present, the high price of commercial $NaBH_4$ has become the biggest obstacle to the success of hydrogen production by sodium borohydride hydrolysis in reality. In addition, the catalysts used for hydrogen production by sodium borohydride hydrolysis often contain precious metals, such as Ru and Pt, which further increase the overall hydrogen production cost.

5.2.2 *Principle of Hydrogen Storage in Light Metal Aluminohydrides*

In aluminohydrides, the four H atoms form a $[AlH_4]^-$ tetrahedron with the Al atom through covalent bonding, and the $[AlH_4]^-$ is then ionically bonded with the metal cation. Typical representatives include $LiAlH_4$ and $NaAlH_4$.

5.2.2.1 $LiAlH_4$ and Its Composite System

$LiAlH_4$ is a white crystalline solid that is relatively stable in dry air at room temperature. But it is extremely sensitive to humid air and proton-containing solvents and can react quickly with them to release hydrogen. $LiAlH_4$ belongs to the monoclinic crystal system, with a space group of $P2_1/c$. The theoretical hydrogen storage mass density of $LiAlH_4$ is 10.5 wt%, and so far, the experimentally measurable hydrogen storage density has reached 7.9 wt%, with a volumetric hydrogen density of 96 kg H_2/m^3.

$$3LiAlH_4 = Li_3AlH_6 + 2Al + 3H_2 \left(5.3wt\%\ H_2, 187-218°C\right) \tag{5.24}$$

$$Li_3AlH_6 = 3LiH + Al + 3/2H_2 \left(2.6wt\%\ H_2, 228-282^{\circ}C\right) \tag{5.25}$$

$$LiH = Li + 1/2H_2 \left(2.6wt\%\ H_2, 370-483^{\circ}C\right) \tag{5.26}$$

The theoretical hydrogen release amounts of the three-step reaction of $LiAlH_4$ are 5.3, 2.6, and 2.6 wt%, respectively. The first decomposition reaction occurs between 187–218 °C, generating the intermediate phase Li_3AlH_6; the second decomposition reaction occurs between 228–282 °C, where the intermediate phase Li_3AlH_6 continues to decompose and release hydrogen, forming LiH and Al; the third reaction occurs between 370–483 °C, where LiH decomposes to release hydrogen, forming Li. Due to the high decomposition temperature of LiH in the third step, its practical application value is relatively small. Nevertheless, since the first step of the $LiAlH_4$ reaction is exothermic, it is difficult to achieve the reverse hydrogen absorption reaction of $LiAlH_4$ from a thermodynamic perspective, and the required reaction conditions are quite harsh. To make Li_3AlH_6 absorb hydrogen at room temperature and complete the phase transition to $LiAlH_4$, it is indispensable to apply a high pressure of over 100 MPa. Such a high hydrogen pressure is difficult to achieve from both technical and safety perspectives. The modification of $LiAlH_4$ is mainly carried out by adding transition metal elements, such as Ti, Fe, Ni, and nanosizing treatment to reduce the hydrogen release temperature of $LiAlH_4$ and improve its reversibility.

Among all the research efforts on the modification of the $LiAlH_4$ hydrogen storage system, doping catalysts is the most studied method. Rafi et al. [91] found that adding 2 mol%TiC to $LiAlH_4$ could reduce the hydrogen release temperature to 85 °C, and heating to 188 °C could release about 6.9 wt% H_2, of which about 5 wt% of H_2 was released in the temperature range of 85–138 °C. DSC tests showed that the activation energies of the three-step reaction of this system were 59, 70, and 99 kJ/mol H_2, respectively. Besides, TiF_3 doping can also significantly improve the hydrogen release performance of $LiAlH_4$ [92, 93]. Adding 2 mol%TiF_3 can reduce the hydrogen release temperature of $LiAlH_4$ to 35 °C. Isothermal hydrogen release results at 140 °C indicate that the sample can release 7.0 wt%H_2 within 80 min, and the activation energies of the two-step hydrogen release have been effectively reduced to 66.76 and 88.21 kJ/mol H_2. In addition, other metal-based catalysts have also been widely studied. For example, adding $CoFe_2O_4$ to the $LiAlH_4$ system can reduce the initial hydrogen release temperature by 90 °C compared to untreated $LiAlH_4$ powder. It can release hydrogen at 65 °C. Under the conditions of 120 °C and 160 min, the sample can produce 6.8 wt% H_2, and its activation energy is also reduced to 52.4 and 86.5 kJ/mol H_2 [94].

By combining metals with carbon materials, $LiAlH_4$ based materials with good catalytic activity and good dispersity can be obtained with improved hydrogen release performance. Studies show that after adding 10 wt% nickel loaded carbon nanoparticles of Ni/C, the dehydrogenation temperature of $LiAlH_4$ can be reduced to 48 °C; Approximately 6.3 wt% of hydrogen can be released within 10 min at the temperature of 140 °C, while under the same condition, $LiAlH_4$ only releases 0.52 wt% of hydrogen. In addition, the activation energy for the first two steps of decomposition can also be reduced to 61.94 and 79.73 kJ/mol [95], respectively.

5.2.2.2 $NaAlH_4$ and Its Composite System

$NaAlH_4$ is in a white crystalline form at room temperature with a density of 1.28 g/cm^3, which has good reversible hydrogen absorption and release properties. When doped with catalysts, it can reversibly absorb and release about 4.5 wt% of hydrogen at temperatures below 100 °C, with no by-products and high purity of hydrogen, what's more, the catalysts are relatively cheap, making it suitable for mid-temperature hydrogen fuel cell powered vehicles (80–200 °C). However, due to the high cost in producing raw materials, there are currently no practical applications of $NaAlH_4$.

At room temperature, $NaAlH_4$ exists in the form of a body-centered tetragonal structure, denoted as α-$NaAlH_4$. Under high pressures, α-$NaAlH_4$ undergoes a phase transition to the orthorhombic β phase, which reduces the volume of α-$NaAlH_4$ by 4%. $NaAlH_4$ can be prepared by mechanical ball milling a mixture of NaH and Al under a H_2 atmosphere. High purity $NaAlH_4$ can be prepared by ball milling for 50 h under a relatively low pressure of 2.5 MPa H_2.

$$NaH + Al + 3/2H_2 = NaAlH_4 \tag{5.27}$$

The mass hydrogen storage density of $NaAlH_4$ is 7.41 wt%, and the hydrogen release conditions are relatively mild, easy to achieve the reverse reaction. The hydrogen release process of $NaAlH_4$ is generally divided into three steps:

$$NaAlH_4 = 1/3Na_3AlH_6 + 2/3Al + H_2\left(36-41\,kJ/mol\,H_2\right) \tag{5.28}$$

$$Na_3AlH_6 = Al + 3NaH + 3/2H_2\left(47\,kJ/mol\,H_2\right) \tag{5.29}$$

$$NaH = Na + 1/2H_2 \tag{5.30}$$

Upon heating from room temperature, $NaAlH_4$ will melt at 180 °C and then start the first step of hydrogen release. The second step of hydrogen release generally occurs above 260 °C, and the third step of hydrogen release requires temperatures higher than 400 °C. Due to the greatly reduced operability of solid hydrogen storage at high temperatures, the practical application of $NaAlH_4$ currently focuses on the first two steps of hydrogen release, which reach a hydrogen storage capacity of 5.6 wt%. Figure 5.17 shows the microstructural evolution of the $NaAlH_4$ hydrogen release process. After $NaAlH_4$ is converted to Na_3AlH_6, due to the change in crystal structure, pores appear between the Na_3AlH_6 unit cells, and aluminum atoms begin to accumulate near these pores. When Na_3AlH_6 is converted to NaH, the aluminum atoms near the pores further accumulate, showing an irregular distribution.

Compared to other solid-state hydrogen storage materials, the thermodynamic performance of $NaAlH_4$ for hydrogen release is already quite good. Therefore, the research in the modification of $NaAlH_4$ mainly focuses on its hydrogen release kinetics and cycling performance, ultimately achieving stable hydrogen absorption and release cycle performance at lower temperatures and applying it to the practical solid-state hydrogen storage scenarios. The catalytic modification of $NaAlH_4$ is generally achieved by doping catalysts, through which the activation energy and hydrogen absorption and release rate of $NaAlH_4$ can be significantly improved. The method of doping catalysts can be divided into "dry doping", "semi-wet doping", and "wet doping" according to the form of the dopant [96]. The principle of this classification is based on the form of the dopants during the doping process. Dry doping refers to the doping of catalysts (such as $TiCl_3$, TiF_3, etc.) into $NaAlH_4$ in a solid form, usually using mechanical ball milling. Wet doping is to suspend $NaAlH_4$

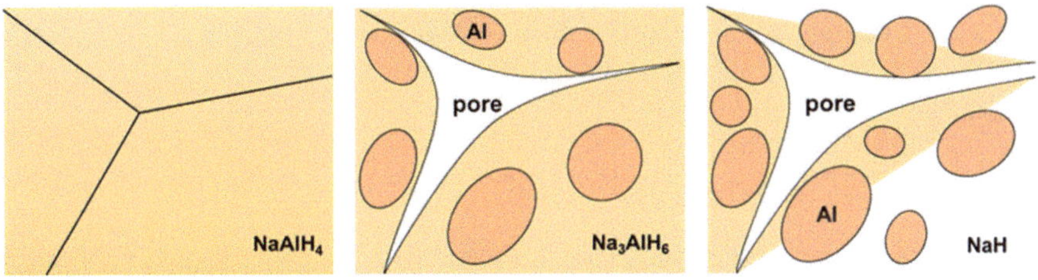

Fig. 5.17 Schematic diagram of the hydrogen release process of $NaAlH_4$

in toluene or dimethyl ether, add Ti alcohol salt and other catalysts for stirring, and finally dry under vacuum to obtain colorless doped $NaAlH_4$ powder. The reversible hydrogen storage capacity can reach 4 wt%, but the hydrogen absorption and release kinetics is poor. Semi-wet doping is a mixture of the previous two methods for doping, and the cycle stability is better than wet doping. Among these three doping methods, dry doping has received the most attention due to its high effectiveness and operational simplicity. Catalyst-doped complex hydrides prepared by ball milling has good hydrogen storage performance and can achieve stable hydrogen absorption and release under suitable conditions. In the modification of $NaAlH_4$, the choice of catalyst is the most important factor. Bogdanovic et al. first introduced Ti as a catalyst for $NaAlH_4$, achieving its reversible hydrogen absorption and release performance. Beside Ti, V, Cr, Nb and other elements adjacent to Ti on the periodic table, also have good catalytic performance. In addition, effective additives that can improve the hydrogen storage performance of $NaAlH_4$ consists of transition metal elements, oxides, fluorides, chlorides, nitrides, etc.

The catalytic mechanism of doping $NaAlH_4$ with such additives mainly include:

1. The additive reacts with $NaAlH_4$ during the ball milling process to form a hydride (such as TiH_X). In the $NaAlH_4$ unit cell, the Al-H bond in AlH is relatively stable, but by doping transition metal hydrides or in-situ generation of transition metal hydrides, the stability of the Al-H bond can be weakened, thereby significantly improving the hydrogen release performance of $NaAlH_4$.
2. The additive generates metal nanoparticles or intermetallic compounds during the ball milling process. These active substances adhere to the surfaces of $NaAlH_4$ and play a good catalytic role in the absorption/release process.
3. The cations in the additives not only change the surface properties of $NaAlH_4$, but also replace the cations in $NaAlH_4$ to change its lattice parameters, leading to the destabilization of $NaAlH_4$ and enabling it to effectively decompose and release hydrogen.

5.2.3 *Principle of Hydrogen Storage in Light Metal Amides*

Light metal amides mainly refer to complex hydrides formed by light metal ions (Li, Na, K, Mg, Ca, etc.) with amino $[NH_2]^-$ or imino $[NH]^{2-}$ ligands. In 2002, Ping Chen et al. [97] reported that Li(Ca)-N-H can be used for reversible hydrogen storage, with a hydrogen storage capacity up to 10.4 wt%, which prospected the research on light metal amides. $LiNH_2$ is obtained by hydrogenating Li_3N, and its reaction formula is:

$$Li_3N + 2H_2 \rightleftharpoons Li_2NH + LiH + H_2 \rightleftharpoons LiNH_2 + 2LiH \tag{5.31}$$

The enthalpy changes of the two-step reversible reactions are 45 and 88.7 kJ/mol H_2, with a complete dehydrogenation temperature above 400 °C. Under the catalytic effect of $TiCl_3$ and VCl_3, the LiH and $LiNH_2$ composite system dehydrogenates about 6 wt% within 150–250 °C and shows good thermodynamic properties and reversibility. Based on the above research, Ping Chen and others explored the reaction mechanism, proposing that the dehydrogenation reaction of $LiNH_2$ and LiH follows an acid-base pair mechanism. Mainly, the H in $LiNH_2$ carries a positive charge, while the H in LiH carries a negative charge, and the strong interaction between the two promotes the breaking of the N-H bond and the metal-H bond, and recombines into H_2.

Although the Li-N-H system has a relatively high hydrogen storage capacity, its thermodynamic performance of hydrogen absorption and release processes is poor. Partially replacing Li elements with the elements having a higher electronegativity, such as Mg and Ca, can effectively improve the hydrogen storage performance of the Li-N-H system [98]. For example, replacing LiH and $LiNH_2$ in the $LiNH_2$-LiH system with MgH_2 and $Mg(NH_2)_2$ can significantly improve the hydrogen storage performance of the Li-N-H system. Compared with the $LiNH_2$-LiH system, the theoretical hydrogen storage capacity of the $Mg(NH_2)_2$-2LiH system is slightly reduced, about 5.6 wt%. Experimental results show that this material can quickly absorb and release more than 5 wt% hydrogen within 250 °C. The enthalpy change of the hydrogen release reaction of $Mg(NH_2)_2$-2LiH is 39–42 kJ/mol H_2, and the entropy change of the hydrogen release reaction is 112–116 J/mol/K H_2. From these, it can be calculated that this system can produce a balance hydrogen pressure of 0.1 MPa at around 75–85 °C, theoretically, this system has the potential for on-board hydrogen storage applications.

From a practical point of view of the $Mg(NH_2)_2$-2LiH system, the concentration of ammonia during the hydrogen release process, the cycle stability of the hydrogen absorption and release, and the effects of water and oxygen on its hydrogen storage performance should be systematically studied [99]. Results show that the concentration of ammonia produced during the hydrogen release process of this system is closely related to its working temperature. The higher the working temperature, the higher the concentration of ammonia produced. After 270 hydrogen absorption and release cycles, the hydrogen storage capacity of this material has been decreased by about 25%. When this material is exposed to air containing saturated water vapor, after 16 cycles of hydrogen absorption and release tests, its capacity and kinetic performance decay patterns are generally consistent with those of unexposed samples, indicating that water and oxygen exert little effects on the hydrogen absorption and release performance of this material. Although thermodynamic calculations show that the $Mg(NH_2)_2$-2LiH system can obtain a balance hydrogen pressure of 0.1 MPa at 80 °C, the initial hydrogen release temperature observed in the experiment is higher than 140 °C, and a considerable hydrogen release rate usually requires a temperature higher than 200 °C.

5.3 Ammonia Boranes and Their Derivatives for Hydrogen Storage

5.3.1 Principle of Hydrogen Storage in Ammonia Boranes

Ammonia borane (NH_3BH_3) is in the form of a crystalline white powder that can exist stably in dry air. It is one of the most common B-N-H compounds, and its theoretical hydrogen storage capacity is as high as 19.6 wt%, making it an ideal hydrogen storage material. In ammonia borane, the H atoms connected to the N atom all exhibit a positive charge; the H atoms connected to the B atom all exhibit a negative charge. As such, there is a strong electrostatic attraction between the positively charged H atoms and the negatively charged H atoms, and this electrostatic interaction is called a double hydrogen bond, denoted as "N-H···H-B". Similar to the classic hydrogen bond, the double hydrogen bond in the ammonia borane molecule has a significant impact on its spatial configuration and physicochemical properties. The reason why ammonia borane molecules can exist stably in a solid state at normal temperatures and pressures is the attractive interaction of the double hydrogen bond of N-H···H-B, which further stabilizes the configuration of the whole ammonia borane molecule. In addition, the thermal decomposition of ammonia borane to produce hydrogen is an exothermic reaction, which is also due to the N and B bonds being converted from coordination bonds to more stable covalent bonds.

Ammonia borane molecule contains three types of proton bonds, B-H, B-N, and N-H, showing high thermal stability. Therefore, its decomposition process requires high energy input, and the required temperature is high so that by-products can be easily formed during the decomposition process. The thermal decomposition process of ammonia borane can be divided into three steps, the reaction are be described as follows:

$$nNH_3BH_3 = [NH_2BH_2]_n + nH_2 \text{ approximately} 110^\circ C \tag{5.32}$$

$$[NH_2BH_2]_n \text{ P } [NHBH]_n + nH_2 \text{approximately} 150^\circ C \tag{5.33}$$

$$[NHBH]_n \text{ P } nBN + nH_2 > 500^\circ C \tag{5.34}$$

The hydrolysis process of ammonia borane is relatively simple, and the reaction formula is as follows:

$$NH_3BH_3 + 2H_2O = NH_{4^+} + BO_{2^-} + 3H_2 \tag{5.35}$$

However, due to its own stability, its hydrolysis process is very slow in the absence of a catalyst. Ammonia borane is relatively stable in aqueous solution and requires an appropriate catalyst to generate hydrogen at room temperature. The efficiency of the dehydrogenation reaction largely depends on the choice of catalysts. In 2006,

Chandra et al. [100] first discovered that Pt, Rh, and Pd-based catalysts showed good catalytic activity in the hydrolysis of ammonia borane to produce hydrogen. Among them, the Pt catalyst shows the best catalytic performance and does not experience obvious deactivation in recycling. For non-precious metal catalysts, commonly used catalysts mainly include Ni, Fe, Co, etc. Among them, Co-based metal catalysts have a relatively high catalytic activity [101]. When the added Co nanoparticle catalyst is 4 mol% of ammonia borane, the hydrolysis process of ammonia borane can be completed within 1.7 min, and the turnover frequency (TOF) value at room temperature is 2.69 min^{-1}. Precious metal catalysts have good catalytic activity, but they are expensive and scarce in nature, making them difficult to obtain and apply on large scale. On the other hand, although non-precious metal catalysts are cheap, their catalytic activity is much lower than that of precious metals. In order to maintain the activity of the catalyst while reducing the use of precious metals, combined precious and non-precious metal catalysts have been developed and they demonstrate good catalytic effects. Mori et al. [102] prepared a series of Ru and Ru-Ni nanoparticles loaded on different inorganic carriers. Compared with the single metal Ru, the TOF [mol $H_2/(mol_{Ru} \cdot min)$] value of the bimetal Ru-Ni increased by 1.5 times. XANES studies confirmed a strong interaction between Ni-Ru, suggesting that the adjacent arrangement of the bimetal Ru-Ni sites could help anchor ammonia borane molecules through electrostatic interactions, thus promoting the hydrolysis of ammonia borane.

In supported metal catalysts, the support material can prevent the formation of impurities on the nanoparticles and suppress the agglomeration of nanoparticles. Moreover, due to its porous structure, the active surface area of metal nanoparticles increases. The support materials can be divided into oxides, carbon materials, and MOFs. Özkar et al. [103] loaded Rh on CeO_2, SiO_2, Al_2O_3, TiO_2, among which Rh^0/CeO_2 showed the best catalytic activity, with a TOF value reaching 2010 mol $H_2/(mol_{Rh} \cdot min)$. Common carbon-based carriers include mesoporous carbon, graphene, graphene oxide, reduced graphene oxide, carbon nanotubes, etc. Wei et al. [104] used SBA-15 as a hard template to synthesize mesoporous carbon-nitrogen materials (MCN) and used this material as a support to load metal Pd and PdNi to prepare a series of ammonia borane hydrolysis catalysts. The pore properties of the MCN support facilitate the mass transfer rates during the reaction process. The resultant catalyst showed good dispersion and the metal particles on the support were small. Due to the interaction between the metal particles and the support interface, the prepared catalyst demonstrated excellent catalytic activity. The TOF value of $Pd_{74}Ni_{26}$/MCN catalyst was as high as 246.8 mol $H_2/(mol_{PdNi} \cdot min)$. MOF materials have highly ordered pore structures and ultra-high specific surface areas, which are beneficial for the dispersion of active metal particles. The metal nanoparticles loaded inside the MOF pores are not easy to agglomerate during the reaction process, making the catalyst to have good stability. Wen et al. [105] used MIL-96(Al) as a support to prepare Ru/MIL-96(Al) catalyst by liquid phase impregnation. The average particle size of Ru particles on MIL-96(Al) was 2 nm. The special structure of the MOF support and the good dispersion of the active metal nanoparticles made the catalyst show good catalytic performance in the hydrolysis of ammonia borane to produce hydrogen, with a TOF value of 231 mol $H_2/(mol_{Ru}\cdot min)$.

In addition, under light conditions, semiconductors form photo-generated electrons and photo-generated hole pairs (e^--h^+), which can migrate to the catalyst surface to participate in redox reactions. Therefore, such charge migration processes can also catalyze the hydrolysis reaction of ammonia borane to release hydrogen. Rej et al. [106] synthesized a core-shell Au-Pd nanostructure, and compared with the dark light, the rate of hydrogen production from the catalytic hydrolysis of ammonia borane under visible light increased by 3 times, reaching 426 mol $H_2/(mol_{cat}\cdot min)$. This is because the surface electric field is enhanced under visible light and hot electrons are concentrated at the edge and apex of the cell, which weakens the uneven surface charge of the B-N bond and facilitates the attack of water molecules.

The future development of hydrogen production from ammonia borane hydrolysis includes the following aspects:

1. Constructing catalysts with lower prices and uneven charge distributions, while simultaneously ensuring their good recyclability and high durability, such as utilizing magnetic samples or integral metal foam catalysts.
2. From the perspective of charge transfer, light induction is beneficial to the hydrolysis reaction of ammonia borane, so it is an environmental friendly strategy to develop semiconductors with catalytic activity for ammonia borane hydrolysis, constructing sensitized systems or heterojunctions, and using sunlight to reduce system energy consumption.
3. The dehydrogenation of ammonia borane is an exothermic reaction. Due to thermodynamic limitations, reversible hydrogen storage is difficult to achieve and the resultant product cannot be rehydrogenated. Therefore, while studying the hydrolysis reaction of ammonia borane, we should also focus on the cheap synthesis methods to regenerate ammonia borane from dehydrogenated products.

5.3.2 Principle of Hydrogen Storage in Ammonia Borane Derivatives

Metal elements M (mainly alkali metals and alkaline earth metals) replace the H atoms connected to the N atom in the ammonia borane (AB) molecule, forming metal ammonia borane (abbreviated as MAB). Compared with AB, MAB has good performance in hydrogen release and suppression the generation of gas impurities. In addition, introducing H-containing groups (NH_3, BH_4, etc.) based on MAB to generate corresponding metal ammonia borane derivatives can also have special hydrogen storage properties. According to the chemical structure, the derivatives of ammonia borane mainly include metal ammonia borane compounds, bimetal ammonia borane compounds, and ammonia-containing derivatives, etc.

Metal ammonia boranes mainly include LiAB, NaAB, KAB, MgAB, AlAB, CaAB, and YAB, etc. There are mainly two methods for the synthesis of MAB: solid-state mechanical ball milling and wet chemical synthesis. In the solid-state synthesis, ball milling can effectively reduce the particle size of the reactants, shorten the material transfer distance, and generate more reactive surfaces. In the wet chemical method, since the reaction is carried out in a liquid, the material transfer is significantly improved, but the disadvantage is the existence of strong M-O bonds between the metal cations and the polar organic solvent, making the solvent difficult to remove and unable to guarantee the purity of the product. Under normal circumstances, MAB can be synthesized by the reaction of the corresponding metal hydride MH_n and AB, and its reaction formula is as follows:

$$MH_n + nNH_3BH_3 = M(NH_2BH_3)_n + nH_2 \tag{5.36}$$

where M = (Li, Na, K, Mg, Ca, Sr). YAB, AlAB, FeA B are generated by the metathesis reaction between MCl_3 and YAB or LiAB or NaAB in tetrahydrofuran (THF) solvent, and their reaction formula is as follows:

$$MCl_3 + 3LiNH_2BH_3 = M(NH_2BH_3)_3 + 3LiCl \tag{5.37}$$

where M = (Y, Al, Fe).

In metal ammonia boranes, due to the replacement of hydrogen atoms by alkali/alkaline earth metals with stronger electron-donating ability, significant changes in the electronic state of nitrogen atoms are incurred, accompanied by changes in the chemical bonding state between B and N. Its B-N bond (about 1.56 Å) is shorter than the one in the ammonia borane molecule (1.58 Å) and the bond length of the M-B is significantly larger than the B-N bond (such as Li-N: 1.98 Å; Na-N: 2.35 Å). The bond length of the N-H and B-H bonds in MAB is much larger than the chemical bonds in the AB [107]. Compared with ammonia borane, the hydrogen release performance of MAB has been significantly improved. Among them, the initial hydrogen release temperature of LiAB is at 91 °C, and the hydrogen release rate is high, about 8 wt% of H_2 can be released after 1 h. KAB releases 6.5 wt% of H_2 at about 80 °C for 3 h, and no NH_3 is generated [108].

The crystal structure of bimetal ammonia boranes is affected by many factors, such as the radius of the metal cation, the number of charges, and the chemical environment of coordinated $[NH_2BH_3]^-$ anion. Different bonding methods and structural configurations may lead to completely different hydrogen release properties [109]. Bimetal cations can provide more significant performance improvement during the thermal decomposition than single metal cations. For example, in $M_2Mg(AB)_4$ (M = Na or K), the small but highly charged Mg^{2+} only forms a tetrahedral coordination with the NH_2 group, while M^+ only forms an octahedral coordination with the BH_3 group. This unique ordered structure arrangement caused by the difference in cation-anion coordination provides direct M-H bonding, improving the induced dehydrogenation process of M-H bonds and thus making the dehydrogenation temperature of $M_2Mg(AB)_4$ significantly lower than the corresponding single metal ammonia boranes. Consequently, the amount of released hydrogen can be significantly increased and the content of byproduct NH_3 can be significantly minimized [110, 111].

Metal ammonia boranes coordinated with ammonia that have been synthesized so far include $MgAB{\cdot}NH_3$, $Li(NH_3)AB$ and $CaAB{\cdot}NH_3$ [112–114]. $MgAB{\cdot}NH_3$ is synthesized by the reaction of MgNH with AB, and H_2 is generated at around 50 °C, with a small amount of NH_3 produced. When the temperature reaches 100, 150, 200 and 300 °C, $MgAB{\cdot}NH_3$ begins to release a large amount of hydrogen, with the amount of released hydrogen being 5.3, 8.4, 9.7, and 11.4 wt%, respectively. $Li(NH_3)AB$ undergoes the dehydrogenation process between 40 and 70 °C, and in the presence of NH_3, 11.18 wt% H_2 can be rapidly released at 60 °C, which has a significant advantage compared to LiAB.

In addition to the above-mentioned metal ammonia borane derivatives, other derivatives, such as LiAB·AB, $LiNH_3BH_4$, $LiNH(BH_3)NH_2BH_3$, etc., which contain both B-H and N-H derivatives [115–117], could show certain advantages in some aspects. For example, LiAB·AB is a derivative formed by LiAB and AB, its dehydrogenation temperature is very low, some derivatives can release 6.0 wt% H_2 at around 57 °C. When it reaches 228 °C, a total of 14 wt% H_2 can be released, while AB needs to reach 500 °C to completely release H_2.

Although metal ammonia boranes and their derivatives have a high hydrogen storage capacity and a relatively low hydrogen release temperature. The irreversible nature of their hydrogen release reaction results in a low overall cycle energy efficiency. Therefore, the ammonia borane system remains in the laboratory research and development stage.

5.4 Physical Adsorbents for Hydrogen Storage

Physical adsorption of hydrogen gas is another way to achieve solid-state hydrogen storage, and microporous materials have been widely used in this field. According to the definition of the International Union of Theoretical and Applied Chemistry (IUPAC), microporous materials typically have pores smaller than 2 nm [118]. These materials have a very large specific surface area, allowing hydrogen gas to be effectively absorbed on the pore walls in the form of molecules, mainly through the weak van der Waals interaction between the solid surface and hydrogen molecules to capture hydrogen gas, thereby achieving a higher storage density than free gas [119]. In 1980, Carpetis et al. [120] systematically tested the performance of different hydrogen adsorption materials under temperature conditions of 65–150 K. Microporous materials adsorb and release hydrogen quickly, and have good reversibility, without the hysteresis phenomenon commonly observed in metal hydrides. However, these materials can only reach a practical value of hydrogen storage at low temperatures so that their application prospect is greatly restricted. In addition, low-temperature hydrogen storage also increases the total cost of hydrogen storage to a certain extent. Therefore, beside improving the maximum hydrogen storage capacity of the microporous material, which is an inherent goal, researchers are also committed to increasing the working temperature of these materials.

Currently, representative physisorption-based hydrogen storage materials include carbon materials, zeolites, metal-organic frameworks (MOFs), covalent organic frameworks (COFs), and porous organic polymers (POPs) (Table 5.4). Porous carbon materials have a high specific surface area, thermal stability, and chemical stability. They are simple to be prepared and can be produced at low costs. The representative material is BPL Carbon [121], which possesses a mass and volume working capacity at 77 K from 8.5 MPa → 0.5 MPa of 1.86 wt% and 16.5 g/L, respectively. Its disadvantages are that the hydrogen adsorption is weak, the pore size and structure are difficult to control, and most of them are non-crystalline. In comparison, zeolites have good hydrogen storage performance due to their high specific surface area, high crystallinity, uniform pore size, high stability, and low cost. For example, the mass hydrogen storage capacity of NaX at 77 K and 1.5 MPa is 1.79 wt% [122]. However, zeolite materials suffer from the similar problems, such as weak adsorption interaction with hydrogen, difficult to control single pore size, limited structure configuration, and low mass hydrogen storage capacity.

Relatively speaking, MOF is a crystalline material composed of metal ions or clusters connected by organic ligands. It is famous for its ultra-high surface area, tunable pore size, and adjustable internal surface properties, and has great potential in gas storage applications [123–125]. It has high crystallinity, ultra-high specific surface area and porosity, high pore size and function control, and contains uncoordinated metal sites, thus it has a good prospect for hydrogen storage applications. For example, the mass and volume hydrogen storage capacities of NU-1501 (Al) in the range of 77 K and 10 MPa to 160 K and 0.5 MPa are 14 wt% and 46.2 g/L, respectively. For Ni_2 (m-dobdc) at a temperature of 298 K, the mass and volume hydrogen storage capacities in the range of 10 MPa to 0.5 MPa are 1.9 wt% and 11 g/L, respectively [126]. However, MOF materials have low hydrogen adsorption capacity at room temperature and poor processability. In addition, covalent organic framework (Covalent Organic Framework, COF) materials and porous organic polymer (Porous Organic Polymer, POP) materials are also widely used studied as hydrogen storage carrier, but they all have problems, such as low adsorption capacity due to lack of strong hydrogen adsorption sites. This section will focus on the application research of porous carbon materials and MOFs as solid-state hydrogen storage materials.

5.4.1 Principle of Hydrogen Storage in Carbon Materials

Carbon-based materials, including activated carbon, carbon nanotubes, etc., due to their special pore structures, are considered suitable as gas storage adsorbents. The related research on using carbon materials to adsorb hydrogen began in the early

Table 5.4 Comparison of the advantages and disadvantages of commonly used physical adsorption hydrogen storage materials

	MOF materials	COF materials	POP materials	Carbon materials	Zeolite materials
Advantages	High crystallinity, ultra-high specific surface area and porosity High pore size and functional controllability, rich in uncoordinated metal sites	High crystallinity, high specific surface area and porosity, chemically controllable structure, structural diversity	Dynamic structure, chemical structure diversity, chemical and water stability	High thermal and chemical stability, high processability, low cost	High crystallinity, high thermal and hydrolysis stability, low cost
Disadvantages	Low hydrogen adsorption capacity at room temperature, poor processability	Lack of hydrogen adsorption sites, difficult to activate, poor machinability	Lack of hydrogen adsorption sites, amorphous, uneven pore size	Lack of hydrogen adsorption sites, uneven pore size, uncontrollable pore structure	Lack of hydrogen adsorption sites, single structure, lack of diversity, low adsorption capacity, poor pore size control
Representative materials	NU-1501 (Al): 14 wt% and 46.2 g/L @77 K and 10 MPa → 160 K and 0.5 MPa; Ni_2 (m-dobdc): 1.9 wt% and 11 g/L @ 298 K and 10 MPa → 0.5 MPa	COF-103: 6.4 wt% and 29.2 g/L @ 77 K and 8.5 MPa → 0.5 MPa	PAF-1: 7 wt% (excess) @ 77 K, 4.8 MPa; PIM-1: 2.6 wt% @ 77 K and 10 MPa	BPL carbon: 1.86 wt% and 16.5 g/L@ 77 K and 8.5 MPa → 0.5 MPa	NaX: 1.79 wt% @77 K and 1.5 MPa

twentieth century [127]. The principle of hydrogen storage in carbon materials is that hydrogen physically adsorbs on the surface of carbon materials with a high specific surface area. Carbon materials demonstrate the characteristics of low density, which is beneficial for weight reduction of equipment. Their chemical and thermal stability during the hydrogen absorption and release processes can improve the reliability of process operation and is beneficial to the application in the field of hydrogen storage. At the same time, carbon materials are abundant in resources, easy to process, and relatively low cost, suitable for industrial and large-scale production.

5.4.1.1 Activated Carbon

There are two ways to activate carbon materials:

1. Physical activation with different oxidizing gases, such as air, O_2, CO_2, steam, or their mixtures.
2. Chemical activation with chemical compounds, such as KOH, NaOH, H_3PO_4, and $ZnCl_2$. It is also possible to combine the chemical activation method (usually with H_3PO_4 or $ZnCl_2$) and the physical activation method (usually with CO_2) to further enhance the porosity and adjust the pore size distribution of resultant activated carbon materials [128].

Many organic products are suitable as raw materials to produce activated carbon. Due to their low cost and abundant yields, woods, sawdust, peats, coconut shells, fruit stones and rice husks are the preferred bio-carbon feedstocks. Carbonized raw materials, such as coal, low-temperature lignite coke, and charcoal, can also be used for the production of activated carbon.

Activated carbon has demonstrated excellent hydrogen storage capacities at low temperatures. In the 1970s, Kidnay and others [129] conducted early research on the use of activated carbon in hydrogen storage. They measured the hydrogen adsorption isotherms of coconut-shell derived activated carbon at a temperature of 76 K and a range of 1–95 bar, and found that 1 kg of activated carbon adsorbed 20.2 g of hydrogen at 25 bar. Carpetis and others [120] tested the F12/350 activated carbon sample, and the maximum usable hydrogen adsorption capacity at 41.5 bar and 65 K was 5.2 wt%, and pointed out that as the temperature decreases, the material's hydrogen storage capacity is also increasing. Especially when it is below 150 K, the increase of hydrogen storage capacity is more significant. Zhou and others [130] tested the hydrogen adsorption capacity of activated carbon AX-21 at 77 K and different pressures and proposed to explain the adsorption capacity of different materials by the changes in packing density and specific surface areas. The loss of the specific surface area of the adsorbent leads to a decrease in the mass storage capacity, while the increase in volumetric density leads to an increase in the release hydrogen amount of the adsorbent in a fixed volume, thereby leading to an increase in the volumetric storage capacity. Jorda'-Beneyto and others [131] increased the hydrogen storage capacity by adding the WSC binder. The refined activated carbon

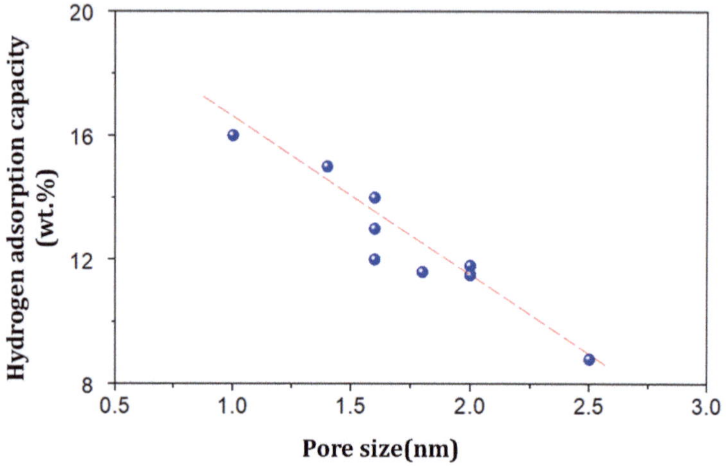

Fig. 5.18 A relationship between the hydrogen absorption capacity and the pore size of carbon materials

monolith was prepared by pyrolysis at 750 °C for 2 h, achieving a high micropore volume and high density while maintaining good mechanical properties. The hydrogen adsorption capacity reached 29.7 g/L at 77 K and 4 MPa. The results show that to achieve a larger hydrogen adsorption amount on a certain volume basis, the material needs to possess a larger micropore volume and a higher stacking density. Sevilla et al. [132] prepared activated carbon from eucalyptus sawdust as a precursor, and under the conditions of 77 K and 2 MPa, the hydrogen adsorption capacity could reach 6.4 wt%. Their results further prove that the storage density of hydrogen is closely related to the pore size of carbon (Fig. 5.18), and pores smaller than 1 nm significantly enhance its hydrogen storage capacity.

Achieving a high hydrogen storage capacity at near room temperature is indeed difficult, and researchers have conducted many studies on the hydrogen storage capacity of activated carbon at room temperature. Gao et al. [133] used steam and KOH activation method to prepare activated viscose-based carbon fibers. Their hydrogen storage capacity increases with the increase of specific surface area and micropore volume. The BET specific surface area and micropore volume of activated viscose-based carbon fibers are 3144 m^2/g and 0.744 m^3/g, respectively. Using a pressure-composition-temperature (PCT) system based on a volumetric method, the hydrogen storage capacity of activated viscose-based carbon fibers at 77 and 298 K and under 4 MPa is measured to be 7.01 and 1.46 wt%, respectively. Fierro et al. [134] prepared activated carbon by chemical activation of smokeless coal and alkali (Na and K) hydroxides, with a surface area of 3220 m^2/g. At 77 K and 4 MPa, a maximum hydrogen storage capacity of 6.0 wt% was obtained; at 298 K and 5 MPa, this maximum value dropped to 0.6 wt%. The poor hydrogen storage capacity of activated carbon at room temperature may be due to the isosteric heat of hydrogen adsorption being 5–8 kJ/mol H_2, which does not allow a large amount of hydrogen molecules to be absorbed at room temperature [128]. Kuchta et al. [135] theoretically showed that in the assumed chemically modified porous carbon, the hydrogen adsorption heat was 15 kJ/mol H_2 or even higher. The pore size was between 0.8–1.1 nm, and the volume storage capacity could reach the target required for practical applications. To achieve a meaningful hydrogen storage capacity at room temperature, the isosteric heat of hydrogen adsorption must be higher. The isosteric heat of hydrogen adsorption on activated carbon can be increased by surface modification, including doping with metal atoms. Alcaiz et al. [136] calculated that H_2 is mainly adsorbed in narrow micropores at room temperature, and micropores larger than 7 Å are ineffective for H_2 adsorption because the H_2 adsorption only occurs in a monolayer. Theoretically, in ideal microporous carbon with a narrow pore window, the maximum adsorption amount of H_2 is about 3.5 wt%.

Hydrogen adsorption capacity, Pore size.

Experimental results also show that doping and micropore size exert a significant impact on the hydrogen storage performance of activated carbon at room temperature. Li and Yang [137] studied the hydrogen storage equilibrium and kinetics on Pt-doped activated carbon (AX-21) containing 5.6 wt% Pt. At room temperature and 10 MPa, the hydrogen storage capacity of the Pt-AX-21 sample (specific surface area of 2518 m^2/g) was 1.2 wt%, which was twice as high as the undoped AX-21 sample (specific surface area of 2880 m^2/g) of about 0.6 wt%. Xia et al. [138] tested the enhanced room-temperature hydrogen storage capacity in activated carbon materials prepared by carbon dioxide activation. These carbon materials have a high specific surface area (2829 m^2/g), a large pore volume (2.34 cm^3/g), and hierarchal pore structure, including a primary micropore with a size between 0.7–1.3 nm and a secondary mesopore with a size between 2–4 nm. One of the best activated carbons showed a hydrogen storage capacity of 0.95 wt% at 298 K and 8 MPa. Data suggested a close relationship between the hydrogen storage capacity and the micropore volume. The authors inferred that the development of micropores, especially a rapid increase of narrow pores with a diameter of about 1.2 nm, seemed to be the key to enhancing the room-temperature hydrogen storage performance of the activated carbon. Geng et al. [139] prepared Pt/Pd doped activated carbon samples using a two-step reduction method (ethylene glycol reduction and hydrogen reduction). Hydrogen adsorption results show that at 298 K, the excess hydrogen adsorption capacity of Pt/Pd doped activated carbon is higher than that of pure activated carbon, which could be attributed to the hydrogen spillover effect. 2.5%Pt and 2.5%Pd co-dopped activated carbon samples at 298 K and 18 MPa showed the highest hydrogen absorption capacity (1.65 wt%). They also pointed out that the particle size and distribution of Pt/Pd catalysts might play a key role in hydrogen adsorption through the spillover effect.

According to current research results [140], this spillover mechanism is defined as: under the conditions of adsorption or formation of active species, the adsorbed or formed active species migrate from the first surface to another surface. The spillover mechanism is realized through three steps:

1. Hydrogen molecules dissociate to hydrogen atoms, which bind with metal atoms.
2. Hydrogen atoms migrate from the metal atom surface to the carbon surface.
3. Hydrogen atoms diffuse on the carbon surface, forming a stable C-H bond after moving a certain distance.

5.4.1.2 Carbon Nanotubes

The structure of carbon nanotubes (CNTs) was discovered by Iijima [141] in 1991. It is of a unique hollow tubular nanostructure, equivalent to rolling a graphite plate with a high aspect ratio, with a pore size of 0.7 nm to several nanometers. The carbon network of the shell layer is closely related to the honeycomb arrangement of carbon atoms in the graphite layer and can be divided into single-walled carbon nanotubes and multi-walled carbon nanotubes. Currently, carbon nanotubes are mainly produced by three methods: arc discharge, laser ablation, and chemical vapor deposition [142]. The first two methods are used to produce carbon nanotubes of better quality but are difficult to work continuously, while chemical vapor deposition is the most common method, but the resultant product has a high density of defects.

Research on the hydrogen storage performance of carbon nanotubes can be traced back to 1997, when Dillon et al. [143] reported a room-temperature hydrogen storage capacity of 5–10 wt% postulated from the thermal desorption tests for carbon nanotubes, which sparked a prosperity using CNT for hydrogen storage applications. Although subsequent results suggested this nominal hydrogen storage capacity was caused by the deposition of metal nanoparticles, these results indeed proved the potential of carbon nanotubes for hydrogen storage applications. In 1999, Ye et al. [144] used the Sieverts' apparatus and measured the adsorption isotherms of hydrogen on purified carbon nanotubes under a low temperature (80 K) and pressure range of 4–8 MPa using a volumetric method. Results showed that the hydrogen adsorption capacity of high-purity carbon nanotube crystals could reach up to 8 wt%. It is believed that when the pressure is higher than 4 MPa, individual single-walled carbon nanotubes will separate, and hydrogen will thus physically adsorb on their exposed surfaces. Nijkamp et al. [145] studied the adsorption of hydrogen on various carbonaceous adsorbents at 77 K within the pressure range of 0–0.1 MPa. The selected samples varied greatly in surface areas. Results showed that the amount of adsorbed hydrogen was found well correlated with the specific surface area, mainly due to the physical adsorption of hydrogen. However, similar to activated carbon, the room temperature hydrogen storage capacity of carbon nanotubes is not satisfied. For example, Nishimiya et al. [146] used the volumetric method to measure the hydrogen isotherms of single-walled carbon nanotubes with an average diameter of 1.32 nm. The maximum hydrogen storage capacities at 295 K & 106.7 kPa and 77 K & 107.9 kPa were 0.932 and 2.37 wt%, respectively. Furthermore, Zhou et al. [147] systematically studied the mechanism of physical adsorption of hydrogen based on the adsorption isotherms collected over a wide temperature and pressure range. They reasoned that the adsorbed hydrogen might be arranged in a single layer on the carbon surface. Therefore, the storage capacity might probably depend on the specific surface area of the carbon. If the interaction between hydrogen and the solid carbon surface remains the van der Waals force, this rule also applies to other organic/inorganic materials. Because the specific surface area of carbon nanotubes is relatively small, it cannot serve as a good adsorbent for hydrogen storage, but super activated carbon has great potential.

By adjusting the surface characteristics and sample properties of carbon nanotubes, the hydrogen adsorption capacity of carbon nanotubes can be improved. Introducing defects can significantly increase the micropore volume, thereby increasing the surface area of CNT. The main methods include ball milling, alkali activation, acid treatment, and bromination. Liu et al. [148] studied the effect of ball milling on the hydrogen adsorption behavior of open short-wall carbon nanotubes. Under a pressure of 8–9 MPa, the hydrogen adsorption amount of carbon nanotubes milled for 10 h at room temperature was 0.66 wt%, about 6 times that of unmilled carbon nanotubes. After 1 h of MgO milling, the hydrogen adsorption amount of carbon nanotubes was 0.69 wt%. The authors suggest that the enhancement of hydrogen adsorption is due to the increase in defects and surface areas of multi-walled carbon nanotubes by ball milling. Chen et al. [149] achieved a huge leap in the hydrogen storage capacity of multi-walled carbon nanotubes from 0.71 to 4.47 wt% at ambient temperature and pressure through KOH activation and high-temperature annealing. Furthermore, Zhang et al. [150] found that using KOH-C redox reactions to form micropores on the walls of multi-walled carbon nanotubes could result in the formation of characteristic micropores at 0.8 nm. More importantly, deep activation could laterally cut the carbon nanotubes, leading to the release of catalyst particles and the formation of medium-size pores. Similarly, enhancing hydrogen adsorption capacity by doping with metal nanoparticles is also a good option. Doping or embedding metal nanoparticles can help maintain the morphology integrity of the carbon nanotubes.

Enhancing hydrogen adsorption through a spillover mechanism has been considered as an effective way to improve the overall hydrogen storage capacity of carbon nanotubes. Lin et al. [151] tested the hydrogen storage capacity of carbon nanotubes before and after doping, with the undoped carbon nanotubes having a hydrogen storage capacity of 0.39 wt%. When nickel nanoparticles are evenly distributed on the surface of carbon nanotubes, up to 1.27 wt% of hydrogen can be stored. Rather et al. [152] prepared multi-walled carbon nanotubes by sputtering and added titanium as a hydrogen spillover catalyst. Under the conditions of 298 K and a balance pressure of 1.6 MPa, the hydrogen storage capacity increased from 0.43 to 2.0 wt%. The increase in the hydrogen storage capacity of the sputtered samples is attributed to the efficient spillover mechanism produced by the good arrangement of titanium nanoparticles on the outer surface of the multi-walled carbon nanotubes. Reyhani et al. [153] used volumetric and electrochemical methods to study the hydrogen storage performance of multi-walled carbon nanotubes modified by Ca, Co, Fe, Ni, and Pd metals. They found that compared with pristine multi-walled carbon nanotubes, the binding energy and hydrogen storage capacity at these metallic sites in multi-walled carbon nanotubes modified by Ca, Co, Fe, Ni, and Pd nanoparticles increased, mainly due to the electron transfer from the metal atoms to carbon atoms. Specifically, Pd nanoparticles at the defect sites will dissociate the H_2 molecules on their surface, thereby resulting in the highest hydrogen storage capacity among the five doped multi-walled carbon nanotubes.

5.4.2 *Principle of Hydrogen Storage in MOF Materials*

Compared with traditional activated carbon and zeolite materials, tuning the functionalities of the MOF pore surfaces can provide many strong hydrogen adsorption sites, thus enabling a higher hydrogen storage capacity. In 2003, Rosi et al. [154] conducted early research on MOF-based hydrogen storage. They tested MOF-5 with a three-dimensional extended porous structure composed of $Zn_4O(BDC)_3$ (H_2BDC = 1,4-benzenedicarboxylic acid). At a low pressure of 2 MPa, it adsorbed 4.5 wt% of hydrogen at 78 K and 1.0 wt% of hydrogen at room temperature. In 2006, Wong et al. measured the hydrogen absorption capacity of a series of MOFs under the saturated pressure of 8 MPa at 77 K, among which MOF-177 (7.5 wt%) showed the highest hydrogen storage capacity. Experimental results also showed that the maximum hydrogen storage capacity in MOFs is closely related to their specific surface area (Fig. 5.19) [155]. Therefore, initial efforts to improve MOF's adsorption performance for H_2 were mainly focused on adjusting the length of the organic linkers to achieve higher porosities. The ideal pore size of MOF materials for hydrogen storage applications should probably match the kinetic diameter of H_2 molecules (~2.9 Å), thereby achieving the strongest interaction between H_2 and MOFs, and maximizing the van der Waals forces [156]. The interaction forces between MOF and hydrogen can be measured by the isosteric heat of adsorption.

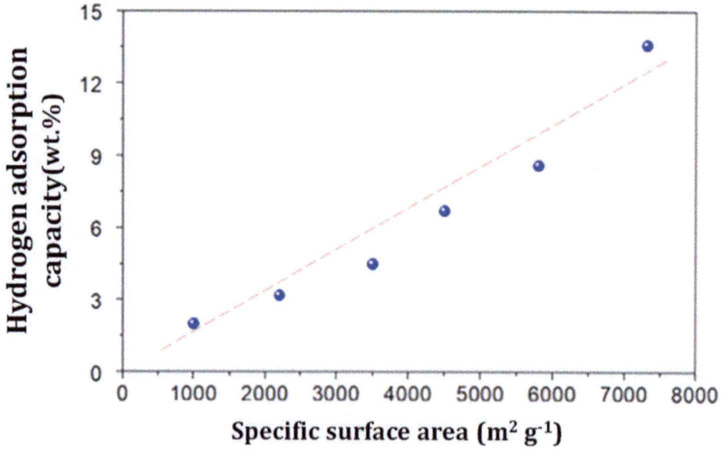

Fig. 5.19 The relationship between the maximum hydrogen storage capacity and surface area of MOF materials

The greater the adsorption heat, the greater the interaction force, and vice versa. Currently, the hydrogen adsorption heat of most MOFs is around 5 kJ/mol H_2, while theoretical calculations indicate that the optimal hydrogen adsorption heat for high-performance MOFs ranges from 15 to 25 kJ/mol H_2 [157]. Currently, methods to enhance the hydrogen storage capacity of MOFs include increasing the unsaturated coordination metal centers, doping metal cations to enhance the charge-induced dipole interaction between MOF and hydrogen molecules, and doping palladium or platinum or other precious metals that can easily dissociate hydrogen molecules to achieve the hydrogen "spillover" mechanism.

Hydrogen adsorption capacity, Specific surface area.

5.4.2.1 Strategies to Improve the Hydrogen Storage Capacity of MOFs

Similar to carbon materials, increasing the heat of H_2 adsorption is another strategy to enhance the hydrogen storage capacity of MOFs. This can be achieved by removing the coordinating solvent molecules from the metal center, forming unsaturated metal sites on the pore surface, and enabling the charge-induced dipole interaction between the open metal sites and H_2 molecules [158]. For example, Xiao et al. [159] tested the hydrogen storage capacity of HKUST-1 at 77 K, with a hydrogen adsorption of 2.27 wt% at 0.1 MPa and 3.6 wt% at 1 MPa. Yong et al. [160] compared the hydrogen storage capacity of both metal vacancy-containing SNU-5 (2.87 wt% at 0.1 MPa; 6.76 wt% at 5 MPa) and metal vacancy-free SNU-4 (2.07 wt% at 0.1 MPa; 4.49 wt% at 5 MPa) at 77 K, indicating that the unsaturated metal sites could contribute to the hydrogen storage. Kapelewski et al. designed a Ni_2(m-dobdc) material with unsaturated Ni^{2+} metal centers, which has a strong MOF-H_2 interaction, with an adsorption heat of 13.7 kJ/mol H_2 and a hydrogen storage capacity of 2.2 wt% at 77 K and 0.1 MPa [157].

Similarly, doping metal atoms (mainly Li^+, Na^+, and K^+) in MOFs, rather than creating open metal sites, can enhance the hydrogen storage capacity of MOFs by increasing the heat of H_2 adsorption through the charge-induced dipole interaction between the metal cations and H_2 molecules. For example, Lim et al. [161] compared the undoped and K^+ doped SNU-200 MOF at 77 K and 0.1 MPa, and the heat of hydrogen adsorption increased from 7.70 to 9.92 kJ/mol H_2, while the corresponding hydrogen storage capacity increased from 1.06 to 1.19 wt%.

Doping with palladium or platinum, precious metals that can easily dissociate hydrogen molecules, can enhance the hydrogen storage capacity of MOFs through the "spillover" mechanism, where dissociated H atoms diffuse from the metal to the MOF pore surface [162]. Li et al. [163] enhanced the hydrogen storage capacity by 74% by growing HKUST-1 on the surface of Pd nanocrystals to form Pd@HKUST-1 nanocomposites. However, Szilágyi et al. [164] argued that the chemical adsorption of hydrogen atoms by the host through the hydrogen spillover mechanism was ensured by a carbon bridge. They introduced catalytic palladium particles into the pores of MOFs, ensuring good contact, thus making the carbon bridge redundant. The addition of palladium nanoparticles indeed increased the hydrogen uptake capacity of MOFs at ambient temperature, but it was found that this was entirely due to the formation of palladium hydrides. Therefore, more reliable experiments and repeatable data are needed to clarify this spillover effect.

5.4.2.2 Progress in MOFs for High-Pressure Hydrogen Storage

For onboard hydrogen storage, the driving range of a vehicle is related to the working capacity of hydrogen, not the absolute hydrogen adsorption capacity, at a chosen temperature and pressure. Generally, the working or available capacity is defined as (Fig. 5.20) [165]: the capacity difference between the supply pressure and release pressure at a specific temperature (pressure swing adsorption process) or the capacity difference between the supply temperature and release temperature at a specific pressure (temperature swing adsorption process). For hydrogen storage in MOF materials, the pressure swing adsorption (PSA) process is generally used. The working capacity of hydrogen is defined as the difference in hydrogen adsorption capacity between the maximum hydrogen charging pressure (about 10 MPa) and the minimum hydrogen release pressure (about 0.5 MPa) at room temperature (296–298 K) [157]. The choice of 10 MPa as the maximum hydrogen charging pressure is because all-metal class I hydrogen storage tanks are relatively cheap and can operate and run safely at this pressure and temperature. Currently, the working capacity of MOF materials under this definition is relatively low (<3.2 wt%), while Ni_2(m-dobdc) material has the highest gravimetric and volumetric hydrogen storage working capacity at 298 K, which are 1.9 wt% and 11 g/L, respectively [157]. Long et al. synthesized the $V_2Cl_{2.8}$(btdd) material, which showed the highest hydrogen adsorption heat (20.6 kJ/mol) [166], through the feedback π-bonding capability enabled by divalent vanadium metal sites. It demonstrated a good room-temperature

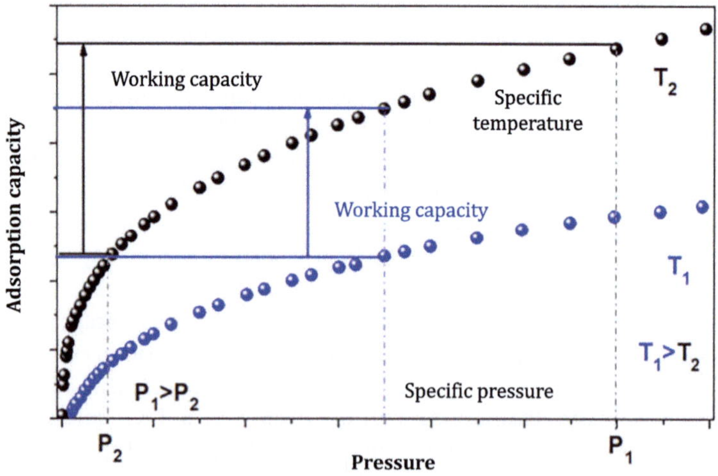

Fig. 5.20 Definition of the hydrogen storage working/available capacity

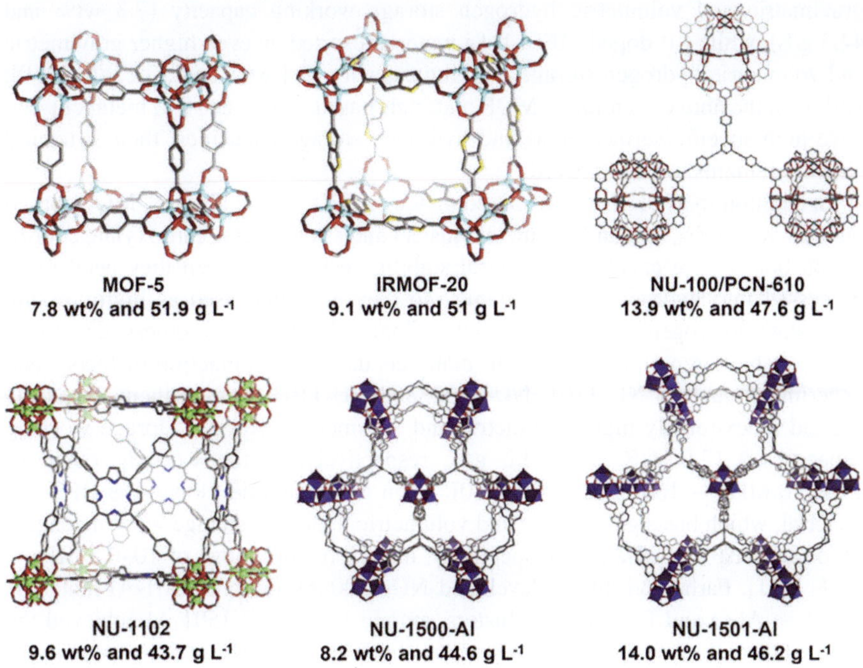

Fig. 5.21 The crystal structure of representative MOFs with the currently highest hydrogen storage working capacity (77 K/10 MPa → 160 K/0.5 MPa) [126]

gravimetric and volumetric hydrogen storage working capacity of 1.5 wt% and 9.8 g/L, respectively. It is worth mentioning that the hydrogen adsorption heat of this MOF material is the only one currently achieved in the optimal hydrogen adsorption heat range of 15–25 kJ/mol H_2 [157].

Adsorption capacity, Working capacity, Specific temperature, Working capacity, Specific pressure, Pressure.

The working capacity of hydrogen at room temperature is relatively low, but by appropriately changing the hydrogen storage operating conditions, such as changing the pressure swing adsorption process operating conditions to 77 K and 10 MPa to 160 K and 0.5 MPa (Fig. 5.21), the working capacity of hydrogen can be greatly increased, which is more conducive to the hydrogen storage tank design and preparation proposed by the DOE Hydrogen Center [126]. The gravimetric and volumetric hydrogen storage working capacity of MOF-5 material is as high as 7.8 wt% and 51.9 g/L (Fig. 5.21) [167]. IRMOF-20, as an extension of MOF-5, has a higher gravimetric and volumetric hydrogen storage working capacity (9.1 wt% and 51 g/L) due to its higher specific surface area [167], while PCN-610 and NU-100 materials have shown an extremely high gravimetric and volumetric hydrogen storage working capacity (13.9 wt% and 47.6 g/L) [168]. Snurr et al., adopted a paradigm combining the machine learning algorithms with grand canonical Monte Carlo (GCMC) simulations, and found that MFU-4l material exhibited a high gravimetric and volumetric hydrogen storage working capacity (7.3 wt% and 44.3 g/L), while Li^+ doped MFU-4l-Li material exerted an even higher gravimetric and volumetric hydrogen storage working capacity (9.4 wt% and 50.2 g/L) [169]. Although the above-mentioned MOF materials based on Cu_2 and Zn_4 metal clusters have high specific surface areas and hydrogen storage capacities, their structural stability remains to be improved.

In addition to MOF materials based on Cu_2 and Zn_4 metal clusters, MOFs based on high-valent Zr_6, Al_3, and Fe_3 metal clusters and strong metal-carboxylate coordination bonds have excellent structural stability, especially when they need to be processed into hydrogen storage tanks, so they are often used as high-pressure solid-state hydrogen storage materials. Gomez-Gualdro and others developed highly stable zirconium-based MOF materials using a combination of theory and experiment, such as NU-1101, NU-1102, and NU-1103. Among them, NU-1103 exerted an extremely high gravimetric and volumetric hydrogen storage working capacity of 12.6 wt% and 43.2 g/L, respectively, at the working range of 77 K/10 MPa → 160 K/0.5 MPa [170]. Chen and others developed the NPF-200 material, which has a gravimetric and volumetric hydrogen storage working capacity of 8.7 wt% and 37.2 g/L, respectively, in the pressure range of 10–0.5 MPa at 77 K [171]. Farha and others developed NU-1500-Al and NU-1501-Al materials based on Al_3O and Fe_3O metal clusters, among which NU-1501-Al achieved the best balance of gravimetric and volumetric hydrogen storage working capacity of 14.0 wt% and 46.2 g/L, respectively (Fig. 5.21) [172]. Despite this, the lower working temperatures, such as 77–160 K, greatly increase its hydrogen storage energy consumption, thus limiting its application in room temperature solid-state hydrogen storage. Table 5.5 summarizes the hydrogen storage performance of common MOFs.

5.4.2.3 Engineering Applications of MOF Materials

From the perspective of practical onboard hydrogen storage applications, gaseous fuels need to be stored at high densities [175]. For example, natural gas is stored in solid thick-walled steel containers at 18–25 MPa at room temperature. The low storage capacity of the storage container, along with its own weight, usually accounts for 90% of the total weight of the system, and the cost of multi-stage complex compression makes this technology impractical [175]. For a driving range of 480–800 km, the onboard gravimetric and volumetric hydrogen storage working capacity targets set by the U.S. Department of Energy are 6.5 wt% and 50 g/L, respectively, which is difficult to achieve by compression alone [176]. In addition, for long-distance onboard hydrogen storage, a higher volumetric hydrogen storage working capacity is needed to address the practical safety and economic barriers in its distribution and utilization, which involves the subsequent processing and granulation applications of MOF materials [177]. Successful packaging and densification of MOF powders into MOF granules determines the downstream processing of MOF materials for hydrogen storage [178, 179]. At present, most calculations of volumetric hydrogen storage working capacity use the crystallographic density of MOFs [126], without

Table 5.5 Hydrogen storage performance of benchmark MOF materials

MOF	BET/ (m^2/g)	Porosity/ (cm^3/g)	Lattice density / (g/cm^3)	Volumetric BET surface area[a]/(m^2/cm^3)	Adsorption heat /(kJ/mol)	Working capacity (adsorption 77 K/10 MPa, desorption 160 K/0.5 MPa)		Working capacity (adsorption 77 K/10 MPa, desorption 77 K/0.5 MPa)		Working capacity (adsorption 296 K/10 MPa, desorption 296 K/0.5 MPa)	
						wt%	g/L	wt%	g/L	wt%	g/L
IRMOF-20 [167]	4070	1.65	0.51	2080		9.1	51.0	5.7	33.4		
HKUST-1 [173]	1980	0.75	0.88	1740	6.5	4.9	46.0	2.5	22.0	1.2	10.8
NU-125 [173]	3230	1.33	0.58	1870	5.1	7.8	49.0	4.8	28.0	1.6	9.5
Cu-MOF-74 [173]	1270	0.47	1.32	1680	5.6	3.0	39.0	1.1	15.0	0.8	10.1
UiO-67 [173]	2360	0.91	0.69	1620	5.8	6.0	41.0	3.4	23.0	1.2	8.0
PCN-250 [173]	1780	0.71	0.90	1595	6.6	4.9	47.0	2.0	18.0	1.2	10.4
CYC U 3Al [173]	2450	1.56	0.48	1170	4.5	8.7	41.0	4.9	24.0	1.8	8.6
NU-1101 [170]	4340	1.72	0.46	1990	5.5	9.1	46.6	6.1	29.0	About 2.0	About 9.3
NU-1103 [170]	6245	2.72	0.30	1860	3.8	12.6	43.2	10.1	33.3	About 3.2	About 9.7
UMCM-9 [168]	5040	2.31	0.37	1860		11.3	47.4	7.3	34.1		
SNU-70 [168]	4940	2.14	0.41	2030	5.1 [174]	10.6	47.9	7.8	34.3		
NU-100 [168]	6050	3.17	0.29	1755	6.1 [174]	13.9	47.6	10.1	35.5	–	–

(continued)

Table 5.5 (continued)

MOF	BET/ (m^2/g)	Porosity/ (cm^3/g)	Lattice density / (g/cm^3)	Volumetric BET surface area[a]/(m^2/cm^3)	Adsorption heat /(kJ/mol)	Working capacity (adsorption 77 K/10 MPa, desorption 160 K/0.5 MPa)		Working capacity (adsorption 77 K/10 MPa, desorption 77 K/0.5 MPa)		Working capacity (adsorption 296 K/10 MPa, desorption 296 K/0.5 MPa)	
						wt%	g/L	wt%	g/L	wt%	g/L
NU-1501-Fe [172]	7140	2.90	0.30	2130	4.0	13.2	45.4	10.8	32.4	About 2.6	About 7.8
NU-1501-Al [172]	7310	2.91	0.28	2060	4.0	14.0	46.2	10.0	28.0	About 3.1	About 8.7
NPF-200 [171]	5830	2.17	0.39	2268	4.5			8.7	37.2		
MFU-4l [169]	3160	1.30	0.56	1766	5.5	7.3	44.3	5.4	33.4	About 1.4	7.8
MFU-4l-Li [169]	4070	1.66	0.48	1950	5.4	9.4	50.2	6.9	35.4	1.8	8.7
Ni2(m-dobdc) [157]	1321	0.56	0.58 (0.366[b])		12.3			About 3.6	About 20.58	About 1.9[c]	11.0[c]
V2Cl2.8(btdd) [166]	1920	1.12	0.64	1229	20.9	–	–	–	–	Approximately 1.5[c]	Approximately 9.8[c]

[a] Based on lattice density
[b] Packaging density
[c] At 298 K

considering the gaps between MOF particles. Using tap density results in a significant decrease in their volumetric hydrogen storage working capacities and increasing the packaging density of MOF materials after processing becomes the key to solving this problem [180]. Due to the low packaging efficiency of powders, the real hydrogen storage capacity of MOF materials cannot reach their theoretical values. Mechanical compression granulation method has been widely used to increase the packaging density of MOF materials, which helps to increase the volumetric hydrogen storage working capacity [181], but it can cause the MOF stability to deteriorate and its structure to collapse.

Long and others, from the University of California, Berkeley, studied the practical hydrogen storage performance of Ni_2(m-dobdc) materials under the temperature swing adsorption (TSA) process operation, and used the packaging density (0.366 g/mL) to calculate the volumetric hydrogen storage working capacity [157]. The volumetric hydrogen storage capacity could reach up to 23 g/L in the temperature range of −75 to 25 °C under the pressure of 10 MPa. Matzger and others at the University of Michigan proposed a strategy by the crystal morphology engineering and grain size control to optimize the real hydrogen storage performance of MOF-5, which had significantly improved the packaging efficiency and volumetric hydrogen storage capacity [182]. Compared with commercial MOF-5 materials (0.189 g/mL), the volumetric hydrogen storage performance of resultant MOF-5 materials has been significantly improved due to the increased packaging density of MOF-5 materials (0.308–0.365 g/mL). It was found that the packaging density of a 7:1 mass ratio mixed materials of MOF-5 (2349) and MOF-5 (808) of different crystal sizes could reach up to 0.380 g/mL. The volumetric hydrogen storage capacity in the working range from 77 K and 10 MPa to 160 K and 0.5 MPa could reach up to 30.5 g/L, exceeding the volumetric hydrogen storage capacity of typical 70 MPa compressed hydrogen systems (25 g/L), and surpassing the volumetric working capacity target of the U.S. Department of Energy in 2020 (30 g/L). This work reveals the key relationship between increasing powder packing density and reducing the collapse of material structure caused by mechanical compression, thus providing a theoretical basis for the development of MOF materials with high specific surface areas and high packaging densities.

In summary, most of the current hydrogen storage research using MOF materials remain focusing on solving scientific problems, such as low MOF room-temperature hydrogen storage capacity and unstable structure, while solving engineering problems, such as low powder packaging density (packaging density) and insufficient mass and heat transfer in hydrogen storage containers to improve the volumetric hydrogen storage capacity and energy utilization efficiency of MOF materials, is equally important for the onboard hydrogen storage research. Future research on the hydrogen storage applications of MOF materials needs to consider these two aspects.

5.5 Solid-State Hydrogen Storage and Transportation Technology and Its Applications

Solid-state hydrogen storage and transportation technology refers to the technology of storing and transporting hydrogen gas through physical/chemical adsorption or forming hydrides, using solid-state hydrogen storage materials as the storage medium, such as hydrogen storage alloys, complex hydrides, MOFs, etc. At present, the most practical value is the hydrogen storage alloy, which is filled in a specific hydrogen storage tank to realize the hydrogen charging and discharging processes. Solid-state hydrogen storage and transportation technology has the following advantages: a high hydrogen storage density with a volumetric hydrogen storage density ≥100 g/L; a lower working pressure, which can directly match the outlet pressure of the water electrolyzers for hydrogen production during hydrogen refueling processes, no need for pressurization equipment; good system safety and no explosion risks; long (>1000 times) reversible hydrogen charging and discharging cycles and repeatable use; simple reaction and high hydrogen release purity.

According to the application scenarios, different systems of hydrogen storage alloys can be chosen:

1. Low-temperature hydrogen storage alloys: hydrogen storage alloys working at temperatures near room temperature, such as $LaNi_5$ series, TiFe series, $TiMn_2$ series, V-based solid solutions, etc., can be collectively called as low-temperature hydrogen storage alloys. These hydrogen storage alloys have relatively low gravimetric hydrogen storage densities (1–3.7 wt%). Because they can release hydrogen at room temperature, they are thus more suitable for hydrogen storage in fixed application scenarios, such as hydrogenation refueling stations, fixed energy storage, and special onboard application scenarios, such as forklifts.
2. High-temperature hydrogen storage alloy: hydrogen storage alloys working at temperatures ≥150 °C, such as Mg series, can be called as high-temperature hydrogen storage alloys. The gravimetric hydrogen storage density of Mg-series alloys is 4–7.6 wt%, which can store and transport hydrogen at normal temperature and pressure. China's total reserves of magnesium resources account for about 22.5% of the world, ranking the first in the world, and the market share of magnesium ingots has maintained above 85% in the past 5 years, with abundant raw material sources and low cost. Therefore, magnesium-based hydrogen storage alloys are suitable for large-scale hydrogen storage and transportation applications, including but not limited to hydrogen metallurgy, large-scale energy storage, hydrogen refueling stations, and other application scenarios.

As early as the 1980s, Japan's Kawasaki Heavy Industries used a 1000 kg mixture of rare earth-nickel-aluminum alloy to make the world's largest solid hydrogen storage tank at the time, with a hydrogen storage capacity of 175 m^3, and successfully used the hydrogen storage alloy container on Toyota's four-stroke engine in 1985, driving a mileage of 200 km on the road. Subsequently, the Japan Chemical Technology Research Institute used a 1200 kg mixture of rare earth AB_5-type alloy

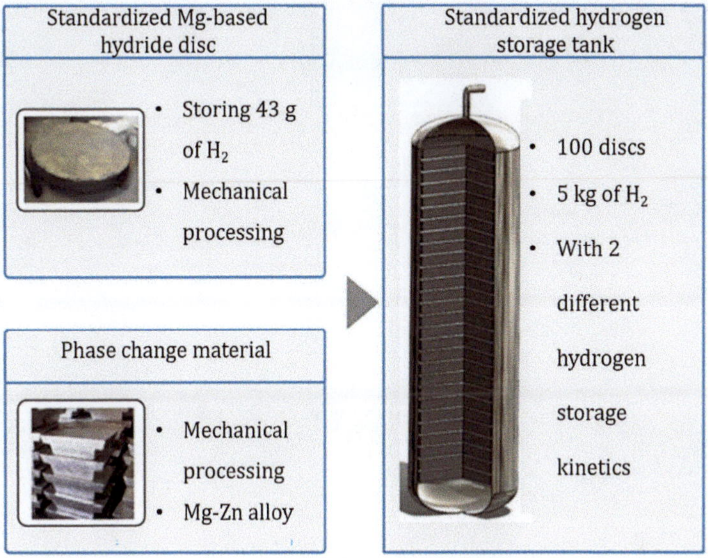

Fig. 5.22 Magnesium-based solid-state hydrogen storage materials and their hydrogen storage tanks produced by McPhy Company

to make a hydrogen storage tank with a storage capacity of 240 m^3. France's McPhy company developed the McStore hydrogen storage system based on magnesium alloy as the hydrogen storage medium around 2010, with a single tank hydrogen storage capacity of up to 5 kg (Fig. 5.22). This system has been currently used as a hydrogen storage medium in the INGRID demonstration project in Italy to achieve the power regulation. In China, the research progress on solid hydrogen storage tanks is generally synchronized with foreign countries. Zhejiang University developed a low-temperature solid hydrogen storage tank with a hydrogen storage capacity of 240 m^3 for hydrogen recovery and purification in the early years; China Grinm Technology Group Co., Ltd. (formerly Beijing General Research Institute for Nonferrous Metals) developed a hydrogen storage system based on AB_2-type hydrogen storage alloys with a hydrogen storage capacity of 40 m^3 in 2012, and successfully coupled with a 5 kW fuel cell system, providing continuous power supplies for communication base stations for nearly 17 h. However, due to the small demand in the hydrogen energy market at that time, solid hydrogen storage and transportation technology did not realize large-scale applications.

Standardized magnesium-based hydrogen storage material disc, Storing 43 g of hydrogen, Mechanical processing.

Phase change materials, Mechanical processing, Mg-Zn alloy.

Standardized hydrogen storage tank, 100 discs, 5 kg of hydrogen, with two different hydrogen absorption and release kinetic properties.

In recent years, with the development of the hydrogen industry, solid-state hydrogen storage technology, due to its high density and high safety features, has been

Fig. 5.23 The H_2One "hydrogen production-hydrogen storage-power generation" integrated system

gradually investigated and applied by many institutions both at home and abroad. The European Union launched the HyCARE project in 2020, which used the TiFe-series hydrogen storage alloy, with an alloy hydrogen storage capacity of 1.42 wt% and a capacity of 1.16 wt% after 250 cycles. The hydrogen storage equipment was coupled with a matching 20 kW PEM hydrogen production equipment and 10 kW PEMFC fuel cell stacks to further improve the overall heat utilization efficiency. In 2020, the University of New South Wales in Australia collaborated with LAVO and launched a 40 kW h hydrogen-fueled battery-based backup power system. The system used a low-temperature hydrogen storage alloy as the hydrogen storage medium, forming a "solar cell-electrolyzed hydrogen production-solid hydrogen storage-fuel cell power generation" integrated system. In 2019, Toshiba Japan developed a fixed "hydrogen production-hydrogen storage-power generation" integrated system called H_2One (Fig. 5.23). This system used a low-temperature hydrogen storage alloy as the hydrogen storage medium and had been successfully applied to the backup power sources in Singapore. In 2015, Australia's Hydrexia company successfully developed a hydrogen storage and transportation equipment based on magnesium alloys, with a designed single vehicle hydrogen storage and transportation capacity of 700 kg, which could be used for large-scale safe storage and transportation of hydrogen.

In recent years, China Grinm Technology Group Co., Ltd. has developed an onboard vehicle hydrogen storage system based on $TiMn_2$ hydrogen storage alloys, with a total hydrogen storage capacity of 17 kg, which has been used in hydrogen fuel cell buses (Fig. 5.24a). At the same time, it has also developed a TiFe hydrogen storage system with a storage capacity of 1000 m^3, which is expected to be used in the Hebei Guyuan Wind Power Hydrogen Production Project as a safe and compact

(a) (b) (c)

a. The world's first low-pressure hydrogen storage alloy fuel cell bus
b. Hydrogen fuel cell-assisted bicycle
c. Large-scale magnesium-based solid-state hydrogen storage prototype device

Fig. 5.24 Domestic solid-state hydrogen storage and transportation applications. (**a**) The world's first low-pressure hydrogen storage alloy fuel cell bus. (**b**) Hydrogen fuel cell-assisted bicycle. (**c**) Large-scale magnesium-based solid-state hydrogen storage prototype device

on-site hydrogen storage buffer. Jiangsu JITRI Advanced Energy Materials Research Institute Co., Ltd. has developed a hydrogen fuel cell-assisted bicycle with a mileage of 80 km based on modified AB_5-type hydrogen storage alloys (Fig. 5.24b), which has been put into industrial park demonstration applications.

Shanghai Jiao Tong University and Hydrexia (China) have jointly developed the first 60 kg-scale magnesium-based solid-state hydrogen storage prototype device in China (Fig. 5.24c), and cooperated with Baowu Clean Energy Co., Ltd. to develop a "solar power generation—water electrolysis hydrogen production—magnesium-based solid-state hydrogen storage/supply" skid-mounted integrated hydrogen energy system, called "Hydrogen Quadriga", which has demonstrated the feasibility of magnesium-based solid-state hydrogen storage technology through its demonstration operation. In summary, solid-state hydrogen storage and transportation technologies for various application scenarios are currently being developed both at home and abroad, but solid-state hydrogen storage and transportation technology is still in the early stage of industrialization, and there are problems to be solved, such as low-cost preparation of materials at large scale, design of large-capacity hydrogen storage tanks, integration of high-temperature waste heat coupling, etc., to achieve efficient and safe hydrogen absorption and release of hydrogen storage alloys.

A solid-state hydrogen storage tank mainly includes solid-state hydrogen storage material, shell, gas pipeline and filter, fin, metal foam, heating pipe and other enhanced heat transfer media, preset spare space, etc.

1. Shell: due to the good sealing and pressure resistance characteristics of the round shell, the shell of the solid-state hydrogen storage tank is mostly cylindrical.

2. Gas pipeline and filter: during the hydrogen charging and discharging process, hydrogen gas flows along the gas pipeline and enters and exits the hydrogen storage tank through the filter. The main function of the filter is to prevent hydrogen storage material particles from being carried away into the gas pipeline by the flow of hydrogen gas. In addition, for large solid-state hydrogen storage tanks, in order to ensure the uniformity of hydrogen gas in the hydrogen storage tank, several filters can be installed in the container to provide a fast flow channel for hydrogen gas, so that the hydrogen pressure in the hydrogen storage tank remains relatively uniform.
3. Fins, metal foams, heating pipes and other enhanced heat transfer media: due to the large contact thermal resistance of the powder bed of the hydrogen storage material, the effective thermal conductivity of the bed body is as low as 1–7 W/(m K), and the exothermic/endothermic attributes of hydrogen absorption/release reactions results in significant temperature fluctuations in the hydrogen storage tank during the hydrogen absorption and release process, which further reduces the hydrogen absorption and release performance of the hydrogen storage tank. Therefore, methods such as fins, metal foams, and heat pipes are often used to enhance the heat transfer performance within the hydrogen storage container.
4. Pre-set spare space: hydrogen storage materials will undergo volume expansion and contraction during the hydrogen absorption and desorption processes. After the long-term cyclic service, the hydrogen storage material particles will pulverize under stress. Under the effect of gravity, small-sized particles will densely accumulate at the bottom of the hydrogen storage tank, leading to a significant stress concentration at the bottom of the tank during the hydrogen absorption process, threatening the structural safety of the hydrogen storage tank. Therefore, generally, 10–20% of the space inside the hydrogen storage tank is reserved to compensate the increased volume of the hydrogen storage materials during hydrogen absorption processes, thus reducing the stress on the container. For hydrogen storage tanks that are longer in the direction of gravity, a multilayer approach can be used to alleviate the self-compaction phenomenon of the hydrogen storage material powder. A typical solid-state hydrogen storage tank design is shown in Fig. 5.25.

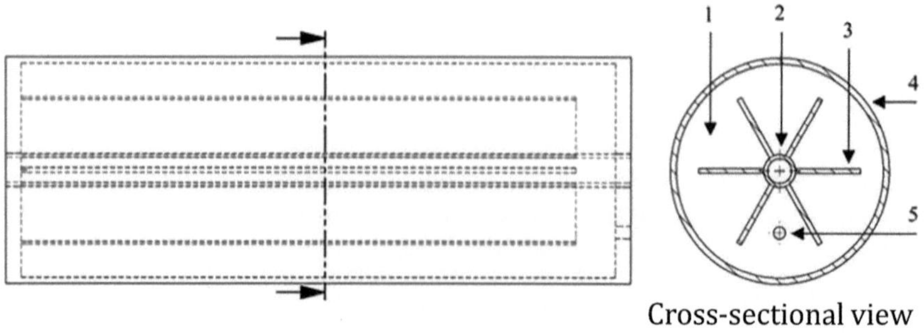

Fig. 5.25 A scheme of typical solid-state hydrogen storage tank design

Solid-state hydrogen storage material, Heating/cooling pipe, Fin, Shell, Gas inlet and outlet, Cross-sectional view.

At present, the design of solid-state hydrogen storage tanks is mostly carried out by combining both theoretical simulations and experiments.

The control equations of the hydrogen absorption and desorption mathematical model in a solid-state hydrogen storage tank mainly include the hydrogen absorption and desorption reaction equilibrium pressure equation and the kinetic equation of the hydrogen storage alloy, as well as the heat and transfer equations describing the tank heat transfer and hydrogen flow processes [183–185]. When constructing the hydrogen absorption and desorption mathematical model in a hydrogen storage tank, the following assumptions are often made: H_2 is an ideal gas; local equilibrium assumption, the pressure, temperature, and hydrogen content in each unit are identical; the radiation heat exchange is ignored; the thermal physical parameters of hydrogen storage alloy, stainless steel and other solids, and hydrogen gas remain constant. For the established mathematical model, it is generally solved on a computer through the finite element method. Self-written programs or commercial software (such as COMSOL Multiphysics, Ansys Fluent, etc.) are adopted to discretize and iteratively solve the mathematical model, and capture the changes in temperature, pressure, hydrogen content, and other variables during the solution process.

5.5.1 Hydrogen Absorption and Desorption Reaction Equilibrium Pressure Equation of Hydrogen Storage Alloys

Through the analysis of the hydrogen absorption and desorption process of the hydrogen storage tank, it is found that the hydrogen absorption and desorption equilibrium pressure of the hydrogen storage material determines whether the hydrogen absorption and desorption reaction can occur. Therefore, it is necessary to consider the PCT curves of the hydrogen storage alloys at different temperatures and obtain the numerical relationships between the equilibrium pressure, temperature, and hydrogen content (H/M). At present, there are mainly the following ways to predict the alloy PCT curve: polynomial fitting + van 't Hoff equation; microscopic parameter method to predict the alloy enthalpy change and PCT curve; statistical thermodynamic theory prediction; phase diagram calculation method. Among them, the first method only needs a PCT curve at a reference temperature and predicts the PCT curve at different temperatures in conjunction with the van 't Hoff equation. The second method considers microscopic structural parameters, such as electronegativity differences and size factors, and predicts the enthalpy change values of a certain alloy, and at the same time predicts the PCT curve by segment processing. The third and fourth methods are based on the principle of equal chemical potential, to construct a thermodynamic database of alloy-hydrogen systems, and then predict the hydrogen absorption and desorption PCT curves of alloys.

Compared with other equilibrium pressure prediction methods, the first method of "polynomial fitting + van 't Hoff equation" has the following advantages: it is simple and easy to obtain the parameters; it has a wide range of applications and accurate prediction results; the predicted curve is smooth and there is no problem existing in the transition area that is difficult to handle in the second to fourth methods, which is very beneficial to enhance the convergence in the solving process. The main disadvantage is that in the model, except for the enthalpy change value, the equilibrium pressure at the reference temperature is fitted by a polynomial, which has no physical meaning. However, the role of the equilibrium pressure equation in the mathematical model is to predict the equilibrium pressure at different temperatures and H/m values. If the predicted equilibrium pressure exerts a sudden change, it is easy to cause singular points in the iterative solution process, which leads to the model's inability to converge, and the calculation thus fails. Therefore, the method of "polynomial fitting + van't Hoff equation" is widely used in mathematical models.

5.5.2 Hydrogen Absorption and Desorption Reaction Kinetic Equation of Hydrogen Storage Alloys

In the mathematical model, it is necessary to construct a description of the instantaneous absorption and desorption rate of the alloy under different pressures, temperatures, and hydrogen contents, to calculate the changes in pressure, temperature, and hydrogen content of each unit within the calculation time interval. The instantaneous absorption and desorption rate is generally calculated based on the alloy density differences and isothermal-isobaric kinetics.

1. Alloy density difference method. The method of calculating the absorption and desorption rate by the alloy density difference was first used by Mayer in 1987, and its equation is:

$$\mathrm{v} = -C_0 \exp\left(-\mathrm{E_a} / \mathrm{R_g T}\right) \mathrm{Ln}\left(\mathrm{P} / \mathrm{P_{eq}}\right)\left(\rho_{MH} - \rho_s\right) \tag{5.38}$$

where, v is the absorption and desorption rate; C_0 is a constant; E_a is the reaction activation energy; ρ_{MH} is the density of the hydride; ρ_s is the density of the alloy phase. This equation has been used in $LaNi_5$, ZrCo and other systems. However, this rate calculation method has significant limitations:

(a) The mechanism of hydrogen absorption and desorption kinetics is unclear, and it cannot describe the hydrogen absorption and desorption mechanism of the alloy.
(b) Insufficient applicability: it cannot be applied to the control step and the situation where the equation parameters change with the pressure or particle size.

Therefore, the method of calculating the absorption and desorption rate by the alloy density difference has application limitations.

2. Isothermal-isobaric kinetic method. Most mathematical models use the isothermal-isobaric kinetic equation to describe the hydrogen absorption and desorption process of the alloy. The hydrogen absorption and desorption kinetic equation is mainly used to describe the hydrogen absorption and desorption rate of the alloy under different temperatures, particle sizes, pressures, and H/m conditions. The general form of the hydrogen absorption and desorption rate equation is:

$$v = k_0 \exp^{(-Ea/RgT)} f\left(P, P_{eq}\right) g(\xi) \tag{5.39}$$

where, k_0 is the intrinsic rate constant; ξ is the reaction fraction, defined as the ratio of the current alloy hydrogen absorption amount to the maximum hydrogen absorption amount; $f(P, P_{eq})$ is the pressure term; $g(\xi)$ is the term related to the reaction fraction. The hydrogen absorption and desorption rate equation generally uses the existing isothermal-isobaric kinetic equation to analyze the reaction fraction-time data during the hydrogen absorption and desorption process, and converts it into a rate equation after obtaining the activation energy and rate constant parameters. At present, the traditional isothermal-isobaric kinetic equations describing the hydrogen absorption and desorption process of the alloy mainly construct the equation composed of the reaction fraction and time, as shown in Table 5.6. These kinetic equations are all semi-empirical models, derived on the basis of certain assumptions. Among them, the Chou model expands the rate constant k, thus revealing the relationship between the rate constant and temperature, pressure, and equilibrium pressure. The similar rate equation can be obtained by taking the partial derivative of the kinetic equation in the table.

The reaction kinetics of alloys is generally tested by the Sievert method (also known as the constant volume method). The test is carried out under isothermal and isobaric conditions, and the hydrogen content of the alloy is calculated through the equation of state, thereby obtaining the curve of the reaction fraction and time. According to different kinetic equations, the reaction fraction is converted into the left-side term of the equation in Table 5.6, and its scatter plot with time is thus made. After linear fitting, the rate-controlling step of the alloy's hydrogen absorption and desorption process under this temperature and pressure condition is judged; fitting the curves under different temperatures and pressures, the expression of the pressure term and the values of the activation energy and rate constant are judged; finally, the kinetic mechanism of the alloy's hydrogen absorption and desorption and the corresponding kinetic equation are obtained, which are used in the mathematical model of the hydrogen absorption and desorption reaction in the solid-state hydrogen storage tank.

Table 5.6 Some representative isothermal-isobaric kinetics equations

Name	Equation	Ratedeterming step
Chou	$1-(1-\xi)^{1/3}=k_0\exp\left(\frac{-E_a}{R_gT}\right)\left(\sqrt{P}-\sqrt{P_{eq}}\right)t$	Surface seepage/ chemical reaction
	$\left[1-(1-\xi)^{1/3}\right]^2=k_0\exp\left(\frac{-E_a}{R_gT}\right)\left(\sqrt{P}-\sqrt{P_{eq}}\right)t$	Diffusion
Jander	$[1-(1-\xi)^{1/3}]^2=kt$	Diffusion
Contracting volume	$1-(1-\xi)^{1/3}=kt$	Interface movement
Ginstling-Brounshtein	$(1-\xi)\ln(1-\xi)+\xi=kt$	Diffusion
Valensi-Carter	$\frac{\omega-\left[1+(\omega-1)\xi\right]-(\omega-1)(1-\xi)^{2/3}}{\omega-1}=kt$	Diffusion
Jonhson-Mehl-Avrami-Kolomogorov (JMAK)	$[-\ln(1-\xi)]^{1/n}=kt$	Nucleation and growth (depends on the value of n)

Note: k is the rate constant; ω is the volume expansion ratio before and after the alloy absorbs and desorbs hydrogen; n is the Avrami index

Because the alloy's hydrogen absorption and desorption process is too complex, its rate and rate-controlling steps will be affected by factors, such as particle size, pressure, temperature, preparation and processing conditions, and catalytic phase. Simply using the density differences to calculate the rate of hydrogen absorption and desorption cannot accurately predict the real rates of hydrogen absorption and desorption of alloys, while using the kinetic equation to analyze the hydrogen absorption and desorption process of alloys under different conditions can reveal the kinetic mechanism of hydrogen absorption and desorption processes of the alloy within a certain range of conditions (including rate-determining steps, activation energy, and rate constants, etc.). Based on the judgment of the mechanism of alloy's hydrogen absorption and desorption processes and the acquisition of kinetic parameters, the kinetic equation can be used to predict the hydrogen absorption and desorption rates under other temperatures, particle sizes, etc., within this range of conditions. However, it is worth noting that since the kinetic equation is semi-empirical, it can only be applicable within the limited range, and the error of extrapolated data beyond this range is relatively large.

5.5.3 Heat Transfer Equation

Due to the complex composition of the solid-state hydrogen storage tank, control equations are generally determined separately for different areas.

1. Powder area and filter area. In the powder area, in addition to the heat conduction between alloy powders, there is also convective heat transfer caused by hydrogen gas flow, so it is necessary to consider the effects of heat conduction and convective heat transfer in the heat transfer equation. The viscosity coefficient of hydrogen gas is extremely low, and the influence of hydrogen gas viscosity dissipation on heat transfer is generally ignored. At the same time, due to the thermal effect of the hydrogen absorption and desorption reaction, a heat source term needs to be introduced in the heat transfer equation. Therefore, the heat transfer equation in the powder area is:

$$\left(\rho C_{P}\right)_{e}\frac{\partial T}{\partial t}+\nabla\left(\rho_{g}C_{pg}uT=\nabla\left(\lambda_{e}\nabla T+S_{T}\right)\right) \tag{5.40}$$

where, $(\rho C_P)_e$ is the product of effective density and specific heat capacity; ρ_g is the density of hydrogen gas; C_{pg} is the specific heat capacity of hydrogen gas; u is the gas flow speed; S_T is the heat source term; λ_e is the effective thermal conductivity of the powder bed.

For the filter area, there is no hydrogen absorption and desorption reaction, so S_T is always 0. For the powder area, the value of S_T is related to the rate of hydrogen absorption and desorption, and its calculation equation is:

$$S_{T}=\pm\frac{\rho_{g}\left(1-\varepsilon\right)M_{H}\Delta H}{M_{H2}M_{g}}v \tag{5.41}$$

where, M_{H2}, M_H and M_s are the relative molar masses of hydrogen gas, hydrogen atoms, and alloy, respectively; ε is the porosity of the powder bed.

The product of effective density and specific heat capacity in the heat transfer equation $(\rho C_P)_e$, is generally obtained by the volume average of the product of density and heat capacity of the alloy and hydrogen:

$$\left(\rho C_{P}\right)_{e}=\varepsilon\rho_{g}C_{pg}+\left(1-\varepsilon\right)\rho_{s}C_{ps} \tag{5.42}$$

where, ρ_s is the solid density; C_{ps} is the specific heat capacity of the solid.

2. Blank area. The blank area only includes the heat transfer of hydrogen, described by the following formula:

$$\rho_g C_{pg} \frac{\partial T}{\partial t} + \nabla(\rho_g C_{pg} uT) = \nabla(\lambda_g \nabla T) \quad (5.43)$$

3. Solid area. For the inner and outer stainless-steel walls and other solid areas, only solid heat conduction exists:

$$\rho_s C_{ps} \frac{\partial T}{\partial t} = \nabla(\lambda_s \nabla T) \quad (5.44)$$

5.5.4 Mass Transfer Equation

Powder area and filter area. For the powder area, it can generally be considered as a porous medium, and Darcy's law can be used to describe the relationship between pressure and gas velocity in the powder bed:

$$u = -\frac{K}{\mu_g} \nabla P \quad (5.45)$$

where, K is the permeability of the powder bed; μ is the viscosity coefficient of the gas.

The change in gas density is described by mass transfer equation:

$$\frac{\partial \varepsilon \rho_g}{\partial t} + \nabla(\rho_g u) = -S_P \quad (5.46)$$

For the filter area, the mass source term S_P is always 0. For the powder area, S_P is related to the rate:

$$S_p = \pm \rho_s (1-\varepsilon) \frac{M_H}{M_s} v \quad (5.47)$$

Blank area. The flow of hydrogen is described by the Navier-Stokes equation:

$$\frac{\partial \rho_g u}{\partial t} + \nabla(\rho_g uu) = -\nabla P + \mu_g \nabla^2 u + \rho_g g \quad (5.48)$$

The change in pressure and gas density is described by:

$$\frac{\partial \rho_g}{\partial t} + \nabla(\rho_g u) = 0 \quad (5.49)$$

Exercises

1. Among the following hydrogen storage metals (alloys), the one with the highest hydrogen storage gravimetric density is ( ).

   A. $LaNi_5$
   B. V
   C. Mg
   D. TiFe

2. Please select the following hydrides belonging to simple metal hydrides ( ).

   A. Mg_2FeH_6
   B. AlH_3
   C. $Mg(BH_4)_2$
   D. $LiAlH_4$

3. The incorrect statement about the hydrolysis hydrogen production equation of solid hydrogen storage materials is ( ).

   A. $NH_3BH_3 + 2H_2O \rightarrow NH_4^{+}+BO_2 + 3H_2$
   B. $MgH_2 + 2H_2O \rightarrow Mg(OH)_2+2H_2$
   C. $NaBH_4 + 2H_2O \rightarrow NaBO_2 + 4H_2$
   D. $LiBH_4 \rightarrow LiH + B + 3/2H_2$

4. The advantages of $LiBH_4$ include ( ).

   A. High hydrogen storage capacity
   B. Hydrogen absorption and release at room temperature
   C. Good reversibility
   D. Fast hydrogen release rate

5. The advantages of AB-type TiFe alloy include ( ).

   A. Strong resistance to impurities
   B. Hydrogen absorption and release at room temperature
   C. Low cost
   D. Long service life

6. The following hydrogen storage materials can reversibly absorb and release hydrogen ( ).

   A. Mg
   B. $LiBH_2NH_3$
   C. Graphene
   D. LiH-$LiNH_2$

7. The enthalpy change of hydrogen absorption for a certain AB_5 alloy is −30 kJ/mol H_2, and the hydrogen absorption plateau pressure at 25 °C is 0.12 MPa. Please calculate the plateau pressure at 40 °C.

8. The main categories of solid-state hydrogen storage materials are ____, ____, ____, and ____.
9. The main components of a solid-state hydrogen storage tank include: _____, _____, _____, _____, _____, and _____.
10. The theoretical mass hydrogen storage density of MgH_2 is ____, and the reversible hydrogen absorption and release reaction equation is _____.
11. The preparation methods of hydrogen storage metals (alloys) include ____, ____, and ____.
12. (1) The three-step hydrogen release reaction of $LiAlH_4$ is ____, ____, and ____.
    (2) The three-step hydrogen release reaction of $NaAlH_4$ is ____, ____, and ____.
13. Briefly describe the main steps of the hydrogen absorption process of metal hydrides.
14. Briefly describe the characteristics of solid-state hydrogen storage and transportation technology.
15. Please write out the reaction equations of MgH_2 and $LiNH_2BH_3$ with liquid water and calculate the mass of hydrogen gas that can be released by complete hydrolysis of 1 kg of each hydrogen storage material.
16. Compare the advantages and disadvantages of metal hydrides, complex hydrides, ammonia boranes and their derivatives, and physical adsorptive hydrogen storage materials.
17. Generally, the volumetric hydrogen storage capacity equals the product of the gravimetric hydrogen storage capacity and the density of MOF materials. According to the definitions of gravimetric and volumetric working capacities of hydrogen storage in this chapter, please answer the following questions:

    A. Complete the following table and select the MOF material with the highest gravimetric and volumetric working capacities of hydrogen storage.
    B. Compare the adsorption heat of the following four materials and discuss how to synthesize MOF materials with high hydrogen storage capacities.

MOF materials	1	2	3	4
Packaging density (g/mL)	0.6	0.8	0.9	0.95
Temperature (K)	298	298	298	298
Adsorption heat (kJ/mol)	7.7	8.4	10.9	18.6
Gravimetric adsorption capacity at 10 MPa pressure (wt%)	6.7	5.5	8.5	10.5
Gravimetric adsorption capacity at 0.5 MPa pressure (wt%)	2.4	1.2	3.6	2.7
Gravimetric hydrogen storage working capacity (wt%)				
Volumetric hydrogen storage working capacity (g/L)				

References

1. Lototskyy MV, Yartys VA, Pollet BG et al (2014) Metal hydride hydrogen compressors: a review. Int J Hydrog Energy 39(11):5818–5851
2. Ni C (2012) Preparation and electrochemical performance of rare earth hydrogen storage electrode alloy for special MH/Ni batteries. Central South University, Changsha
3. Huang H, Huang K, Liu S (2008) Progress in research on hydrogen storage technology and its key materials. New Chem Mater 36(11):4–43
4. Züttel A (2003) Materials for hydrogen storage. Mater Today 6(9):24–33
5. Lai QW, Paskevicius M, Sheppard DA et al (2015) Hydrogen storage materials for mobile and stationary applications: current state of the art. ChemSusChem 8(17):2789–2825
6. Cai Y, Xu J, Hu F et al (2018) Hydrogen storage technology and materials. Chemical Industry Press, Beijing
7. Zhang Q, Zou J, Ren L et al (2020) Progress in research on core-shell structured nano-magnesium-based composite hydrogen storage materials. Mater Sci Technol 28(3):10
8. Liu T, Wang C, Wu Y (2014) Mg-based nanocomposites with improved hydrogen storage performances. Int J Hydrog Energy 39(26):14262–14274
9. Jia Y, Sun C, Shen S et al (2015) Combination of nanosizing and interfacial effect: future perspective for designing Mg-based nanomaterials for hydrogen storage. Renew Sust Energ Rev 44:289–303
10. Zou J, Long S, Chen X et al (2015) Preparation and hydrogen sorption properties of a Ni decorated Mg based Mg@ Ni nano-composite. Int J Hydrog Energy 40(4):1820–1828
11. Zou J, Zeng X, Ying Y et al (2013) Study on the hydrogen storage properties of core-shell structured Mg-RE(RE=Nd, Gd, Er) nano-composites synthesized through arc plasma method. Int J Hydrog Energy 38(5):2337–2346
12. Zou J, Zeng X, Ying Y et al (2012) Preparation and hydrogen sorption properties of a nano-structured Mg based Mg-La-O composite. Int J Hydrog Energy 37(17):13067–13073
13. Li Q, Lin Q, Jiang L et al (2002) Progress in the synthesis of magnesium-based hydrogen storage alloys by hydriding combustion method. Rare Metals 26(5):386–390
14. Zhang H, Zheng X, Liu Y et al (2016) Progress in the application of rare earth elements in hydrogen storage materials. J Chin Rare Earth Soc 34(1):10
15. Han X (2016) Research on rare earth AB5 type low cobalt and cobalt-free hydrogen storage alloys. Northeastern University, Shenyang
16. Chen S, Zhang N, Kan H et al (2017) Current status and prospects of research on LaNi5 series hydrogen storage alloys. New Chem Mater 45(9):26–28
17. Noritake TA, Towata M, Seno S et al (2002) Chemical bonding of hydrogen in MgH_2. Appl Phys Lett 81(11):2008–2010
18. Aguey-Zinsou KF, Ares-Fernandez JR (2010) Hydrogen in magnesium: new perspectives toward functional stores. Energy Environ Sci 3(5):526–543
19. Andreasen A (2008) Hydrogen refuelling properties of Mg-Al alloys. Int J Hydrog Energy 33(24):7489–7497
20. Vajo JJ et al (2004) Altering hydrogen storage properties by hydride destabilization through alloy formation: LiH and MgH_2 destabilized with Si. J Phys Chem B 108(37):13977–13983
21. Si TZ, Zhang JB, Liu DM et al (2013) A new reversible Mg_3Ag-H_2 system for hydrogen storage. J Alloys Compd 581:246–249
22. Zhong HC, Wang H, Liu JW et al (2011) Altered desorption enthalpy of MgH_2 by the reversible formation of Mg(In) solid solution. Scr Mater 65(4):285–287
23. Cui J, Wang H, Liu J et al (2013) Remarkable enhancement in dehydrogenation of MgH_2 by a nano-coating of multi-valence Ti-based catalysts. J Mater Chem A 1(18):5603
24. Skripnyuk VM, Rabkin E (2012) Mg_3Cd: a model alloy for studying the destabilization of magnesium hydride. Int J Hydrog Energy 37(14):10724–10732

25. Ouyang L, Tong Y, Dong H et al (2006) Hydrogen storage performance of Mg_3La compound and the effect of Ni addition on its performance. In Proceedings of the 7th National Hydrogen Energy Conference. [S. l.]: [s. n.]
26. Reilly JJ, Wiswall RH (1968) Reaction of hydrogen with alloys of magnesium and nickel and the formation of Mg_2NiH_4. Inorg Chem 7(11):2254–2256
27. Didisheim JJ, Zolliker P, Yvon K et al (1984) Dimagnesium iron (II) hydride, Mg_2FeH_6, containing octahedral FeH_6^{4-} anions. Inorg Chem 23(13):1953–1957
28. Gennari FC, Castro FJ, Gamboa JJA (2002) Synthesis of Mg_2FeH_6 by reactive mechanical alloying: formation and decomposition properties. J Alloys Compd 339(1–2):261–267
29. Saita I, Saito K, Akiyama T (2005) Hydriding combustion synthesis of $Mg_2Ni_{1-x}Fe_x$ hydride. J Alloys Compd 390(1–2):265–269
30. Pozzo M, Alfe D (2009) Hydrogen dissociation and diffusion on transition metal (=Ti, Zr, V, Fe, Ru, Co, Rh, Ni, Pd, Cu , Ag)-doped Mg(0001) surfaces. Int J Hydrog Energy 34(4):1922–1930
31. Cui J, Liu JW, Wang H et al (2014) Mg-TM (TM: Ti, Nb, V, Co, Mo or Ni) core-shell like nanostructures: synthesis, hydrogen storage performance and catalytic mechanism. J Mater Chem A 2(25):9645–9655
32. Zhu W, Panda S, Lu C et al (2020) Using a self-assembled two-dimensional MXenes-based catalyst (2D-Ni@Ti_3C_2) to enhance hydrogen storage properties of MgH_2. ACS Appl Mater Interfaces 12(45):50333–50343
33. Bhatnagar A, Pandey SK, Vishwakarma A et al (2016) Fe_3O_4@Graphene as a superior catalyst for hydrogen de/absorption from/in MgH_2/Mg. J Mater Chem A 4(38):14761–14772
34. Lototskyy M, Denys R et al (2018) An outstanding effect of graphite in nano-MgH_2-TiH_2 on hydrogen storage performance dagger. J Mater Chem A 6(23):10740–10754
35. Liu Y, Zou J, Zeng X et al (2015) Study on hydrogen storage properties of Mg-X (X= Fe, Co, V) nano-composites co-precipitated from solution. RSC Adv 5(10):7687–7696
36. Liu Y, Zou J, Zeng X et al (2014) Hydrogen storage properties of a Mg-Ni nanocomposite coprecipitated from solution. J Phys Chem C 118(32):18401–18411
37. Xia G, Tan Y, Chen X et al (2015) Monodisperse magnesium hydride nanoparticles uniformly self-assembled on graphene. Adv Mater 27(39):5981–5988
38. Zhu W, Ren L, Lu C et al (2021) Nanoconfined and in situ catalyzed MgH_2 self-assembled on 3D Ti_3C_2 MXene folded nanosheets with enhanced hydrogen sorption performances. ACS Nano 15(11):18494–18504
39. Li W, Li C, Ma H et al (2007) Magnesium nanowires: enhanced kinetics for hydrogen absorption and desorption. J Am Chem Soc 129(21):6710–6711
40. Bsenberg U, Doppiu S, Mosegaard L et al (2007) Hydrogen sorption properties of MgH_2-$LiBH_4$ composites. Acta Mater 55(11):3951–3958
41. Tayeh T, Awad AS, Nakhl M et al (2014) Production of hydrogen from magnesium hydrides hydrolysis. Int J Hydrog Energy 39(7):3109–3117
42. Huang M, Ouyang L, Wang H et al (2015) Hydrogen generation by hydrolysis of MgH_2 and enhanced kinetics performance of ammonium chloride introducing. Int J Hydrog Energy 40(18):6145–6150
43. Yang B, Zou J, Huang T et al (2019) Enhanced hydrogenation and hydrolysis properties of core-shell structured Mg-MO_x (M=Al, Ti and Fe) nanocomposites prepared by arc plasma method. Chem Eng J 371:233–243
44. Mao J, Zou J, Lu C et al (2017) Hydrogen storage and hydrolysis properties of core-shell structured Mg-MF_x (M=V, Ni, La and Ce) nano-composites prepared by arc plasma method. J Power Sources 366:131–142
45. Song Q (2020) Discussion on fuel cell technology of underwater unmanned vehicles. Ship Sci Technol 42(23):150–154
46. Li Y, Zhang L, Han S (2020) Research and progress on the structure and hydrogen storage performance of rare earth-magnesium-nickel superlattice alloys. J Yanshan Univ 44(3):323–330

47. Kadir K, Sakai T, Uehara I (1997) Synthesis and structure determination of a new series of hydrogen storage alloys; RMg_2Ni_9 (R=La, Ce, Pr, Nd, Sm and Gd) built from $MgNi_2$ Laves-type layers alternating with AB_5 layers-ScienceDirect. J Alloys Compd 257(1–2):115–121
48. Kadir K, Sakai T, Uehara I (1999) Structural investigation and hydrogen capacity of YMg_2Ni_9 and $(Y_{0.5}Ca_{0.5})(MgCa)Ni_9$: new phases in the AB_2C_9 system isostructural with $LaMg_2Ni_9$. J Alloys Compd 287(1–2):264–270
49. Kohno T, Yoshida H, Kawashima F et al (2000) Hydrogen storage properties of new ternary system alloys: La_2MgNi_9, $La_5Mg_2Ni_{23}$, La_3MgNi_{14}. J Alloys Compd 311(2):L5–L7
50. Guo X, Wang S, Liu X et al (2011) Laves phase hydrogen storage alloys for super-high-pressure metal hydride hydrogen compressors. Rare Metals 30(003):227–231
51. Zotov TA, Sivov RB, Movlaev EA et al (2011) IMC hydrides with high hydrogen dissociation pressure. J Alloys Compd 509(supp-S2):S839–S843
52. Shilov AL, Padurets LN, Kost ME (1985) Thermodynamics of hydrides of intermetallic compounds of transition metals. Zhurnal Fizicheskoi Khimii 59(8):1857–1875
53. Kinaci A, Aydinol MK (2007) Ab initio investigation of FeTi-H system. Int J Hydrog Energy 32(13):2466–2474
54. Thompson P, Pick MA, Reidinger F et al (1978) Neutron diffraction study of β iron titanium deuteride. J Phys F 8(4):L75–L80
55. Zhao D, Shang H, Li Y et al (2017) Research on the application of titanium-iron-based hydrogen storage alloys in the field of vehicle energy storage. Rare Metals 41(5):515–533
56. Sandrock GD, Goodell PD (1980) Surface poisoning of $LaNi_5$, FeTi and (Fe, Mn) Ti by O_2, Co and H_2O. J Less-Common Metals 73(1):161–168
57. Hirata T (1985) Hydrogen absorption and desorption properties of $FeTi_{1.14}O_{0.03}$ in impure hydrogen containing CO, CO_2 and oxygen. J Less Common Metals 107(1):23–33
58. Szajek A, Jurczyk M, Jankowska E (2003) The electronic and electrochemical properties of the TiFe-based alloys. J Alloys Compd 348(1–2):285–292
59. Yan Y, Yan K, Chen Y et al (2004) Research progress on vanadium-based solid solution type hydrogen storage alloys. Rare Metals 28(4):6
60. Pei P, Zhang P, Zhang B et al (2006) V series hydrogen storage alloys and their alloying. Mater Rev 20(10):123–127
61. Hao L, Chen Y, Yan Y et al (2007) Influence of Ni or Mn on hydrogen absorption-desorption performance of V-Ti-Cr-Fe alloys. Mater Sci Eng A 459(1–2):204–208
62. Yan Y (2007) Study on the structure and hydrogen absorption and desorption behavior of V-Ti-Cr-Fe hydrogen storage alloy. Sichuan University, Chengdu
63. Züttel A, Rentsch S, Fischer P et al (2003) Hydrogen storage properties of $LiBH_4$. J Alloys Compd 356-357:515–520
64. Kollonitsch J, Fuchs O, Gábor V (1954) New and known complex borohydrides and some of their applications in organic syntheses. Nature 173(4394):125–126
65. Mosegaard L, Møller B, Jørgensen JE et al (2007) Intermediate phases observed during decomposition of $LiBH_4$. J Alloys Compd 446:301–305
66. Benzidi H, Garara M, Lakhal M et al (2018) Vibrational and thermodynamic properties of $LiBH_4$ polymorphs from first-principles calculations. Int J Hydrog Energy 43(13):6625–6631
67. Züttel A, Borgschulte A, Orimo SI (2007) Tetrahydroborates as new hydrogen storage materials. Scr Mater 56(10):823–828
68. Orimo S, Nakamori Y, Kitahara G et al (2005) Dehydriding and rehydriding reactions of $LiBH_4$. J Alloys Compd 404:427–430
69. Mauron P, Buchter F, Friedrichs O et al (2008) Stability and reversibility of $LiBH_4$. J Phys Chem B 112(3):906–910
70. Pinkerton FE, Meisner GP, Meyer MS et al (2005) Hydrogen desorption exceeding ten weight percent from the new quaternary hydride $Li_3BN_2H_8$. J Phys Chem B 109(1):6–8
71. John JV, Sky LS, Florian M (2005) Reversible storage of hydrogen in destabilized $LiBH_4$. J Phys Chem B 109(9):3719–3722

72. Nakamori Y, Miwa K, Ninomiya A et al (2006) Correlation between thermodynamical stabilities of metal borohydrides and cation electronegativites: first-principles calculations and experiments. Phys Rev B 74(4):045126
73. Liu H, Jiao L, Zhao Y et al (2014) Improved dehydrogenation performance of $LiBH_4$ by confinement into porous TiO_2 micro-tubes. J Mater Chem A 2(24):9244–9250
74. Surrey A, Minella CB, Fechler N et al (2016) Improved hydrogen storage properties of $LiBH_4$ via nano-confinement in micro-and mesoporous aerogel-like carbon. Int J Hydrog Energy 41(12):5540–5548
75. Ngene P, Verkuijlen MHW, Zheng Q et al (2011) The role of Ni in increasing the reversibility of the hydrogen release from nanoconfined $LiBH_4$. Faraday Discuss 151:47–58
76. Thiangviriya S, Utke R (2015) $LiBH_4$ nanoconfined in activated carbon nanofiber for reversible hydrogen storage. Int J Hydrog Energy 40(11):4167–4174
77. Lei Z, Sun W, Song L et al (2018) Enhanced hydrogen storage properties and reversibility of $LiBH_4$ confined in two-dimensional Ti_3C_2. ACS Appl Mater Interfaces 10(23):19598–19604
78. Shao J, Xiao X, Fan X et al (2014) Low-temperature reversible hydrogen storage properties of $LiBH_4$: a synergistic effect of nanoconfinement and nanocatalysis. J Phys Chem C 118(21):11252–11260
79. Zhang X, Zhang L, Zhang W et al (2021) Nano-synergy enables highly reversible storage of 9.2 wt% hydrogen at mild conditions with lithium borohydride. Nano Energy 83:105839
80. Avorotynska O, El-Kharbachi A, Deledda S et al (2016) Recent progress in magnesium borohydride $Mg(BH_4)_2$: fundamentals and applications for energy storage. Int J Hydrog Energy 41(32):14387–14403
81. Paskevicius M, Pitt MP, Webb CJ et al (2012) In-situ X-ray diffraction study of gamma-$Mg(BH_4)_2$ decomposition. J Phys Chem C 116(29):15231
82. Yang J, He F, Ping S et al (2012) Reversible dehydrogenation of $Mg(BH_4)_2$-LiH composite under moderate conditions. Int J Hydrog Energy 37(8):6776–6783
83. Liu Y, Yang Y, Zhou Y et al (2012) Hydrogen storage properties and mechanisms of the $Mg(BH_4)_2$-$NaAlH_4$ system. Int J Hydrog Energy 37(22):17137–17145
84. Yu XB, Guo YH, Sun DL et al (2010) A combined hydrogen storage system of $Mg(BH_4)_2$-$LiNH_2$ with favorable dehydrogenation. J Phys Chem C 114(10):4733–4737
85. Sun N, Xu B, Zhao S et al (2015) Influences of Al, Ti and Nb doping on the structure and hydrogen storage property of $Mg(BH_4)_2$(001) surface—a theoretical study. Int J Hydrog Energy 40(33):10516–10526
86. Lai Q, Aguey-Zinsou KF (2017) Destabilisation of $Ca(BH_4)_2$ and $Mg(BH_4)_2$: via confinement in nanoporous Cu_2S hollow spheres. Sustain Energy Fuels 1(6):1308–1319
87. Chong L, Zou J, Zeng X et al (2013) Mechanisms of reversible hydrogen storage in $NaBH_4$ through NdF_3 addition. J Mater Chem A 1(12):3983–3991
88. Chong L, Zou J, Zeng X et al (2014) Effects of LnF_3 on reversible and cyclic hydrogen sorption behaviors in $NaBH_4$: electronic nature of Ln versus crystallographic factor. J Mater Chem A 3(8):4493–4500
89. Chong L, Zeng X, Ding W et al (2015) $NaBH_4$ in "graphene wrapper:" significantly enhanced hydrogen storage capacity and regenerability through nanoencapsulation. Adv Mater 27(34):5070–5074
90. Christian M, Aguey-Zinsou KF (2013) Synthesis of core-shell $NaBH_4$@ M(M=Co, Cu, Fe, Ni, Sn) nanoparticles leading to various morphologies and hydrogen storage properties. Chem Commun 49(60):6794–6796
91. Rafi-ud-Din, Zhang L, Ping L et al (2010) Catalytic effects of nano-sized TiC additions on the hydrogen storage properties of $LiAlH_4$. J Alloys Compd 508(1):119–128
92. Liu SS, Sun LX, Zhang Y et al (2009) Effect of ball milling time on the hydrogen storage properties of TiF_3-doped $LiAlH_4$. Int J Hydrog Energy 34(19):8079–8085
93. Zang L, Cai J, Zhao L et al (2015) Improved hydrogen storage properties of $LiAlH_4$ by mechanical milling with TiF_3. J Alloys Compd 647:756–762

94. Li Z, Zhai F, Wan Q et al (2014) Enhanced hydrogen storage properties of $LiAlH_4$ catalyzed by $CoFe_2O_4$ nanoparticles. RSC Adv 4(36):18989–18997
95. Zang L, Liu S, Guo H et al (2017) In situ synthesis of 3D flower-like nanocrystalline Ni/C and its effect on hydrogen storage properties of $LiAlH_4$. Chem Asian J 18:350–357
96. Luo H (2018) The influence of Ti, Y doping on the in-situ preparation and reversible hydrogen absorption and desorption properties of $NaAlH_4$. Guangxi University, Nanning
97. Chen P, Xiong ZT, Luo JZ et al (2002) Interaction of hydrogen with metal nitrides and imides. Nature 420:302–304
98. Luo W (2004) ($LiNH_2$-MgH_2): a viable hydrogen storage system. J Alloys Compd 381(1–2):284–287
99. Liu Y, Hu J, Wu G et al (2008) Formation and equilibrium of ammonia in the $Mg(NH_2)_2$-2LiH hydrogen storage system. J Phys Chem C 112(4):1293–1298
100. Chandra M, Qiang X (2007) Room temperature hydrogen generation from aqueous ammonia-borane using noble metal nano-clusters as highly active catalysts. J Power Sources 168(1):135–142
101. Liu P, Xiaojun G, Kang K et al (2017) Highly efficient catalytic hydrogen evolution from ammonia borane using the synergistic effect of crystallinity and size of noble-metal-free nanoparticles supported by porous metal-organic frameworks. ACS Appl Mater Interfaces 9(12):10759–10767
102. Mori K, Miyawaki K, Yamashita H (2016) Ru and Ru-Ni nanoparticles on TiO_2 support as extremely active catalysts for hydrogen production from ammonia-borane. ACS Catal 6:3128–3135
103. Akbayrak S, Tonbul Y, Özkar S (2016) Ceria supported rhodium nanoparticles: superb catalytic activity in hydrogen generation from the hydrolysis of ammonia borane. Appl Catal B Environ 198:162–170
104. Wei W, Lu ZH, Yan L et al (2018) Mesoporous carbon nitride supported Pd and Pd-Ni nanoparticles as highly efficient catalyst for catalytic hydrolysis of NH_3BH_3. ChemCatChem 10(7):1620–1626
105. Wen L, Su J, Wu X et al (2014) Ruthenium supported on MIL-96: an efficient catalyst for hydrolytic dehydrogenation of ammonia borane for chemical hydrogen storage. Int J Hydrog Energy 39(30):17129–17135
106. Rej S, Hsia CF, Chen TY et al (2016) Facet-dependent and light-assisted efficient hydrogen evolution from ammonia borane using gold-palladium core-shell nanocatalysts. Angew Chem Int Ed 128(25):7338–7342
107. Wu C, Wu G, Xiong Z et al (2010) Stepwise phase transition in the formation of lithium amidoborane. Inorg Chem 49(9):4319–4323
108. Himashinie V (2010) Potassium(I) amidotrihydroborate: structure and hydrogen release. J Am Chem Soc 132(34):11836–11837
109. Chen X, Feng Y, Gu Q et al (2013) Synthesis, structures and hydrogen storage properties of two new H-enriched compounds: $Mg(BH_4)_2(NH_3BH_3)_2$ and $Mg(BH_4)_2(NH_3)_2(NH_3BH_3)$. Dalton Trans 42(40):14365–14368
110. Jepsen LH, Skibsted J, Jensen TR (2013) Investigations of the thermal decomposition of MBH_4-$2NH_3BH_3$, M= Na, K. J Alloys Compd 580:S287–S291
111. Chen W, Huang Z, Wu G et al (2015) New synthetic procedure for $NaNH_2(BH_3)_2$ and evaluation of its hydrogen storage properties. Sci China Chem 58(1):169–173
112. Chua YS, Wu G, Xiong Z et al (2010) Synthesis, structure and dehydrogenation of magnesium amidoborane monoammoniate. Chem Commun 46(31):5752–5754
113. Chua YS, Wu G, Xiong Z et al (2009) Calcium amidoborane ammoniate—synthesis, structure, and hydrogen storage properties. Chem Mater 21(20):4899–4904
114. Xia G, Yu X, Guo Y et al (2010) Amminelithium amidoborane $Li(NH_3)NH_2BH_3$: a new coordination compound with favorable dehydrogenation characteristics. Chem Eur J 16(12):3763–3769

115. Wu C, Wu G, Xiong Z et al (2010) $LiNH_2BH_3 \cdot NH_3BH_3$: structure and hydrogen storage properties. Chem Mater 22(1):3–5
116. Fu H, Yang JZ, Wang XJ et al (2014) Preparation and dehydrogenation properties of lithium hydrazidobis (borane)($LiNH(BH_3)NH_2BH_3$). Inorg Chem 53(14):7334–7339
117. Johnson SR, David WIF, Royse DM et al (2009) The monoammoniate of lithium borohydride, $Li(NH_3)BH_4$: an effective ammonia storage compound. Chem Asian J 4(6):849–854
118. McCusker LB, Liebau F, Engelhardt G (2001) Nomenclature of structural and compositional characteristics of ordered microporous and mesoporous materials with inorganic hosts(IUPAC recommendations 2001). Pure Appl Chem 73(2):381–394
119. Simonyan VV, Johnson JK (2002) Hydrogen storage in carbon nanotubes and graphitic nanofibers. J Alloys Compd 330:659–665
120. Carpetis C, Peschka W (1980) A study on hydrogen storage by use of cryoadsorbents. Int J Hydrog Energy 5(5):539–554
121. Furukawa H, Yaghi OM (2009) Storage of hydrogen, methane, and carbon dioxide in highly porous covalent organic frameworks for clean energy applications. J Am Chem Soc 131(25):8875–8883
122. Langmi HW, Walton A, Al-Mamouri MM et al (2003) Hydrogen adsorption in zeolites A, X, Y and RHO. J Alloys Compd 356–357:710–715
123. Xu W, Tu B, Liu Q et al (2020) Anisotropic reticular chemistry. Nat Rev Mater 5(10):764–779
124. Trickett CA, Helal A, Al-Maythalony BA et al (2017) The chemistry of metal-organic frameworks for CO_2 capture, regeneration and conversion. Nat Rev Mater 2:17045
125. Siegelman RL, Kim EJ, Long JR (2021) Porous materials for carbon dioxide separations. Nat Mater 20(8):1060–1072
126. Chen Z, Kirlikovali KO, Idrees KB et al (2022) Porous materials for hydrogen storage. Chem 8(3):693–716
127. Mcbain JW XCIX (1909) The mechanism of the adsorption (“sorption”) of hydrogen by carbon. Lond Edinb Dublin Philos Mag J Sci 18(108):916–935
128. Sevilla M, Mokaya R (2014) Energy storage applications of activated carbons: supercapacitors and hydrogen storage. Energy Environ Sci 7(4):1250–1280
129. Kidnay AJ, Hiza MJ (1967) High pressure adsorption isotherms of neon, hydrogen, and helium at 76K. In: Advances in cryogenic engineering. Springer, Boston, pp 730–740
130. Zhou L, Zhou Y, Sun Y (2004) Enhanced storage of hydrogen at the temperature of liquid nitrogen. Int J Hydrog Energy 29(3):319–322
131. Jordá-Beneyto M, Lozano-Castelló D, Suárez-García F et al (2008) Advanced activated carbon monoliths and activated carbons for hydrogen storage. Microporous Mesoporous Mater 112(1):235–242
132. Sevilla M, Fuertes AB, Mokaya R (2011) High density hydrogen storage in superactivated carbons from hydrothermally carbonized renewable organic materials. Energy Environ Sci 4(4):1400–1410
133. Gao F, Zhao DL, Li Y et al (2010) Preparation and hydrogen storage of activated rayon-based carbon fibers with high specific surface area. J Phys Chem Solids 71(4):444–447
134. Fierro V, Szczurek A, Zlotea C et al (2010) Experimental evidence of an upper limit for hydrogen storage at 77 K on activated carbons. Carbon 48(7):1902–1911
135. Kuchta B, Firlej L, Pfeifer P et al (2010) Numerical estimation of hydrogen storage limits in carbon-based nanospaces. Carbon 48(1):223–231
136. Alcañiz-Monge J, Román-Martínez MC (2008) Upper limit of hydrogen adsorption on activated carbons at room temperature: a thermodynamic approach to understand the hydrogen adsorption on microporous carbons. Microporous Mesoporous Mater 112(1):510–520
137. Li Y, Yang RT (2007) Hydrogen storage on platinum nanoparticles doped on superactivated carbon. J Phys Chem C 111(29):11086–11094
138. Xia K, Hu J, Jiang J (2014) Enhanced room-temperature hydrogen storage in super-activated carbons: the role of porosity development by activation. Appl Surf Sci 315:261–267

139. Geng Z, Wang D, Zhang C et al (2014) Spillover enhanced hydrogen uptake of Pt/Pd doped corncob-derived activated carbon with ultra-high surface area at high pressure. Int J Hydrog Energy 39(25):13643–13649
140. Rostami S, Pour AN, Izadyar M (2018) A review on modified carbon materials as promising agents for hydrogen storage. Sci Prog 101(2):171–191
141. Iijima S (1991) Helical microtubules of graphitic carbon. Nature 354(6348):56–58
142. Li W, Wang H, Ren Z et al (2008) Co-production of hydrogen and multi-wall carbon nanotubes from ethanol decomposition over Fe/Al_2O_3 catalysts. Appl Catal B Environ 84(3):433–439
143. Dillon AC, Jones KM, Bekkedahl TA et al (1997) Storage of hydrogen in single-walled carbon nanotubes. Nature 386(6623):377–379
144. Ye Y, Ahn CC, Witham C et al (1999) Hydrogen adsorption and cohesive energy of single-walled carbon nanotubes. Appl Phys Lett 74(16):2307–2309
145. Nijkamp MG, Raaymakers J, van Dillen AJ et al (2001) Hydrogen storage using physisorption-materials demands. Appl Phys A 72(5):619–623
146. Nishimiya N, Ishigaki K, Takikawa H et al (2002) Hydrogen sorption by single-walled carbon nanotubes prepared by a torch arc method. J Alloys Compd 339(1):275–282
147. Zhou L, Zhou Y, Sun Y (2006) Studies on the mechanism and capacity of hydrogen uptake by physisorption-based materials. Int J Hydrog Energy 31(2):259–264
148. Liu F, Zhang X, Cheng J et al (2003) Preparation of short carbon nanotubes by mechanical ball milling and their hydrogen adsorption behavior. Carbon 41(13):2527–2532
149. Chen CH, Huang CC (2007) Hydrogen storage by KOH-modified multi-walled carbon nanotubes. Int J Hydrog Energy 32(2):237–246
150. Zhang X, Chen W (2015) Mechanisms of pore formation on multi-wall carbon nanotubes by KOH activation. Microporous Mesoporous Mater 206:194–201
151. Lin KY, Tsai WT, Yang TJ (2011) Effect of Ni nanoparticle distribution on hydrogen uptake in carbon nanotubes. J Power Sources 196(7):3389–3394
152. Rather SU, Hwang SW (2016) Comparative hydrogen uptake study on Titanium-MWCNTs composite prepared by two different methods. Int J Hydrog Energy 41(40):18114–18120
153. Reyhani A, Mortazavi SZ, Mirershadi S et al (2011) Hydrogen storage in decorated multi-walled carbon nanotubes by Ca, Co, Fe, Ni, and Pd nanoparticles under ambient conditions. J Phys Chem C 115(14):6994–7001
154. Rosi NL, Eckert J, Eddaoudi M et al (2003) Hydrogen storage in microporous metal-organic frameworks. Science 300(5622):1127–1129
155. Wong-Foy AG, Matzger AJ, Yaghi OM (2006) Exceptional H_2 saturation uptake in microporous metal-organic frameworks. J Am Chem Soc 128(11):3494–3495
156. Suh MP, Park HJ, Prasad TK et al (2012) Hydrogen storage in metal-organic frameworks. Chem Rev 112(2):782–835
157. Kapelewski MT, Runevski T, Tarver JD et al (2018) Record high hydrogen storage capacity in the metal-organic framework Ni_2(m-dobdc) at near-ambient temperatures. Chem Mater 30(22):8179–8189
158. He T, Pachfule P, Wu H et al (2016) Hydrogen carriers. Nat Rev Mater 1(12):16059
159. Xiao B, Wheatley PS, Zhao X et al (2007) High-capacity hydrogen and nitric oxide adsorption and storage in a metal-organic framework. J Am Chem Soc 129(5):1203–1209
160. Lee YG, Moon HR, Cheon YE et al (2008) A comparison of the H_2 sorption capacities of isostructural metal-organic frameworks with and without accessible metal sites: [{Zn_2(abtc)(dmf)$_2$}$_3$] and [{Cu_2(abtc)(dmf)$_2$}$_3$] versus [{Cu_2(abtc)}$_3$]. Angew Chem Int Ed 47(40):7741–7745
161. Lim DW, Chyun SA, Suh MP (2014) Hydrogen storage in a potassium-ion-bound metal-organic framework incorporating crown ether struts as specific cation binding sites. Angew Chem Int Ed 53(30):7819–7822
162. Allendorf MD, Hulvey Z, Gennett T et al (2018) An assessment of strategies for the development of solid-state adsorbents for vehicular hydrogen storage. Energy Environ Sci 11(10):2784–2812

163. Li G, Kobayashi H, Taylor JM et al (2014) Hydrogen storage in Pd nanocrystals covered with a metal-organic framework. Nat Mater 13(8):802–806
164. Szilágyi PÁ, Callini E, Anastasopol A et al (2014) Probing hydrogen spillover in Pd@MIL-101(Cr) with a focus on hydrogen chemisorption. Phys Chem Chem Phys 16(12):5803–5809
165. Mason JA, Oktawiec J, Taylor MK et al (2015) Methane storage in flexible metal-organic frameworks with intrinsic thermal management. Nature 527(7578):357–361
166. Jaramillo DE, Jiang HZH, Evans HA et al (2021) Ambient-temperature hydrogen storage via Vanadium(ii)-dihydrogen complexation in a metal-organic framework. J Am Chem Soc 143(16):6248–6256
167. Ahmed A, Liu Y, Purewal J et al (2017) Balancing gravimetric and volumetric hydrogen density in MOFs. Energy Environ Sci 10(11):2459–2471
168. Ahmed A, Seth S, Purewal J et al (2019) Exceptional hydrogen storage achieved by screening nearly half a million metal-organic frameworks. Nat Commun 10(1):1568
169. Chen Z, Mian MR, Lee SJ et al (2021) Fine-tuning a robust metal-organic framework toward enhanced clean energy gas storage. J Am Chem Soc 143(45):18838–18843
170. Gómez-Gualdrón DA, Wang TC, García-Holley P et al (2017) Understanding volumetric and gravimetric hydrogen adsorption trade-off in metal-organic frameworks. ACS Appl Mater Interfaces 9(39):33419–33428
171. Zhang X, Lin RB, Wang J et al (2020) Optimization of the pore structures of MOFs for record high hydrogen volumetric working capacity. Adv Mater 32:1907995
172. Chen Z, Li P, Anderson R et al (2020) Balancing volumetric and gravimetric uptake in highly porous materials for clean energy. Science 368(6488):297–303
173. Connolly BM, Madden DG, Wheatley AEH et al (2020) Shaping the future of fuel: monolithic metal-organic frameworks for high-density gas storage. J Am Chem Soc 142(19):8541–8549
174. Huang Y, Cheng Y, Zhang J (2021) A review of high density solid hydrogen storage materials by pyrolysis for promising mobile applications. Ind Eng Chem Res 60(7):2737–2771
175. Pirngruber GD, Llewellyn PL (2011) Metal-organic frameworks: applications from catalysis to gas storage. Wiley-VCH Verlag GmbH & Co. KGaA, Weinheim
176. Kundu T, Shah BB, Bolinois L et al (2019) Functionalization-induced breathing control in metal-organic frameworks for methane storage with high deliverable capacity. Chem Mater 31(8):2842–2847
177. Hu Z, Wang Y, Shah BB et al (2019) CO_2 capture in metal-organic framework adsorbents: an engineering perspective. Adv Sustain Syst 3(1):1800080
178. Hu Z, Kundu T, Wang Y et al (2020) Modulated hydrothermal synthesis of highly stable MOF-808(Hf) for methane storage. ACS Sustain Chem Eng 8:17042–17053
179. Peterson GW, Decoste JB, Glover TG et al (2013) Effects of pelletization pressure on the physical and chemical properties of the metal-organic frameworks $Cu_3(BTC)_2$ and UiO-66. Microporous Mesoporous Mater 179:48–53
180. Suresh K, Aulakh D, Purewal J et al (2021) Optimizing hydrogen storage in MOFs through engineering of crystal morphology and control of crystal size. J Am Chem Soc 143(28):10727–10734
181. García-Holley P, Schweitzer B, Islamoglu T et al (2018) Benchmark study of hydrogen storage in metal-organic frameworks under temperature and pressure swing conditions. ACS Energy Lett 3(3):748–754
182. Prasad TK, Suh MP (2012) Control of interpenetration and gas-sorption properties of metal-organic frameworks by a simple change in ligand design. Chem Eur J 18(28):8673–8680
183. Lin X, Zhu Q, Leng H et al (2019) Numerical analysis of the effects of particle radius and porosity on hydrogen absorption performances in metal hydride tank. Appl Energy 250:1065–1072
184. Lin X, Yang H, Zhu Q et al (2020) Numerical simulation of a metal hydride tank with $LaNi_{4.25}Al_{0.75}$ using a novel kinetic model at constant flows. Chem Eng J 401:126115
185. Lin X, Xie W, Zhu Q et al (2021) Rational optimization of metal hydride tank with $LaNi_{4.25}Al_{0.75}$ as hydrogen storage medium. Chem Eng J 421:127844

Appendixes

Appendix A: Abbreviations

ACC Activated carbon
ANSI American National Standards Institute
ASME American Society of Mechanical Engineers
ASU Air separation unit
ATR Autothermal reforming
BZ Benzene
CCS Carbon capture and storage
CDP Proton conductive membrane
CHE Cyclohexane
CLAS Chemical looping ammonia synthesis
CNT Carbon nano tube
COF Covalent-organic framework
CSA Canadian Standards Association
CTAB Cetyl trimethyl ammonium bromide
DAC Direct air capture
DBT Dibenzyltoluene
DE Decomposition
DEC Decahydronaphthalene
DOE Department of Energy
ECAS Electrochemical ammonia synthesis
FTIR Fourier transform infrared spectroscopy
GHSV Gas hourly space velocity
H18-DBT Perhydro-dibenzyltoluene
H2G Hydrogen to gas
H2H Hydrogen to hydrogen
H2P Hydrogen to powder
H2T Hydrogen to heat

J. Zou, *Hydrogen Storage and Transportation*,
https://doi.org/10.1007/978-981-96-2875-9

HE Heat exchanger
HER Hydrogen evolution reaction
HEV Hybrid electrical vehicle
HS Hydrogen storage
HT High temperature
IEA International Energy Agency
ISO International Organization for Standardization
LCOS Levelized cost of energy
LH_2 Liquid hydrogen
LOHC Liquid organic hydrogen carriers
LTLower temperature
MCH Methyl cyclohexane
MH Metal hydride
MOF Metal-organic framework
MSLV Multifunctional steel layered vessel
MSR Methanol steam reforming
MWCNT Multiwalled carbon nanotube
NAP Naphthalene
NEC N-ethylcarbazole
NIST National Institute of Standard and Technology
NRR Nitrogen reduction reaction
OER Oxygen evolution reaction
P2H Power to hydrogen
PAS Plasma ammonia synthesis
PCSA Photochemical synthesis of ammonia
PCT Pressure-composition-temperature
PEM Proton exchange membrane
PEMFC Proton exchange membrane fuel cell
PNEC Perhydro-N-ethylcarbazole
POP Porous polymer
POX Partial oxidation
PPC P-doped porous carbon
rGO Reduced graphene oxide
TOF Turnover frequency
TOL Toluene
XRD X-Ray diffraction

Appendix B: Symbols

Symbol	Unit	Meaning
P	Pa	Pressure
V	m^3	Volume
n	mol	Molar mass of substance
R	J/(mol K)	Universal gas constant
T	K	Absolute temperature
Z	—	Compressibility factor
A	$m^6 \cdot Pa/mol^2$	Dipole interaction or repulsion constant
B	m^3/mol	Volume occupied by hydrogen molecules
C	K/Pa	Coefficient of the equation of state for hydrogen
K	—	Equilibrium constant
H	J	Enthalpy of the gas
U	J	Internal energy of the gas
μ	K/Pa	Joule-Thomson coefficient
ΔH	kJ/mol	Enthalpy change of reaction
ΔS	J/mol	Entropy change of reaction
F	—	Degrees of freedom
p	—	Number of phases
v	1/s	Reaction rate
k_B	J/K	Boltzmann constant
E_a	kJ/mol	Activation energy of reaction
S	MPa	Minimum yield strength
δ	mm	Nominal wall thickness
D	mm	Nominal diameter
F_f	—	Design factor
E_f	—	Axial joint factor
T_f	—	Temperature reduction factor
H_f	—	Material performance factor
ρ	kg/m^3	Density
H/M	—	Hydrogen content
k	1/s	Rate constant
ξ	—	Reaction Fraction
ω		Volume expansion ratio before and after hydrogen absorption and desorption
C_p	J/(kg K)	Heat capacity at constant pressure
u	m/s	Flow rate
M	g/mol	Relative molar mass
ε	—	Porosity
μ_g	Pa s	Gas viscosity coefficient
S_T	J/(s m^3)	Heat source term
S_P	kg/(s m^3)	Mass y term

MIX
Papier aus verantwortungsvollen Quellen
Paper from responsible sources
FSC® C105338

If you have any concerns about our products,
you can contact us on
ProductSafety@springernature.com

In case Publisher is established outside the EU,
the EU authorized representative is:
Springer Nature Customer Service Center GmbH
Europaplatz 3, 69115 Heidelberg, Germany

Printed by Libri Plureos GmbH
in Hamburg, Germany